10th Scientific Conference Air Protection in Theory and Practice 2017

E3S Web of Conferences Volume 28 (2018)

Zakopane, Poland
18 - 21 October 2017

Editors:

Marianna Czaplicka
Oskar Czechowski
Janina Fudala
Magdalena Jablonska-Czapla
Joanna Kyziol-Komosinska
Grzegorz Majewski
Katarzyna Juda-Rezler
Wioletta Rogula-Kozlowska
Izabela Sowka

ISBN: 978-1-5108-5570-0

Printed from e-media with permission by:

Curran Associates, Inc.
57 Morehouse Lane
Red Hook, NY 12571

Some format issues inherent in the e-media version may also appear in this print version.

Printed by Curran Associates, Inc. (2018)

For additional information, please contact EDP Sciences – Web of Conferences
at the address below.

EDP Sciences – Web of Conferences
17, Avenue du Hoggar
Parc d'Activité de Courtabœuf
BP 112
F-91944 Les Ulis Cedex A
France

Phone: +33 (0) 1 69 18 75 75
Fax: +33 (0) 1 69 28 84 91

contact-edps@webofconferences.org

Additional copies of this publication are available from:

Curran Associates, Inc.
57 Morehouse Lane
Red Hook, NY 12571 USA
Phone: 845-758-0400
Fax: 845-758-2633
Email: curran@proceedings.com
Web: www.proceedings.com

TABLE OF CONTENTS

X^{th} Scientific Conference
Air Protection in Theory and Practice

18-21st October 2017, Zakopane (Poland)

Editors:

Marianna Czaplicka, Oskar Czechowski, Janina Fudała, Magdalena Jabłońska-Czapla, Joanna Kyzioł-Komosińska, Grzegorz Majewski, Katarzyna Juda-Rezler, Wioletta Rogula-Kozłowska, Izabela Sówka

Organizers:

Institute of Environmental Engineering Polish Academy of Sciences in cooperation with the scientific network AIRCLIM-NET

AIRCLIM-NET

PREFACE

For many years, the problem of air pollution has been perceived as an important risk factor for human health, decreasing the quality of life and worsening the condition of the natural environment. Recently, more and more attention has been paid not only to air pollution on a local scale, but also to the assessment of transboundary transport of air pollutants and global phenomena, such as depletion of ozone layer or the greenhouse effect. The development of instrumental methods of analysis, as well as new measurement methods dedicated to the control of pollutants immission and emission, enabled the detection of ultrafine aerosol particles and even the trace concentrations of gaseous pollutants, which in the long term allows us to assess the risks associated with the presence of toxic micro-contaminants in the atmosphere. The progress in the use of effective methods for dust removal and utilization of gaseous pollutants in the sector of power generation and industry, combined with increasingly restrictive legal regulations in the field of atmosphere protection, led to a significant reduction of emissions of major air pollutants. Nevertheless, a number of problems, resulting e.g. from the conventional combustion of solid fuels or air pollution in cities, still remain unresolved. Meanwhile, the current state of knowledge allows to generate more and more new statistics on the current or forecasted number of cardiovascular and respiratory diseases caused by inhalation of polluted air. According to the above, all actions aimed at exchanging information and experiences regarding the widespread air protection against pollution are perfectly justified. An example of such activities was the Xth edition of the scientific conference entitled "Air Protection in Theory and Practice", held in Zakopane (Poland), on 18-21st October 2017. The conference was organized by the Institute of Environmental Engineering Polish Academy of Sciences located in Zabrze and gathered 70 delegates from the whole country and abroad (38 oral presentations and 26 posters were presented).

The aim of the conference was to present the scientific achievements and to exchange views on the following topics: air pollution and climate change, emissions of dust and gas pollution, the spread and transformation of air pollutants, methods used for emission reduction, alternative energy sources and cleaner energy technologies, legal aspects of emission control and problems connected with air quality management. The conference enabled the exchange of experiences and gave an opportunity to the discussions on the coordination of further research related to air protection and application activities.

The Organizing Committee would like to express the warmest thanks to all Authors of the conference publications for the effort put into their elaboration, and to the Reviewers for the reliable evaluation of the works.

On behalf of the Organizing Committee
Chairman, Prof. Dsc PhD. Eng. Marianna Czaplicka

edp sciences

Web of Conferences

Statement of Peer review

In submitting conference proceedings to *Web of Conferences*, the editors of the proceedings certify to the Publisher that

1. They adhere to its **Policy on Publishing Integrity** in order to safeguard good scientific practice in publishing.
2. All articles have been subjected to peer review administered by the proceedings editors.
3. Reviews have been conducted by expert referees, who have been requested to provide unbiased and constructive comments aimed, whenever possible, at improving the work.
4. Proceedings editors have taken all reasonable steps to ensure the quality of the materials they publish and their decision to accept or reject a paper for publication has been based only on the merits of the work and the relevance to the journal.

Title, date and place of the conference: X-th Scientific Conference Air Protection in Theory and Practice, 18-21'st October 2017, Zakopane

Proceedings editor(s): Janina Fudala

Date and editor's signature 12.12.2017r

17, avenue du Hoggar - PA de Courtaboeuf – BP 112 - 91944 Les Ulis Cedex A (France)
Tél. : 33 (0)1 69 18 75 75 - Fax : 33(0)1 69 07 45 17 – www.edpsciences.org

The Use of Principal Component Analysis for Source Identification of $PM_{2.5}$ from Selected Urban and Regional Background Sites in Poland

Barbara Błaszczak[1,*]

[1] Institute of Environmental Engineering, Polish Academy of Sciences, 34 M. Skłodowska-Curie St., 41-819 Zabrze, Poland

Abstract. The paper reports the results of the measurements of water-soluble ions and carbonaceous matter content in the fine particulate matter ($PM_{2.5}$), as well as the contributions of major sources in $PM_{2.5}$. Daily $PM_{2.5}$ samples were collected during heating and non-heating season of the year 2013 in three different locations in Poland: Szczecin (urban background), Trzebinia (urban background) and Złoty Potok (regional background). The concentrations of $PM_{2.5}$, and its related components, exhibited clear spatiotemporal variability with higher levels during the heating period. The share of the total carbon (TC) in $PM_{2.5}$ exceeded 40% and was primarily determined by fluctuations in the share of OC. Sulfates (SO_4^{2-}), nitrates (NO_3^-) and ammonium (NH_4^+) dominated in the ionic composition of $PM_{2.5}$ and accounted together ~34% (Szczecin), ~30% (Trzebinia) and ~18% (Złoty Potok) of $PM_{2.5}$ mass. Source apportionment analysis, performed by PCA-MLRA model (Principal Component Analysis – Multilinear Regression Analysis), revealed that secondary aerosol, whose presence is related to oxidation of gaseous precursors emitted from fuel combustion and biomass burning, had the largest contribution in observed $PM_{2.5}$ concentrations. In addition, the contribution of traffic sources together with road dust resuspension, was observed. The share of natural sources (sea spray, crustal dust) was generally lower.

1 Introduction

Air pollution by particulate matter (PM) is one of the most serious environmental problems of all developed and developing countries [1]. Much attention is paid to fine PM factions – $PM_{2.5}$ (particles with aerodynamic diameter ≤ 2.5 μm), mainly due to their harmful impact on human health [2, 3]. What's more, fine particulates are capable of long-term persistence in the air, affect climate, ecological conditions and visibility [4, 5].
$PM_{2.5}$ are complex mixtures, originating either directly from natural and anthropogenic sources, such as biomass burning, vehicle exhaust and industry, or as secondary pollutants through heterogeneous chemical reactions in the atmosphere [6-8]. Carbonaceous matter (elemental carbon – EC, organic carbon – OC) and inorganic ions, have the highest

* Corresponding author: barbara.blaszczak@ipis.zabrze.pl

contribution to PM mass [9, 10]. The latter, especially sulfates (SO_4^{2-}), nitrates (NO_3^-) and ammonium (NH_4^+), are an important constituent of the secondary inorganic aerosol (SIA) [11, 12]. This component greatly affects the concentration and composition of fine PM, both in areas remote from emission sources, as well as in urban areas [10-14].
The PM concentrations and chemical composition can vary significantly in time and space, which depend on the main emission sources and chemical reactions occurring in the atmosphere [12, 15]. Therefore, it is necessary to further study the chemical characteristics and sources of $PM_{2.5}$ in as many sites as possible. The research on a spatial and seasonal variation of PM constituents in different types of measurement sites is particularly important for Central and Eastern Europe where knowledge about PM chemical compositions is still very limited [16].
Characterizing and evaluating the contribution of the different aerosol sources can constitute the basis for planning efficient strategies of air quality management [4]. Commonly used methods of source apportionment are receptor models (RMs), which use the chemical speciation of aerosols collected at the receptor to infer the contribution of the possible sources to the ambient concentration of particulate matter and their chemical composition [7]. Several receptor models have been devised and widely used, such as chemical mass balance [17], positive matrix factorization [18] and principal component analysis [17, 19].
Taking into account the above considerations, the aim of the following study was to analysed concentrations and chemical compositions of $PM_{2.5}$ samples from 3 different measurement sites in Poland. The preliminary assessments of the $PM_{2.5}$ sources were also performed and its relative contribution to observed $PM_{2.5}$ concentrations were determined.

2 Materials and methods

2.1 PM2.5 sampling and chemical analysis

Research material were $PM_{2.5}$ samples from 3 air quality monitoring stations pertinent to Regional Inspectorate for Environmental Protection (RIEP), located in Trzebinia (urban background), Szczecin (urban background) and Zloty Potok (rural background) [Fig. 1]. The study included 2 periods of 2013 year, representing the heating (I-III) and non-heating (V-VII) season. 24 hours PM samples were collected on quartz fiber filters with the use of low-volume samplers. Mass concentration of $PM_{2.5}$ was determined by the gravimetric method, according to the standard PN-EN 14907:2006b. Ambient air quality – Standard gravimetric measurement method for the determination of the $PM_{2.5}$ mass fraction of suspended particulate matter.
The methodology followed was already presented in details in previous papers [eg. 10] and is briefly described below. A 1.5-cm^2 piece was cut out from each filter and was analysed for the OC and EC concentrations. Carbonaceous matter content was determined by thermal-optical carbon analyser with FID detection – Model 4L Main Oven Assembly, manufactured by Sunset Laboratory Inc. In this study, the analysis was performed using the "EUSAAR_2" protocol, which has been developed as a proposed standard method for European measurement stations under the EUSAAR (European Supersites for Atmospheric Aerosol Research) project [20]. The remaining part of the filter was used to analyse water-soluble inorganic ions: Cl^-, NO_3^-, SO_4^{2-}, Na^+, NH_4^+, K^+, Ca^{2+}, and Mg^{2+}. The ion content was determined by ion chromatograph manufactured by Herisau Metrohm AG (Switzerland), equipped with conductivity detector. Chromatographic separation was performed on anion- and cation-exchange columns (respectively Metrosep A Supp 3 or 5 and Metrosep C3 or C4).

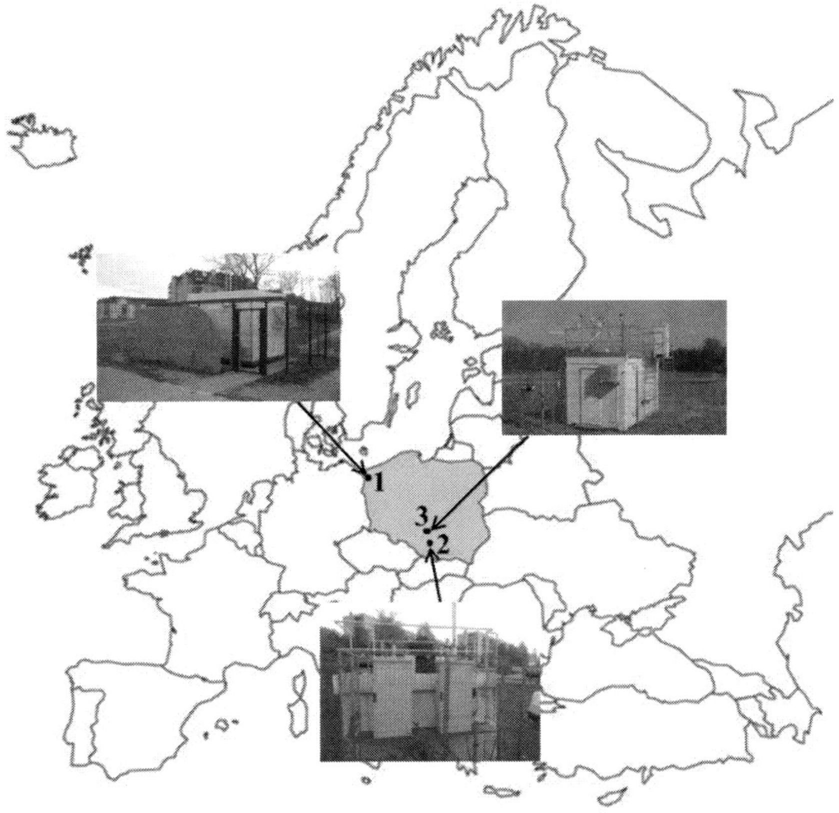

Fig. 1. Location of measurement sites: 1) Szczecin (14°39'48''E; 53°22'51''N); 2) Trzebinia (19°28'41''E; 50°09'23''N); 3) Złoty Potok (19°27'29''E; 50°42'38''N)

2.2 Principal Component Analysis/Multiple Linear Regression (PCA-MLR)

A preliminary source identification study of the water soluble ionic species and the carbonaceous matter was carried out by the principal component analysis coupled with multilinear regression analysis (PCA-MLRA) [21]. The PCA-MLR is an important receptor model that has been applied in several studies [17, 19, 22]. This model does not require information on source profiles. The source categories can be identified according to the PM ambient dataset, and the contribution of each source category can be also estimated [22].
All the computations were performed using the software package Statistica 12.0 (Stat Soft company). PCA was performed separately for both measuring periods. The necessity of separate treatment of heating and non-heating seasons results from the differences in the meteorological situation and changes in the emission profile (especially increased activity of anthropogenic sources during heating season) which have together a significant impact on PM concentrations and chemical composition.
Eleven parameters were included in the analysis (NO_3^-, SO_4^{2-}, NH_4^+, Na^+, Cl^-, K^+, Mg^{2+}, Ca^{2+}, EC, SOC i POC). The exceptions were $PM_{2.5}$ samples from Złoty Potok, where Ca^{2+} and Mg^{2+} ions were not taken into account, due to a large number of results below the detection limit. Bearing in mind the chemical behaviour of organic carbon [11], the secondary organic carbon (SOC) and primary organic carbon (POC) were considered as independent variables. The method dor the determination of SOC and POC content in $PM_{2.5}$ was specified in previous works [11].

In this study, the decision of selecting the number of principal components relied on Kaiser criterion (factors having an eigenvalue greater than 1.0) [21]. The interpretation of the principal components was based on the variables with factor loadings with absolute values greater than 0.6. The contribution of each source to the PM burden was quantitatively assessed by the means of MLRA procedure proposed in [21].

3 Results and discussion

3.1 Characterization of $PM_{2.5}$ and major components

The daily mean concentrations of $PM_{2.5}$, as the concentration of its related components, ranged in wide limits of values and exhibited spatial and seasonal variations. Much higher $PM_{2.5}$ concentrations were found during the heating period than the non-heating one, with the values respectively: 24.4 and 8.7 μg/m^3 (Szczecin), 39.2 and 16.7 μg/m^3 (Trzebinia), 34.0 and 16.3 μg/m^3 (Złoty Potok). These differences can be attributed to unfavourable meteorological conditions in winter (lower wind speeds and mixing heights, frequent temperature inversions) and, what is more important, an increase in the activity of local emission sources of PM, especially biomass burning and fossil fuel combustion in domestic furnaces and in small local heating plants [10-12].
The concentrations of total carbon (TC) in $PM_{2.5}$, averaged for a whole measuring period, equalled 6.65 μg/m^3, 11.03 μg/m^3 and 10.05 μg/m^3, respectively for Szczecin, Trzebinia and Złoty Potok. The total carbon (TC) content in $PM_{2.5}$ mass exceeded generally 40% and was primarily determined by fluctuations in the share of OC, which was generally above 30% of the $PM_{2.5}$ mass. On average, water-soluble ions contribute above one-third by weight to $PM_{2.5}$, with the exception of a rural background station in Złoty Potok (averaged contribution ~20%). At all measurement sites, ionic composition of $PM_{2.5}$ was dominated by SO_4^{2-}, NO_3^-, and NH_4^+. Combined share of the above mentioned ions was (on average) ~34%, ~30% and ~18% of $PM_{2.5}$ mass, respectively in Trzebinia, Szczecin and Zloty Potok. The SIA mass contributions to $PM_{2.5}$ at all sites were similar during heating and non-heating period, opposite to TC contribution, which was quite higher during heating period.
The obtained results were compared with the values observed at selected urban and regional background sites in Europe [Fig. 1.]. It should be mentioned that concentrations of main components in analysed $PM_{2.5}$ samples were comparable for Trzebinia and Zloty Potok sites from Southern Poland – one of the regions of Poland with the highest degree of urbanization and pollution of all the components of the environment [11].
The combined share of secondary inorganic ions in $PM_{2.5}$ from investigated areas was lower (eg. Birmingham [14], Corso Firenze [24], Melpitz [27]) or comparable (Madrid [25], Barcelona [26], Rotterdam [23]), conversely to TC shares. Mean seasonal contribution of TC in $PM_{2.5}$ from Szczecin and Trzebinia sites were relatively higher than those observed in urban background sites specified in Fig. 1. In the case of regional background site in Złoty Potok the higher TC share in $PM_{2.5}$, in comparison with other similar sites, was particularly evident. The dominance of carbonaceous matter content in $PM_{2.5}$ mass from Polish measurement sites is due to the unique position of coal as an energy source, especially in power generation sector, and the wide use of biomass (mainly in rural areas) in the low-efficient furnaces, which results in the combustion conditions favorable the emissions of substances rich in organic carbon. Moreover, the contribution of industrial emission (e.g. cocking plants, iron works, waste incineration plants) to particulate pollution in Poland is also high against other European countries [10-12, 19].

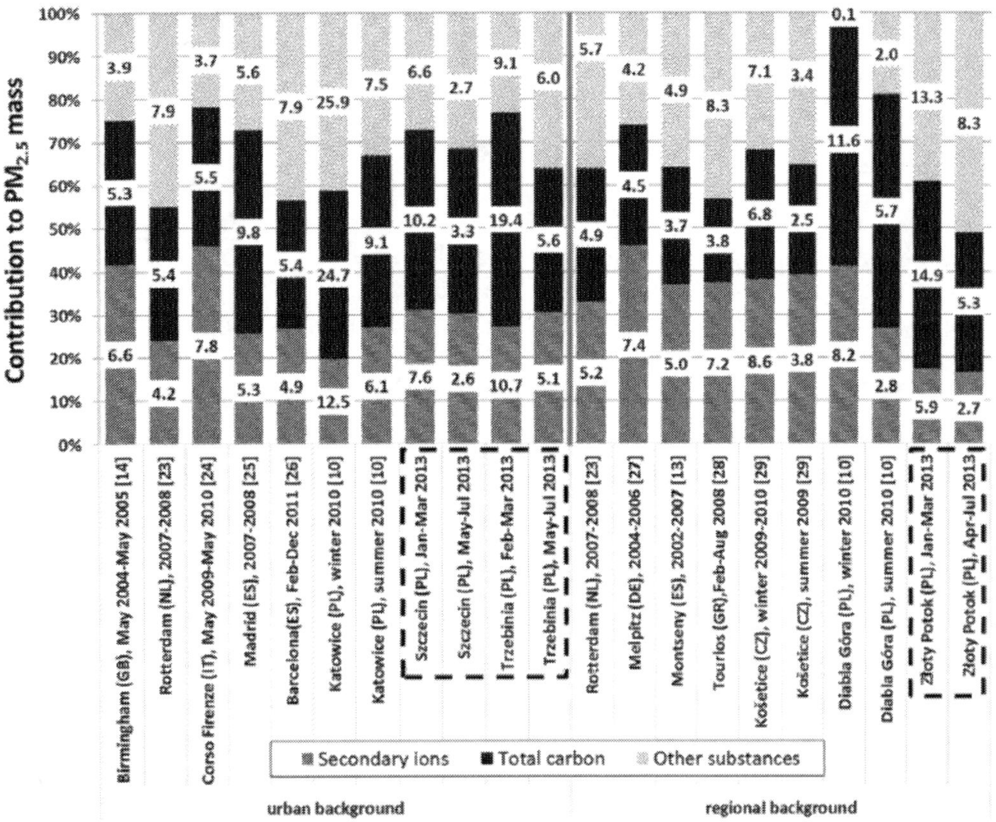

Fig. 1. Contributions of secondary inorganic ions and total carbon to fine particulate matter mass from selected urban and regional background sites in Europe
(the grey line on the graph separates urban and regional background sites; 3 sites under the consideration of this study are distinguished by the black dotted lines)

3.2 Source identification of $PM_{2.5}$

Table 1 shows a number of principal components (PC) for $PM_{2.5}$ samples from analysed measurement sites. The main statistical parameters of extracted components – initial eigenvalue, and variance explained (%), as well as the contribution of these components to observed $PM_{2.5}$ concentrations, was also presented.
In the case of Szczecin site, PCA-MLRA model allowed to extract 3 and 4 principal components, respectively in the heating and non-heating period, explaining up to ~76% and ~86% of the total variance. During the heating period, the main contribution in the $PM_{2.5}$concentrations showed sea spray (PC3: Na^+, Cl^-), with the average value ~46%. The first component (NO_3^-, SO_4^{2-}, NH_4^+, and SOC) (~28%), suggested the role of secondary aerosols, which results from the oxidation of gaseous precursors emitted from fuel combustion processes [19]. A comparable contribution (average: ~26%) was noted for vehicle emissions along with road dust resuspesion (PC2: EC, POC, Ca^{2+}, Mg^{2+}). During the non-heating period, the highest share in the observed $PM_{2.5}$ concentration (~47%) was identified for secondary aerosol (PC2 : SO_4^{2-}, NH_4^+, SOC). The combustion of solid fuels, identified by PC3 (EC, POC), equalled on average ~28%. The share of secondary nitrate (PC4: NO_3^-), as well as mixed natural sources (sea salt and mineral dust) (PC1: Cl^-, Na^+, K^+, Ca^{2+}, Mg^{2+}) was significantly lower.

Table 1. Varimax rotated factor loadings for $PM_{2.5}$ ambient dataset in Szczecin, Trzebinia and Złoty Potok

Szczecin	**Heating season**			**Non-heating season**			
No. of components	**PC 1**	**PC 2**	**PC 3**	**PC 1**	**PC 2**	**PC 3**	**PC 4**
Components	**NH_4^+, SO_4^{2-}, NO_3^-, SOC**	**Mg^{2+}, Ca^{2+}, EC, POC**, *K^+*	**Na^+, Cl^-**	**Cl^-, K^+, Mg^{2+}, Ca^{2+}, Na^+**	**NH_4^+, SOC, SO_4^{2-}**	**EC, POC**	**NO_3^-**
Initial eigenvalue	5.53	1.57	1.24	4.94	1.82	1.67	0.99
Variance explained	50.30	14.28	11.27	44.92	16.56	15.21	9.00
Contribution [%]	**27.83**	**25.73**	**46.45**	**10.02**	**47.28**	**28.56**	**9.11**
Trzebinia	**Heating season**			**Non-heating season**			
No. of components	**PC 1**	**PC 2**	**PC 3**	**PC 1**	**PC 2**	**PC 3**	
Components	**NH_4^+, SO_4^{2-}, NO_3^-,** *K^+*	**EC, POC**	**SOC, Cl^-, Na^+,** *Ca^{2+}, Mg^{2+}, K^+*	**EC, POC, K^+, Na^+**	**NH_4^+, SO_4^{2-}, Ca^{2+},** *SOC*	**NO_3^-, Cl^-**	
Initial eigenvalue	5.08	1.68	1.62	3.42	1.95	1.46	
Variance explained	46.16	15.28	14.73	31.08	17.73	13.30	
Contribution [%]	**48.23**	**9.06**	**42.71**	**67.84**	**23.45**	**8.71**	
Złoty Potok	**Heating season**			**Non-heating season**			
No. of components	**PC 1**	**PC 2**		**PC 1**	**PC 2**	**PC 3**	
Components	**NH_4^+, SO_4^{2-}, NO_3^-, SOC, K^+**	**Cl^-, Na^+, EC, POC**		**EC, POC, NH_4^+,** *NO_3^-*	**SOC, SO_4^{2-},** *Na^+*	**Cl^-, Na^+,** *NO_3^-*	
Initial eigenvalue	5.98	1.09		3.42	1.58	1.27	
Variance explained	66.40	12.13		37.97	17.51	14.12	
Contribution [%]	**44.28**	**55.72**		**51.63**	**38.28**	**2.38**	

*Note: The variables with factor loadings greater than 0.6 are in bold

At Trzebinia site, 3 principal components were identified both in the heating and non-heating period, explaining ~76% and ~62% of the total variance, respectively. During the heating period, the dominant sources of $PM_{2.5}$ were associated with the combustion of fuels, represented by PC1 (~48%) and PC3 (~43%). PC1 was high loaded with NO_3^-, SO_4^{2-}, NH_4^+ and K^+, and therefore might showed a combined contribution of secondary inorganic aerosol and biomass burning. PC3 (SOC, Cl^-, Na^+, K^+, Ca^{2+}, Mg^{2+}) would pointed out the role of municipal combustion sources (presence of fly ashes particles, typically enriched in alkaline metals). Moreover, the contribution of primary emissions (combustion of fuels eg. in vehicle engines) (PC2: EC, POC) was found. During the non-heating period, the largest contributor in observed $PM_{2.5}$ concentrations (~68%) was a vehicle traffic accompanied by road dust resuspension (P1: EC, POC, Na^+, K^+). Secondary aerosol (PC2:

SO_4^{2-}, NH_4^+, Ca^{2+}, SOC), originating from transformations of gaseous PM precursors emitted in coal combustion processes, has played a minor role (average: ~23%). Factor 3 (NO_3^-, Cl^-) could represent open burning processes as well as wood combustion and/or waste incineration.
With regard to the regional background station in Złoty Potok PCA-MLRA model allowed to extract 2 and 3 principal components, respectively for the heating and non-heating period, explaining ~56% and ~70% of the total variance. During the heating one, the large source of $PM_{2.5}$ was secondary aerosol formation after gaseous emission from different sources (PC1: NH_4^+, SO_4^{2-}, NO_3^-, SOC; ~44%). The dominant source of $PM_{2.5}$ was traffic sources (road salt and emissions from car engines, often outdated and with high emission rates), represented by PC2 (Na^+, Cl^-, EC, POC ; ~56%). A significant role of traffic sources, along with secondary ammonium nitrate, in the observed $PM_{2.5}$ was found also during the non-heating season – PC1 (NO_3^-, NH_4^+, EC, POC; ~52%). A slightly lower share was recorded for secondary aerosol (PC2: SO_4^{2-}, SOC), with an average contribution ~38%. A small proportion was found for factor 3 (NO_3^-, Na^+, Cl^-), which could present the contribution of wood burning, or, what is more likely, the shares of particles of natural origin, which have been changing in the atmosphere during transportation.

4 Conclusion

$PM_{2.5}$ samples were collected at two urban background (Szczecin, Trzebinia) and one regional background (Złoty Potok) measurement sites and analysed for the main chemical components (OC, EC, NO_3^-, SO_4^{2-}, NH_4^+, Na^+, Cl^-, K^+, Mg^{2+}, Ca^{2+}). Much higher $PM_{2.5}$ concentrations, as well as concentrations of carbonaceous matter and water-soluble ions, were recorded during the heating period, due to the changes in the emission profile and unfavourable meteorological conditions. Comparing the obtained results with the data recorded on different urban and rural background sites across Europe, it was found that SIA contribution in $PM_{2.5}$ mass from investigated areas is lower or comparable, while the share of total carbon was relatively high. In the latter case, it is due to the existing structure of energy consumption in Poland, with a dominance of fossil fuels and the wide use of biomass (especially in rural areas) in the low-efficient furnaces.
PCA-MLRA model allowed for sources identification of $PM_{2.5}$ from investigated areas. The chemical composition of $PM_{2.5}$ appeared to be influenced mainly by secondary aerosols, emitted from fossil fuel combustion, vehicle exhaust, as well as biomass burning. The seasonal variability of secondary aerosol contribution was quite different for selected sites. For urban background site in Szczecin, higher share of this $PM_{2.5}$ source was identified during the heating period. Opposite situation was noted in the case of Trzebinia and Złoty Potok sites, localised in Southern Poland – one of the regions of Poland with the highest degree of urbanization and pollution of all the components of the environment.
A large contribution of traffic sources along with road dust resuspension was also identified by PCA-MLRA model, especially in the case of Złoty Potok site. Natural sources (sea spray, crustal dust) had, in general, a minor contribution to the observed $PM_{2.5}$ composition. The exception was Szczecin site, localised in the West Pomeranian Province – one of the regions of Poland with the lowest emissions of PM and its gaseous precursors.

This work was supported by the Polish National Science Centre partly under PRELUDIUM funding scheme 2nd edition, Project no. UMO-2011/03/N/ST10/05542. The results presented in this work will be included in PhD thesis of the author. The author would like also to thank Prof. Dr Eng. Katarzyna Juda-Rezler, the dissertation supervisor, and Dr Eng. Krzysztof Klejnowski, the auxiliary supervisor, for comprehensive assistance during the realisation of the work, as well as Dr Eng Magdalena Reizer for help in the interpretation of the results and Dr Eng Barbara Mathews for conducting the analysis of $PM_{2.5}$ chemical composition.

References

1. EEA Report No 5/2015. Air Quality in Europe – 2015 report (Luxembourg: Publications Office of the EU, 2015)
2. K.-H Kim, E. Kabir, S. Kabir, Environ. Int. **74** (2015)
3. C.A. Pope, D.W. Dockery, D.W., J. Air Waste Manage. Assoc. **56** (2006)
4. E. Diapouli, M. Manousakas, S. Vratolis, V. Vasilatou, Th. Maggos, D. Saraga, Th. Grigoratos, G. Argyropoulos, D. Voutsa, C. Samara, K. Eleftheriadis, Atmos Environ **164** (2017)
5. G. Majewski, W. Rogula-Kozłowska, P.O. Czechowski, A. Badyda, A. Brandyk, Atmosphere, **6**(8) (2015)
6. W. Rogula-Kozłowska, B. Kozielska, K. Klejnowski, S. Szopa, *Arch Environ Prot* **39**(1) (2013)
7. C.A. Belis, F. Karagulian, B.R. Larsen, P.K. Hopke, Atmos Environ **69** (2013)
8. G. Majewski, W. Rogula-Kozłowska, Theor Appl Climatol **125**(1-2) (2016)
9. J.C. Chow, D.H. Lowenthal, L.-W.A. Chen, X. Wang, J.G. Watson, *Air Qual Atmos Health* **8** (2015)
10. W. Rogula-Kozłowska, K. Klejnowski, P. Rogula-Kopiec, L. Ośródka, E. Krajny, B. Błaszczak, B. Mathews, *Air Qual Atmos Health* **7** (2014)
11. B. Błaszczak, W. Rogula-Kozłowska, B. Mathews, K. Juda-Rezler, K. Klejnowski, P. Rogula-Kopiec, Aerosol Air Qual Res **16**(10) (2016)
12. B. Błaszczak, K. Juda-Rezler, W. Rogula-Kozłowska, M. Reizer, B. Mathews, K. Maciejewska, K. Klejnowski, Environ Eng Sci **34**(4) (2017)
13. J. Pey, N. Pérez, S. Castillo, M. Viana, T. Moreno, M. Pandolfi, J.M. López-Sebastián, A. Alastuey, X. Querol, Atmos Res **94** (2009)
14. J. Yin, R.M. Harrison, Atmos Environ **42** (2008)
15. K. Dimitrou, E. Remoundaki, E. Mantas, P. Kassomenos, Atmos Environ **116** (2015)
16. EMEP Status report 1/2016. Transboundary particulate matter, photo-oxidants, acidifying and eutrophying components. Norwegian Meteorological Institute (2016)
17. G.-L. Shi, F. Zeng, X. Li, Y.-C. Feng, Y.-Q. Wang, G.-X. Liu, T. Zhu, Atmos Environ **45** (2011)
18. M. Manousakas, H. Papaefthymiou, E. Diapouli, A. Migliori, A.G. Karydas, I. Bogdanovic-Radovic, K. Eleftheriadis, Sci Total Environ **574** (2017)
19. M. Reizer, K. Juda-Rezler, Air Qual Atmos Health, **9**(5) (2016)
20. P. Prati, A. Zucchiatti, F. Lucarelli, P.A. Mandò, Atmos Environ **34** (2000)
21. G.D. Thurston, J.D. Spengler, Atmos Environ **19** (1985)
22. J. Zhou, Z. Xing, J. Deng, K. Du, Atmos Environ **135**, 20-30 (2016)
23. M. Keuken, H.D. van der Gon, K. van der Valk, Sci Total Environ **408** (2012)
24. E. Cuccia, D. Massabò, V. Ariola, M. Bove, P. Fermo, A. Piazzalunga, P. Prati, P., Atmos Environ **67** (2013)
25. P. Salvador, B. Artíñano, M. Viana, A. Alastuey, X. Querol, Atmos Environ **57** (2012)
26. N. Pérez, J. Pey, C. Reche, J. Cortés, A. Alastuey, X. Querol, Sci Total Environ **571** (2016)
27. G. Spindler, A. Grüner, K. Müller, S. Schlimper, H. Herrmann, J Atmos Chem **70** (2013)
28. St. Pateraki, V.D. Assimakopoulos, A. Bougiatioti, G. Kouvarakis, N. Mihalopoulos, Ch. Vasilakos, Sci Total Environ **424** (2012)
29. J. Schwarz, M. Cusack, J. Karban, E. Chalupníčková, V. Havránek, J. Smolík, V. Ždímal, Atmos Res **176-177** (2016)

Dust emission from wet, low-emission coke quenching process

Bogusław Komosiński[1], *Bartłomiej* Bobik[1], *Tomasz* Konieczny[1], and *Ewelina* Cieślik[1]
[1]Polish Academy of Sciences, Institute of Environmental Engineering, M. Skłodowskiej-Curie 34, 41-819 Zabrze, Poland

Abstract: Coke plants, which produce various types of coke (metallurgical, foundry or heating), at temperatures between 600 and 1200°C, with limited access to oxygen, are major emitters of particulates and gaseous pollutants to air, water and soils. Primarily, the process of wet quenching should be mentioned, as one of the most cumbersome. Atmospheric pollutants include particulates, tar substances, organic pollutants including B(a)P and many others. Pollutants are also formed from the decomposition of water used to quench coke (CO, phenol, HCN, H_2S, NH_3, cresol) and decomposition of hot coke in the first phase of quenching (CO, H_2S, SO_2) [1]. The development of the coke oven technology has resulted in the changes made to different types of technological installations, such as the use of baffles in quench towers, the removal of nitrogen oxides by selective NO_x reduction, and the introduction of fabric filters for particulates removal. The BAT conclusions for coke plants [2] provide a methodology for the measurement of particulate emission from a wet, low-emission technology using Mohrhauer probes. The conclusions define the emission level for wet quenching process as 25 g/Mg_{coke}. The conducted research was aimed at verification of the presented method. For two of three quench towers (A and C) the requirements included in the BAT conclusions are not met and emissions amount to 87.34 and 61.35 g/Mg_{coke} respectively. The lowest particulates emission was recorded on the quench tower B and amounted to 22.5 g/Mg_{coke}, therefore not exceeding the requirements.

1 Legal basis for the measurement of particulate matter and gaseous pollutants emission

1.1 Directive 2010/75/EU of the European Parliament and of the Council of 24 November2010 on industrial emissions (integrated pollution prevention and control)

EU Member States, and therefore Poland, were obliged to transpose the provisions of the Directive by 7 January 2013. As a consequence, Poland has prepared amendments to a

[1]Corresponding author: boguslaw.komosinski@ipis.zabrze.pl

number of legislative acts by changing the existing rules on integrated permits, emission standards for installations and land protection. The most important changes were introduced in the Law on Environmental Protection and some other laws. As a result of the legislation introduced in 2011-2013, the existing Community legislation included in the IED [5] has been harmonized and consolidated, replacing six industrial standards [6-11]:

a) Directive 2008/1/EC of the European Parliament and of the Council of 15 January 2008 concerning integrated pollution prevention and control,
b) Directive 2000/76/EC of the European Parliament and of the Council of 4 December 2000 on the incineration of waste,
c) Directive 2001/80/EC of the European Parliament and of the Council of 23 October 2001 on the limitation of emissions of certain pollutants into the air from large combustion plants,
d) Directive 1999/13/EC on the limitation of emissions of volatile organic compounds due to the use of organic solvents in certain activities and installations,
e) Directives 78/176/EWG, 82/186/EWG, 92/112,EWG,
f) Regulation (EC) No 166/2006 of the European Parliament and of the Council of 18 January 2006 concerning the establishment of a European Pollutant Release and Transfer Register.

National legislation imposes:

a) integrated permits,
b) permission for the introduction of dusts and gases,
c) Ordinance of the Minister of the Environment of 4 November 2014 on emission standards for certain types of installations, sources of fuel combustion and devices for incineration or co-incineration of waste.

New directives and regulations aim to improve the prevention and control of pollution caused by industrial activities. They also aim to better the environment as a result of the reduction of industrial emissions and the achievement of the objectives set out in the thematic strategies.

1.2 Periodic measurements of particulates emission from coke quench towers in the light of the BAT conclusions

In point 51 of the BAT conclusions for coke oven plants [2], the emission reduction takes place using conventional wet, low-emission quenching technology. The permissible emission, determined as an average during the sampling period, is 25 g/Mg_{coke}. The Mohrhauer method, in accordance with the VDI 2303 standard, is the measurement method.

2 Coke quench tower

The final stage of coke production is its wet quenching. After hot coke is pushed from the oven, the quench car travels by rail to a quench tower. The lower part of the quench tower consists of the quench car chamber, and the upper one of the quench chimney. The large steam plume appears from spraying water onto hot coke. The stack effect of the tower and the air suction holes directs the air, water and particles into the atmosphere.

15 to 25 Mg of coke is pushed from the oven(depending on the type), and this amount is subjected to quenching. The water is sprayed onto hot coke from the nozzles, and after quenching the car leaves the tower [12, 13].

To remove particulates and tar substances from the steam plume and to unify the stream, baffles are installed inside the quench towers.

Fig. 1. Coke quench tower.

2.1 Measurement method

The methodology for manual determination of particulate emissions from wet quenching using Mohrhauer probes (Mohrhauer's non-isokinetic method), was developed on the basis of German standard VDI 2303 *"Limiting of dust emissions in the extinguishing of coke; coking plants and gas works"* [3], in accordance with the BAT Conclusions for Coke Oven Plants [2].

Specialized measuring equipment and laboratory facilities are required to determine the concentration of dust discharged into the atmosphere from the quench tower. Materials from which the measuring instruments should be made are the basis for proper measurements.

Non-isokinetic Mohrhauer method is used to measure particulates emission from quench towers, from which the steam plume has the parameters given in Table 1. It is adapted for exhaust gases containing significant amounts of water vapor. This method is dedicated to conventional quench towers with minimized particulates emission.

Table 1. Steam plume parameters.

Parameter	Unit	Range
H_2O	kg H_2O/kg$_{dry\ gas}$	0.8-2.0
Water vapor emission	Kg/Mg$_{coke}$	250-500
Dust emission	g/Mg$_{coal}$	20-800
Flue gas velocity	m/s	0-12

Fig. 2. Top of a quench tower with Mohrhauer's probes installed.

Mohrhauer method consists of capturing a part of the steam plume by the probe. Vapor enters the probe through the square-shaped inlet port. Vapor leaves the probe through the outlet with a fabric (e.g. gauze) with minimal flow resistance. The dust deposited in the probe is transferred quantitatively to the containers after removing the back cover of the probe. The deposition of particulates in the probe is caused by a change of flow direction, centrifugal force, decreasing vapor flow rate, and filter effect of the fabric [14]. Measurements are carried out at the top of a quench tower during the quenching process. In order to achieve satisfactory accuracy, it is recommended to carry out a measurement for 6 - 10 coke quenchings. Total time of measurement may vary as it depends on the production efficiency of a coke plant – usually there are 4-5 quenchings per hour, each lasting for about 5 minutes. Particulates settled on walls of a probe are flushed with water and poured into a condensate container. To determine the amount of dust deposited in the probes, the dust is quantitatively removed from the condensate, dried and weighed using paper filters. Filters are conditioned. The conditioning of filters is carried out at 105°C for 2 hours. Subsequently, the filters are cooled in a silica gel desiccator and weighed on a weight with accuracy of 0.001 g. The conditioning and weighing process is repeated until a constant filter mass is obtained.

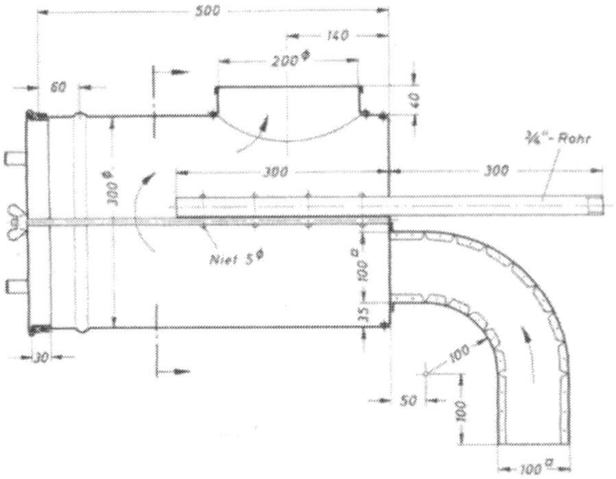

Fig. 3. Mohrhauer's probe [3].

Emission is calculated from the following formula:

$$E_p = \frac{b \cdot c}{d \cdot e \cdot f} \tag{1}$$

- E_p – emission, g/Mg_{coke},
- b – mass of precipitated dust, g,
- c – area of the quench tower outlet, m^2,
- d – area of the probe's inlet, m^2,
- e – number of coke quenchings,
- f – weight of coke pushed from the oven, Mg.

3 Results

Measurements were carried out on three quench towers that were adapted for the use of probes. There were 6 independent measurements of 8 coke quenchings on the two towers and 3 independent measurements of 6 coke quenchings on the third tower.

Tables 2, 3 and 4 show the emission in grams of dust per tonne of quenched coke, as well as the exceedances of the permissible value of 25 g/Mg_{coke} as established by the BAT conclusions. The research carried out indicates that the highest particulates emission was recorded on the quench tower A - 98 g/Mg_{coke}, which exceeded the allowable limit by 3.9. Although the tower is one of the most modern, the exceedances may be due to the type of coke produced, distribution of quenching water, or the pollution of water [14, 15]. Requirements were also exceeded in the case of the quench tower C – the average emission amounted to 61.35g/Mg_{coke}, exceeding the limit by 2.5.

The lowest particulates emission was recorded on the quench tower B, whose construction (including the battery) was completed in 2012. The emission was 22.5 g/Mg, not exceeding the requirement included in the BAT conclusions. The emission values for the towers A, B and C are shown in Figure 5 and presented in Table 2, 3 and 4.

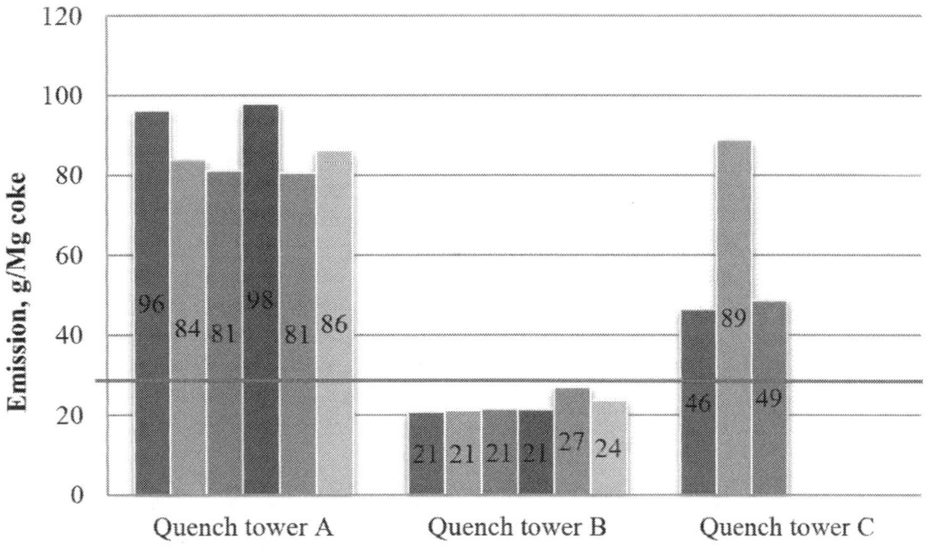

Fig. 4. Particulates emission from the analyzed towers.

Table 2. Results for the quench tower *A*.

Quench tower *A*								
Number of quenchings	8	--						
Mass of coke	196.8	Mg						
Area of the quench tower outlet	89.1	m^2						
Area of the probe's inlet	0.06	m^2						
Measurement number			1	2	3	4	5	6
Emission	g/Mg_{coke}		95.16	83.9	81.24	97.9	80.58	85.27
Average emission	g/Mg_{coke}		**87.34**					
Exceedance of the permissible value of 25 g/Mg_{coke}			3.8	3.4	3.2	3.9	3.2	3.4

Table 3. Results for the quench tower *B*.

Quench tower *B*								
Number of quenchings	8	--						
Mass of coke	136	Mg						
Area of the quench tower outlet	49.82	m^2						
Area of the probe's inlet	0.06	m^2						
Measurement number			1	2	3	4	5	6
Emission	g/Mg_{coke}		20.66	21.14	21.37	21.32	26.87	23.62
Average emission	g/Mg_{coke}		**22.5**					
Exceedance of the permissible value of 25 g/Mg_{coke}			0.8	0.8	0.9	0.9	1.1	0.89

Table 4. Results for the quench tower *C*.

Quench tower *C*					
Number of quenchings	6	--			
Mass of coke	123	Mg			
Area of the quench tower outlet	89.1	m^2			
Area of the probe's inlet	0.06	m^2			
Measurement number			1	2	3
Emission	g/Mg_{coke}		46.43	88.91	48.62
Average emission	g/Mg_{coke}		**61.35**		
Exceedance of the permissible value of 25 g/Mg_{coke}			1.9	3.6	1.9

4 Conclusions

During the measurements it was determined that for two of the three quench towers analyzed, the emission limit of 25 g/Mg_{coke} was exceeded. Average emissions for towers A and C amounted 87.34 and 61.35 g/Mg_{coke} respectively. Due to the same coke quenching technology and the same measurement method, variations in the results may be due to the type of coke produced, water distribution technology and contamination of water. It was also found that the requirements presented in the BAT conclusions were met only in the case of the quench tower B, which was built in 2012. It should be noted that the need to meet the requirements included in the BAT conclusions will force further improvements in coke quenching technology as well as further emission measurements.

References

1. J. Hehlmann, B. Marcinkiewicz, J. Kapała: *Mokre gaszenie koksu w wieżach z wypełnieniem komórkowym,* Arch. of Envir. Prot. **25**, 33 (1999)
2. 2012/135/EU: Commission Implementing Decision of 28 February 2012 establishing the best available techniques (BAT) conclusions under Directive 2010/75/EU of the European Parliament and of the Council on industrial emissions for iron and steel production (notified under document C(2012) 903)
3. VDI 2303 *Limiting of dust emissions in the extinguishing of coke; coking plants and gas works* (1966)
4. Polish Standard PN-64/Z-01001 (1964)
5. Directive 2010/75/EU of the European Parliament and of the Council of 24 November 2010 on industrial emissions (integrated pollution prevention and control)
6. Directive 2008/1/EC of the European Parliament and of the Council of 15 January 2008 concerning integrated pollution prevention and control
7. Directive 2000/76/EC of the European Parliament and of the Council of 4 December 2000 on the incineration of waste
8. Directive 2001/80/EC of the European Parliament and of the Council of 23 October 2001 on the limitation of emissions of certain pollutants into the air from large combustion plants
9. Directive 1999/13/EC on the limitation of emissions of volatile organic compounds due to the use of organic solvents in certain activities and installations
10. Directives 78/176/EWG
11. Regulation (EC) No 166/2006 of the European Parliament and of the Council of 18 January 2006 concerning the establishment of a European Pollutant Release and Transfer Register
12. K. Klejnowski, *Najlepsze dostępne techniki (BAT). Wytyczne dla branży koksowniczej,* (Wydawnictwo Instytutu Chemicznej Przeróbki Węgla, Zabrze 2006)
13. M. Tatara, A. Tatara, W. Hummer, B. Komosiński, *Proces gaszenia koksu metodą mokrą, stan obecny i perspektywy rozwoju w świetle zmian wymogów BAT* (Cokemaking Conference 2010, Instytut Chemicznej Przeróbki Węgla, Zakopane 2010)
14. *Particulates' emission from wet coke quenching process,* Research Procedure PB23, edition 2 from 01.02.2016
15. EPA *National Emission Standards for Hazardous Air Pollutants (NESHAP) for Coke Ovens: Pushing, Quenching, and Battery Stacks – Background Information for Proposed Standards. Final Report,* (2001)
16. W. L. Filho, I. Butorina *Approaches to Handling Environmental Problems in the Mining and Metallurgical Regions*, NATO Sci. Ser. **20,** 86 (2002)

E3S Web of Conferences **28**, 01003 (2018) https://doi.org/10.1051/e3sconf/20182801003
Air Protection in Theory and Practice

Re-entrained road dust PM10 emission from selected streets of Krakow and its impact on air quality

Marek Bogacki[1,*], *Marian* Mazur[1], *Robert* Oleniacz[1], *Mateusz* Rzeszutek[1], and *Adriana* Szulecka[1]

[1] AGH University of Science and Technology, Department of Environmental Management and Protection, A. Mickiewicza 30 Av., 30-059 Krakow, Poland

Abstract. Scientific research studies conducted in various parts of the world confirm that PM10 concentrations in urban air depend to a great extent on the resuspension processes of the dust deposited on the road surface. The paper presents the results of the study related to the determination of the re-entrained PM10 emissions from four selected streets of Krakow (Southern Poland) together with the assessment of its impact on air quality. Examined streets are characterised by different traffic intensity (from 500 to over 20 000 vehicles per day) and individual vehicle structure. Dust material sampling and estimation of the PM10 emission were conducted according to the U.S. EPA methodology (AP 42 Fifth Edition). Two variants of sample collection were applied: from the road surface including the area at the curb (4 streets) and from the road surface alone (1 street). The estimates of resuspended road dust emission as well as the reference values derived from the U.S. EPA guidelines were used to assess the impact of this emission on the PM10 levels in the air at the location of one of the analysed streets. This assessment was conducted using the CALINE4 mathematical model. The study showed that the PM10 emissions from the re-entrained road dust can be responsible for up to 25 % in the winter and 50 % in the summer of the total PM10 concentrations in the air near the roads.

1 Introduction

The PM10 fine dust concentrations in the air are determined by the spatio-temporal distribution of its emission as well as the atmospheric propagation factors, which depend on, inter alia, the orography and local meteorological conditions. In urban and industrial regions the exceedances of permitted PM10 levels are commonly observed. This situation frequently results from the presence of numerous low emission sources situated in a small area, usually densely built, which impede the dispersion effectiveness of air pollutants. Particularly susceptible to excessive air pollutant concentrations in the air are the areas located within the street canyons [1-4]. This can be explained by the occurrence of specific

* Corresponding author: bogacki@agh.edu.pl

air circulation conditions that are not conducive to unhindered dispersion of pollutants emitted from the road transport [5-6].

PM10 dust in the air can be of the anthropogenic origins, but it can also result from physiochemical reactions between different substances present in the air (secondary organic and inorganic aerosols) [7]. Numerous studies on the PM10 chemical composition conducted for the areas within street canyons point out that the particulate matter does not solely originate from fuel combustion in motor vehicles, but constitutes a mixture of dusts of different origins and individual chemical compositions [8]. In addition to the organic and elementary carbon, the silicates, carbonates, metal oxides, phosphates, secondary organic and inorganic aerosols can be found in dust particles. However the chemical composition, particle size and morphology are determined strictly by the source of origin [9]. Literature studies indicate that particles of less than 100 nm in diameter are mainly emitted from internal combustion engines and are based on the organic and elementary carbon [10]. Particles with a diameter of 2.5-10 μm and more are emitted primarily from the road surfaces, where the dust originating from industry and energy sector is deposited together with the particulate matter resulting from abrasion of the road surface, car tires and brake pads as well as particles caused by the wind erosion processes and dust loading brought from construction sites, unpaved roads, etc. [7, 11-13]. Passing motor vehicles create a vacuum in the space between the road and car chassis, triggering the entrainment of the dust from the road to the air. Those particles are primarily of the organic origins, including soil dust and organic parts, and the products of abrasion of the vehicle consumables [14].

Many studies conducted worldwide indicate that the resuspended dust emission constitutes a significant share in the PM10 concentrations in the air near the transport routes [6, 8, 14-20]. Research carried out in the US city of Tucson (Arizona) in the winter showed that the fraction of particulate matter from the fuel combustion in car engines and the share of re-entrained road dust ranged from 4-40 % and 30-50 % of the total PM10 concentration in the air, respectively [21]. On the other hand, the study conducted in Barcelona (Spain) [7] pointed out that the ratio of PM10 emissions from automotive engines to PM10 resuspended dust emissions from roads is equal to only 1.2. However, the same emission ratio for PM2.5 particulate matter is equal to 4.

During the evaluation of the PM10 sources contribution in the background levels in the vicinity of roads, it is of significant importance to unambiguously describe the nature of the street. In the paper [6] the results of measurements carried out for two types of roads were showed. The first type was represented by an urban road located in a street canyon (buildings height of 10-20 m), characterized by heavy traffic (about 20 000 vehicles·day^{-1}) and frequent traffic jams (Zürich, Switzerland) and the second type was represented by a motorway with a traffic density of approx. 50 000 vehicles·day^{-1} (Reiden, canton of Lucerne, Switzerland). Obtained results indicated that the level of PM10 concentrations in the air within the street canyon is influenced in 41% by the dust emitted from car engines, in 38 % by the particles re-entrained from the road surface and in 21% by the particles originating from brake pads wear. In the case of the motorway, however, these shares are 41, 56 and 3 %, respectively.

Similar studies executed in different countries in the world have shown unequivocally the significant influence of the dust emitted from the road pavements on air quality, especially in urban areas. Assuming that by the year 2020 the PM10 emission from the internal combustion engines will decrease by 30-45 % but the other emission will remain the same, it should be expected that the resuspended dust emissions share will increase to about 80-90 % (with the current number of road vehicles assumed). In the case of an increase in the number of cars, the share of re-entrained road dust emissions will be even higher [22]. Thus, the actions aiming at the reduction of PM10 emissions into the air cannot be limited merely to the establishment of more restrictive PM10 emission standards from internal combustion engines, but should also involve decision making that would lead to radical decrease in the re-entrained road dust emissions by introducing more efficient and intensified road-cleaning techniques [23-25]. Only a major reduction in re-entrained road dust emission will allow many urbanized areas to meet the current European Union standards specifying daily and annual permitted levels of PM10 concentrations [26].

Krakow is one of many Polish cities where PM10 concentrations in the air are exceeded on a regular basis, which is observed especially at the traffic air quality monitoring stations [27]. The aim of the research described in the paper was to estimate the level of re-entrained road dust emissions from four selected streets of Krakow and its impact on air quality using one of the analysed streets as an example.

2 Research methodology

2.1 Area of research

Experimental campaign was conducted on 10.10.2015 from 0:00 to 1:00 in the four selected sections of Krakow streets, characterized by different traffic intensity during the day. In the study two one-lane streets were considered: Lea (500-5 000 vehicles·day^{-1}) and Reymonta (5 000-10 000 vehicles·day^{-1}) as well as two two-lane streets: Armii Krajowej (10 000-20 000) and Krasinskiego (over 20 000 vehicles·day^{-1}). All of the analysed streets are located in the centre of Krakow. Two of them (i.e. Lea St. and Krasinskiego Av.) are situated within the street canyons.

2.2 Dust sampling methodology

Samples of dust material from selected road sections were collected according to the American Environmental Protection Agency (EPA) guidelines [28, 29]. Two variants of sampling were considered: (1) from the road surface and the area at the curb (all streets), and (2) only from the road surface (Reymont St.). Three samples of dust material were collected from each street. Total weight of samples collected from every street exceeded 0.5 kg. The sampling areas, perpendicular to the axis of the road, were strip-shaped with a width of 100 cm. Samples were collected with a brush and a dustpan. Dust sampling was carried out 15 minutes after the road traffic had been temporally suspended. The most recent rainfall occurred in the period of 26-27.09.2015, while the last street cleaning took place at night between 30.09 and 1.10.2015.

2.3 Physical examination of the collected samples

The dust samples collected from each street were unified and weighed together with the relative humidity and granulometric distribution analysis performed. Determination of the granulometric composition of the dust for the particle size above 0.071 mm was carried out using the sieving methodology in accordance with the Polish Standard PN-R-04032. The granulometric composition of the dust fraction below 0.071 mm (the smallest fraction received after sifting on sieves) in the range of 0.00002 mm to 2 mm was determined using Mastersizer 2000 (Malvern Instruments) laser diffraction particle size analyser. The wet method was used for the granulometric analysis with propan-2-ol dispersion agent. Analysis was performed for each sample twice, once without ultrasound and once with ultrasound application to assist the agglomerates separation within the analysed dust.

2.4 Methodology for estimating re-entrained road dust emission

Re-entrained particulate matter emissions from the analysed road sections were determined in accordance with the U.S. EPA guidelines [28]. In the first step, the emission factors for resuspended road dust was determined based on the measured value of the road silt load (sL) in g·m^{-2} and the average mass of the vehicles on the road in Mg·vehicles^{-1}. The dust emission factor was calculated based on the road type, depending on the average daily traffic expressed in vehicles·day^{-1}. The emission factor unit is g·km^{-1}·vehicles^{-1}. The average vehicle weight was calculated on the basis of the vehicle structure information derived from periodic measurements of the traffic intensity and vehicle structure carried out for the analysed sections of considered roads. The emission rate was quantified as a function of time, since the traffic intensity is strongly dependent on the time of the day.

According to the methodology applied in the study, three aspects were taken into consideration when estimating the road dust emission rates: traffic intensity in vehicles·h^{-1} or vehicles·day^{-1}, length of the road section in km and the number of hours in the year during which the atmospheric precipitation intensity reached at least 0.245 mm·h^{-1}. Only the results of calculations of hourly maximum emission rates in kg·h^{-1} were presented in the study.

2.5 Methodology for estimating the impact of resuspended PM10 on air quality

The assessment of the influence of the re-entrained road dust PM10 emission on air quality was conducted for one of the roads considered in the study (i.e. Krasinskiego Av.) located within the street canyon. Calculations were carried out using the CALINE4 dispersion model for predicting pollutant concentrations from road transport [30]. Two variants of re-entrained dust emissions were adopted. In option 1, the value of resuspended dust emission was determined on the basis of the actual road silt load (sL in g·m^{-2}), whereas in variant 2 the emission value was estimated based on the reference values specified in the U.S. EPA methodology [28].

3 Results and discussion

Performed physical analysis verified low humidity of the collected samples. The moisture content in relation to the dry mass of the samples ranged from 0.256 to 1.693 %, with an average value of 0.626 % and a standard deviation of 0.377 %. Analysis of the granulometric composition of the samples (Table 1) showed, that the dust deposited on the analysed roads mainly consisted of coarse particles of 75-500 μm and 50-2000 μm in diameter. On average, they contribute to 55 and 32 % of the analysed samples mass, respectively. The fraction of fine particles below 75 μm, that can be lifted by travelling vehicles as a result of turbulences, ranges from 1.73 to 4.10 %. The analysis results lead to a conclusion that the largest dissimilarities in the granulometric distributions obtained with and without ultrasound application were observed for the particle size up to 10 μm. Differences between percentage share of the 0-2.5 μm and the 2.5-10 μm particles for both variants are equal to 28.9-54.5 (with ultrasound application) and 43.6-75.6 % (without ultrasound), which indicates that in the extreme cases the effect of ultrasound application increased the share of the 0-2.5 μm particles even by 71.1 % and the 2.5-10 μm particles by 56.4 %. Therefore, the results of the granulometric analysis of the dust determined with the application of ultrasound were used for the emission calculations.

Table 1. The results of granulometric distribution analysis of the dust collected from the analysed streets performed with and without ultrasound application.

Street name	Analysis with (+) or without (-) ultrasound application	Particle fraction shares [% w/w]					
		0-2.5 µm	2.5-10 µm	10-75 µm	75-500 µm	500-2000 µm	>2000 µm
Lea St.	-	0.05	0.18	3.02	68.31	26.68	1.76
	+	0.09	0.24	3.07	68.15	26.69	1.76
Reymonta St.	-	0.03	0.10	1.46	52.54	39.58	6.28
	+	0.09	0.22	1.41	52.39	39.60	6.29
Reymonta St.*	-	0.04	0.12	1.57	32.72	31.97	33.58
	+	0.10	0.26	1.53	32.53	31.98	33.59
Krasinskiego Av.	-	0.07	0.29	2.55	66.68	29.01	1.39
	+	0.25	0.66	2.30	66.27	29.12	1.40
Armii Krajowej St.	-	0.10	0.35	3.54	59.36	35.26	1.39
	+	0.18	0.50	3.43	59.21	35.28	1.39

* sample collected only from the road surface

One of the main objectives of the study was to determine the specific parameter values (silt load – sL) for the analysed road sections characterizing the road dust loading of particle size up to 75 µm. In the Fig. 1 the sL parameter values for the analysed roads are presented, which were determined based on the weight and granulometric distribution of the samples. The values of sL parameter for the considered streets are characterised by large discrepancies, which result from the type of the road, different traffic intensity as well as other factors, i.e. undergoing constructions near the road. The lowest value of the sL parameter of 0.1 $g \cdot m^{-2}$ determined for Reymonta St.* is connected to the fact that the corresponding sample was collected only from the road surface excluding the area at the curb.

Calculated values of the sL parameter for the analysed roads and its reference values derived from the U.S. EPA guidelines [28] were used to estimate both measured and reference values of the PM10 emission rates (Fig. 2). On the basis of the determined emission factors and the actual traffic conditions on roads during peak hours, the actual and reference PM10 emission rates from the considered roads were calculated (Fig. 3).

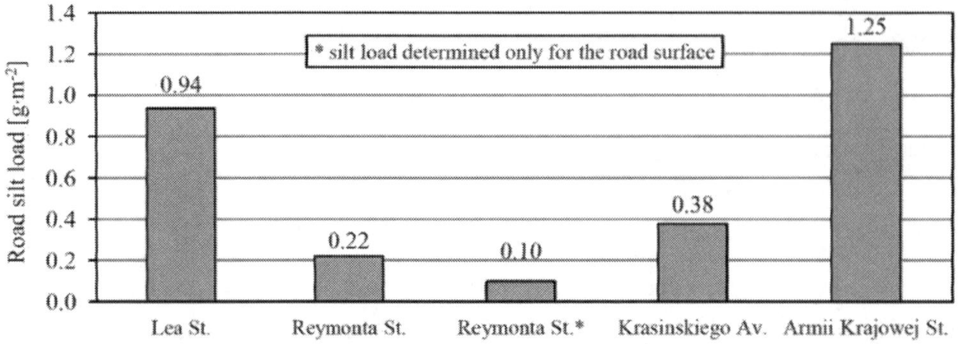

Fig. 1. Actual road silt load (sL) calculated for considered streets (Reymonta St.* - sL calculated based on the dust collected only from the road surface)

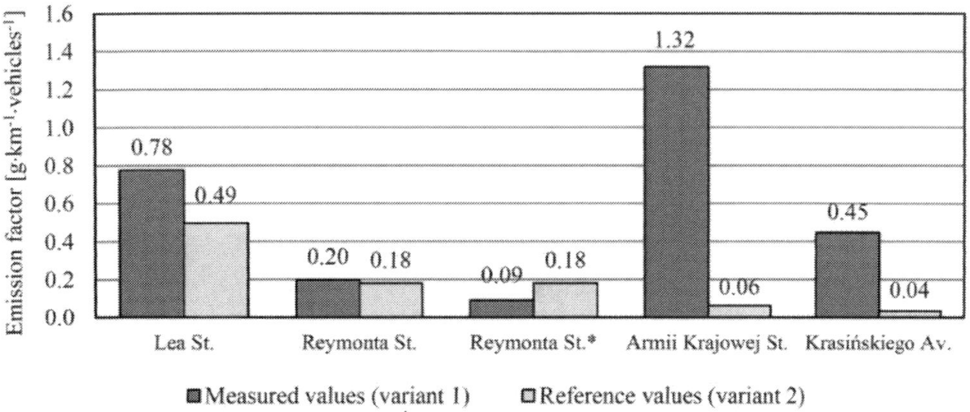

Fig. 2. Measured and reference values of the PM10 emission factors from the analysed streets (Reymonta St.* - emission factor calculated based on the dust collected only from the road surface)

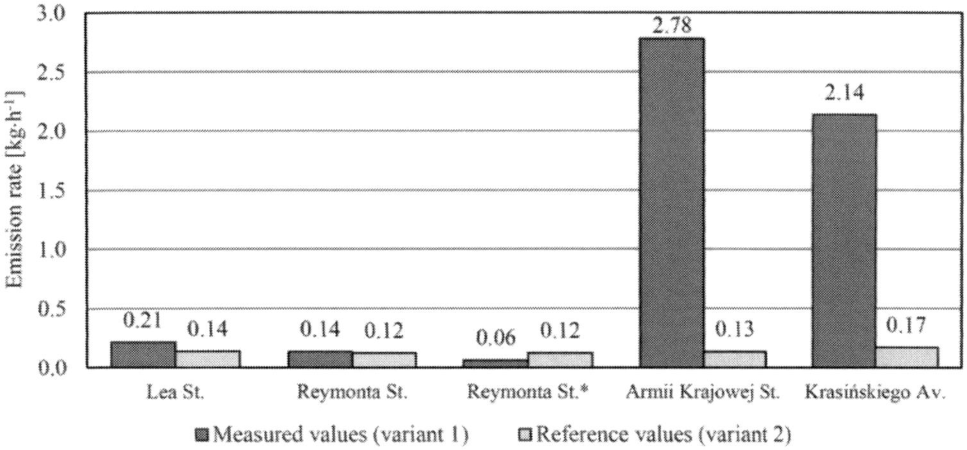

Fig. 3. Comparison of the maximum 1-hour PM10 dust emission rates from the analysed streets based on the measured and reference values of the sL silt load parameter (Reymonta St.* - emission rate calculated based on the dust collected only from the road surface)

Comparative analysis of the actual and reference hourly emission rates showed that in most cases the values of the actual emissions were higher than the reference data. The only exception is the Reymonta St.*, where the actual value of the resuspended PM10 emission was lower than the one obtained from the reference silt load values. The largest discrepancy was found in the case of Armii Krajowej St. where the actual emission is more than 20 times higher than the reference value. Emission rate of this magnitude resulted from additional pollution source on the Armii Krajowej St. in the form of vehicles from the nearby construction site.

Calculated the actual and reference maximum 1-hour PM10 emission values were used to predict the effect of this emission on the modelled PM10 air concentrations across the cross-section in the Krasinskiego street canyon. The concentrations based on actual emission rates are presented in the Fig. 4, while those based on the reference emissions are shown in the Fig. 5. In addition, for both cases the influence of different atmospheric stability conditions on the calculated concentrations was analysed. The presented data were determined for the three Pasquill-Gifford-Turner (PGT) dispersion parameters corresponding to the following atmospheric stability conditions: extremely unstable (PGT = 1), neutral (PGT = 4) and extremely stable (PGT = 7).

The results show that the highest PM10 concentration is concentrated directly above the roadway axes. Moreover, it is proportional to the traffic intensity of the corresponding roadway. On the other hand, within the area of a green belt separating two roadways a drastic decrease in the maximum 1-hour PM10 concentrations is observed. This results from the fact that the modelled wind direction parallel to the road axis is dominant in street canyons. In the adopted calculation scenario, a small deviation from the standard wind direction (σ_{WD}) of 15° was applied taking into account that the wind in this area is incapable of large deviations due to the continuous barrier of buildings creating the canyon.

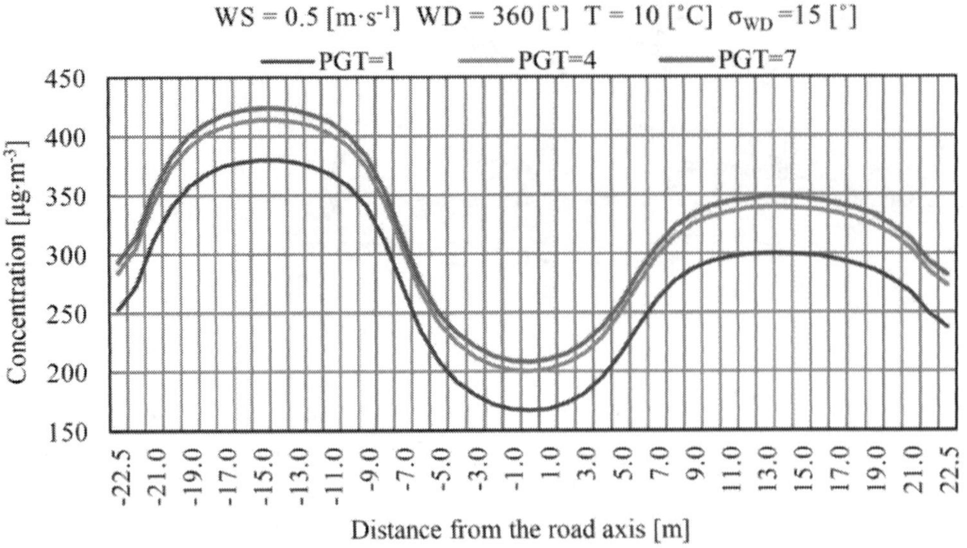

Fig. 4. Distribution of the maximum 1-hour PM10 concentrations in the Krasinskiego street canyon cross-section for different atmospheric stability conditions and the actual emission rates (based on the actual road silt load - sL)

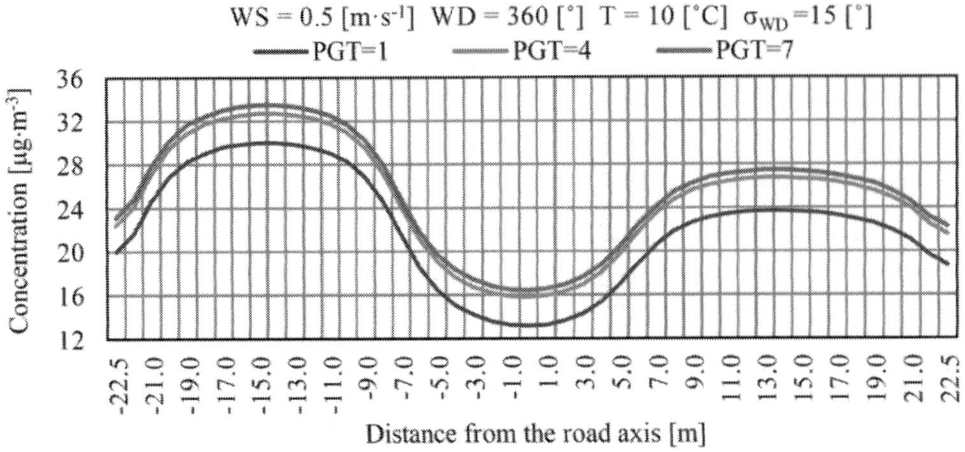

Fig. 5. Distribution of the maximum 1-hour PM10 concentrations in the Krasinskiego street canyon cross-section for different atmospheric stability conditions and the reference emission rates (based on the reference road silt load - sL)

Comparative analysis of the PM10 emission calculation results obtained based on the reference and the actual road silt load showed considerably high discrepancies in the modelled PM10 concentrations. Within the green belt area between the roadways where the air quality monitoring station is located, determined PM10 levels for the actual emission rates range from 170 to 210 $\mu g \cdot m^{-3}$, whereas the concentrations for the reference emission values range from 13 to 17 $\mu g \cdot m^{-3}$, with the largest values occurring during extremely stable atmospheric conditions. At the roadway axes, the concentrations calculated for both variants are approx. 2 times higher. Comparison of the modelling results and measured PM10 concentrations at the analysed air quality monitoring station during non-heating period indicates that the outcomes based on the road dust emission rate calculations using the reference value of the sL parameter are more reliable. The results of the analysis confirm that the sample collection of the deposited dust for the purpose of assessing the impact of resuspended road dust emissions on the PM10 concentrations in the air should be limited only to the road surface (excluding area at the curb).

Comparison of the modelling results with PM10 air concentrations observed at this location (traffic station) in 2014-2016 during different periods of the year showed that the road dust re-entrainment processes from the Krasinskiego Av. surface in the summer can be responsible for 40-50 %, while in the winter even up to 25 % of the total PM10 concentrations in the air. Moreover, the outcomes pointed to a significant influence of wind speed on the PM10 concentrations in street canyons. Maximum 1-hour concentrations at wind speed from 0.5 to 5.0 $m \cdot s^{-1}$ range from 31 to 3 $\mu g \cdot m^{-3}$, but only a slight increase in wind speed, e.g. from 0.5 to 1.0 $m \cdot s^{-1}$ leads to decrease in maximum 1-hour PM10 concentrations in the air by up to 60 %. This tendency is observed as well at urban background stations located in Krakow [31, 32].

4 Conclusions

Conducted research proved that dust particles resuspended from roads may constitute an important share in the total PM10 air concentrations in the vicinity of transport routes, especially in urban areas that are insufficiently ventilated (e.g. inside street canyons). These observations are consistent with findings in the study discussing the same considered street canyon in 2012 [33]. In addition, they pointed to the possible major impact of the fine dust emitted from the road surface in the Krasinskiego Av. in Krakow on the observed PM10 concentrations in the air.

Conclusions formulated in the study allowed to verify the applicability of the sL parameter reference value presented in the U.S. EPA guidelines [28] for considered road types in Krakow. Positive verification of the PM10 emissions from roads based on the assumed sL values confirmed that the appropriate practice of dust sample collection for the purpose of evaluating the impact of this emission on air quality is to exclude areas near the curb and focus on the road surface alone. Nonetheless, the PM10 emissions calculated on the basis of the dust material collected from both the road surface and the area at the curb accurately represent the situation corresponding to the high PM10 concentrations (peaks) observed periodically at the traffic air quality monitoring stations resulting from the mechanical street cleaning events. However, this statement requires further experimental research and analysis on such episodes.

Reduction of the re-entrained road dust emissions is therefore an important challenge facing municipalities. Decrease in the amount of particles deposited on the roads can be attained, for example, as a result of intense atmospheric precipitation, but the preventive measures undertaken by urban services responsible for maintaining the cleanliness of the streets are crucial. As denoted in the work [25], these actions should involve regular dusting and cleaning of the streets, pavements and curbs, as well as the introduction of legal regulations and control measures to prevent further street pollution caused by vehicles leaving unpaved roads, construction sites or transporting uncovered dusty/powdery cargoes.

The paper has been prepared within the scope of the AGH UST statutory research no. 11.11.150.008. The authors would like to thank the authorities and municipal services of the city of Krakow for their assistance in carrying out the research.

References

1. I. Düring, A. Lohmeyer, L. Zippack, Auswertung der Messungen des BLUES während der Abspülmassnahme im Bereich der Messstation Neuenlander Straße in Bremen (2005)
2. W. Bächlin, I. Düring, A. Lohmeyer, L. Zippack, Auswertung der Messungen des BLUME während der Abspülmaßnahme am Abschnitt Frankfurter Allee 86 (2004)
3. I. Düring, T. Hoffmann, A. Lohmeyer, E. Nitzsche, Auswertung der Messungen des Blume während der Verbesserten Strassenreinigung am Abschnitt Frankfurter Allee 86 (2007)
4. G. Omstedt, B. Bringfelt, C. Johansson, Atmos. Environ. **39**, 6088 (2005)
5. C.-H. Chang, R.N. Meroney, J. Wind Eng. Ind. Aerodyn. **91**, 1141 (2003)
6. N. Bukowiecki, P. Lienemann, M Hill, M. Furger, A. Richard, F. Amato, A.S.H. Prévôt, U. Baltensperger, B. Buchmann, R. Gehrig, Atmos. Environ. **44**, 2330 (2010)
7. F. Amato, S. Nava, F. Lucarelli, X. Querol, A. Alastuey, J.M. Baldasano, M. Pandolfi, Sci. Total. Environ. **408**, 4309 (2010)
8. F. Carraz, K.G. Taylor, S. Stainsby, D. Robertson, North West Geography **6**, 1 (2006)
9. P. Pant, R.M. Harrison, Atmos. Environ. **77**, 78 (2013)
10. M. Marjamäki, J. Keskinen, Particulates, Characterisation of Exhaust Particulate Emissions from Road Vehicles, Deliverable 2, Vehicle exhaust particulates characterisation, properties, instrumentation and sampling requirements (EC, DG TrEn, 2001)
11. F. Amato, X. Querol, C. Johansson, C. Nagl, A. Alastuey, Sci. Total. Environ. **408**, 3070 (2010)
12. M. Abu-Allaban, J.A. Gillies, A.W. Gertler, R. Clayton, D. Proffitt, Atmos. Environ. **37**, 5283 (2003)
13. A. Thorpe, R.M. Harrison, Sci. Total. Environ. **400**, 270 (2008)
14. F. Amato, O. Favez, M. Pandolfi, A. Alastuey, X. Querol, S. Moukhtar, B. Bruge, S. Verlhac, J.A.G. Orza, N. Bonnaire and others, Atmos. Environ. **129**, 114 (2016)
15. F. Amato, A. Karanasiou, P. Cordoba, A. Alastuey, T. Moreno, F. Lucarelli, S. Nava, G. Calzolai, X. Querol, Environ. Sci. Technol. **48**, 8069 (2014)
16. F. Amato, A. Karanasiou, T. Moreno, A. Alastuey, J.A.G. Orza, J. Lumbreras, R. Borge, E. Boldo, C. Linares, X. Querol, Atmos. Environ. **61**, 580-587 (2012)
17. F. Amato, M. Pandolfi, T. Moreno, M. Furger, J. Pey, A. Alastuey, N. Bukowiecki, A.S.H. Prevot, U. Baltensperger, X. Querol, Atmos. Environ. **45**, 6777 (2011)
18. M. Crippa, P.F. DeCarlo, J.G. Slowik, C. Mohr, M.F. Heringa, R. Chirico, L. Poulain, F. Freutel, J. Sciare, J. Cozic and others, Atmos. Chem. Phys. **13**, 961 (2013)
19. O. Favez, H. Cachier, J. Sciare, Y. Le Moullec, Atmos. Environ. **41**,7969 (2007)
20. F. Shou-bin, T. Gang, L. Gang, H. Yu-hu, Q. Jian-ping, C. Shui-yuan, Atmos. Environ. **43**, 6003 (2009)
21. J. Chow, J. Watson, D. Lowenthal, C. Frazier, B. Hinsvark, L. Pritchett, Wintertime PM10 and PM2.5 chemical compositions and source contributions in Tucson, Arizona,

In: J. Chow, D. Ono (eds.), *Transactions, PM10 Standards and Nontraditional Particulate Controls* (Air Waste Management Assoc., Pittsburgh, Pennsylvania, 1992)

22. M. Rexeis, S. Hausberger, Atmos. Environ. **43**, 4689 (2009)
23. R. Pitt, R. Bannerman, R. Sutherland, The Role of Street Cleaning in Stormwater Management (Water World and Environmental Resources Congress, American Society of Civil Engineers, Salt Lake City, 2004)
24. G. Ionescu, T. Apostol, E.C. Rada, M. Ragazzi, V. Torretta, U.P.B. Sci. Bull., Series D 75, 2, 175 (2013)
25. B. Benito-López, M.R. Moreno-Enguix, J. Solana-Ibañez, Waste Manage. **31**, 1099 (2011)
26. X. Querol X, A. Alastuey, T. Moreno, M.M. Viana, S. Castillo, J. Pey, S. Rodríguez, B. Artiñano, P. Salvador, M. Sánchez and others, Atmos. Environ. **42**, 3964-3979 (2008)
27. M. Rzeszutek, A. Szulecka, Ekonomika Transportu i Logistyka, **59**, 339 (2016)
28. U.S. EPA, AP 42 Fifth Edition, I, 13, Miscellaneous Sources, 13.2.1 Paved Roads (2011)
29. U.S. EPA, AP 42 App. C1, Procedures For Sampling Surface/Bulk Dust Loading (1993)
30. P.E. Benson, CALINE4 - a dispersion model for predicting air pollutant concentrations near roadways (California Department of Transportation, Sacramento, Report No. FHWA/CATL-84/15, 1989)
31. R. Oleniacz, M. Bogacki, M. Rzeszutek, A. Kot, Meteorological factors affecting air quality in Krakow, In: J. Konieczyński (ed.), *Ochrona powietrza w teorii i praktyce, Tom* 2 (IPIŚ PAN, Zabrze, 2014)
32. R. Oleniacz, M. Bogacki, A. Szulecka, M. Rzeszutek, M. Mazur, JCEEA, **33**, 315 (2016)
33. M. Bogacki, M. Rzeszutek, K. Heba, JCEEA, **33**, 63 (2016)

Monitoring of PAHs in the natural protected areas in non-heating season using Norway spruce (*Picea abies* (L.) Karst) needles

Jacek Borgulat[1,*], *Tomasz* Staszewski[1], *Włodzimierz* Łukasik

[1]Institute for Ecology of Industrial Areas, Kossutha 6, 40-844 Katowice, Poland

Abstract. This paper presents the results of the screening investigation of the environmental burden of polycyclic aromatic hydrocarbons (PAHs) in the natural protected areas in non-heating season. Current year spruce needles were used as a bioacummulative indicator. The total exposure at 15 PAHs (2-6 rings) and carcinogenic potential of these compounds were taken as parameters describing the hazard level. Sampling, in a uniform way, was made in the Silesian Voivodeship landscape parks and the selected Polish national parks as well as in the reference sites with characteristic pattern of PAHs emission, namely in industrialized urban areas and near the expressway. The presence of PAHs, including carcinogenic ones, was shown in all the localities; their content in needles sampled in the natural protective areas was slightly diversified and ranged from 145 to 339 ng/g^{-} (d. w.). Higher differences were observed for carcinogenic potential index (CP) which varied from 2.2 to 18.2. The analysis of PAHs profiles has shown that despite the seasonal lack of intensive emissions from domestic heating, in the majority of natural protective areas pyrogenic sources have also been identified. Profiles of PAHs found in Karkonoski and Ojcowski NPs suggest their petrogenic origin probably due to the intensive tourism (emissions from car engines) in those areas.

Keywords: Polycyclic aromatic hydrocarbons (PAHs), carcinogenic PAHs, *Picea abies*, biomonitoring, non-heating time

1 Introduction

PAHs in ambient air are produced by incomplete combustion of coal, oil, petroleum, wood and plastics. PAHs contain from 2 to 7 rings, while the most toxic of them contain from 4 to 7 rings [1]. The adverse effects of these compounds on human health are connected with metabolic conversion into mutagenic, carcinogenic and teratogenic

* Corresponding author: j.borgulat@ietu.pl

substances. They easily bond with RNA and DNA and as the result they disrupt the replication processes, which may lead to tumour formation [2]. PAHs with a number of rings ≥ 4 are absorbed on particulate matter in the atmosphere, while PAHs with fewer rings may persist in a gas phase [3]. PAHs are strongly absorbed by organic matter in soil and are unavailable for plant uptake [4]. Therefore, PAH deposition on leaves is a function of their concentration in the air [5]. Thus leaves are a convenient passive sampler for the monitoring PAH content in the air. Particularly the coniferous, including spruce, are good a candidate for PAHs bioindicator due to all-year exposure to air pollutants and chemical structure of needles; their cuticle possesses strong lipophilic properties affecting the accumulation of airborne PAHs [6–9]. The waxy surface of needles not only traps particulates but it can also intercept vapor phase contaminants [10]. The amount and sorption properties of cuticle also depends on age and time of year (mainly due to ambient temperature) [11].

The content of PAHs in assimilatory organs increases together with the time of exposure [12,13]. The leaves collected in winter contain higher amount of PAHs when compared to the ones sampled in spring – summer time. On one hand it is due to a greater deposition of these contaminants on leaf surface resulting from a higher emissions of several pollutants in winter, mainly due to domestic heating [14] and on the other hand due to cumulative properties of leaves surface. Commonly the investigation of PAHs deposit is made using the one- and second-year needles collected in summer and winter time and the PAHs ratios have been used by many authors [18–21] to identify sources of PAHs.

Monitoring benzo[a]pyrene (BaP) concentration in the air, which has been carried out in Poland in 98 stations in towns across the country, showed that the permissible level of this compound
(1 ng/m^3) [15] was exceeded in 92 sites. It was also found that concentration of BaP in the air was several times higher in autumn - winter time when compared to the non-heating time [16]. For example in Katowice at annual concentration amounting to 5.1 ng/m^3, the seasonal values were 8.3 ng/m^3 and 1.1 ng/m^3, respectively [17].

The primary objective of this study was to assess levels of PAH deposited on the current year spruce needles in non-heating time in the remote areas of national and landscape parks and compare them to the amount and profile of PAHs in needles from town areas and from the vicinity of the expressway. Moreover, the attempt was made to identify sources of PAHs and to determine their carcinogenic potential.

2 Materials and methods

2.1 Sampling sites characterisation

Sampling sites are presented in Figure 1. Sampling was conducted in September with new growth spruce needles taken from six trees and the averaged samples were analysed. In the protected areas the samples were taken in the middle of the forest. In urban locations, the samples were taken about 2 km from the town centre and about 50 meters from the expressway running through the rural area.

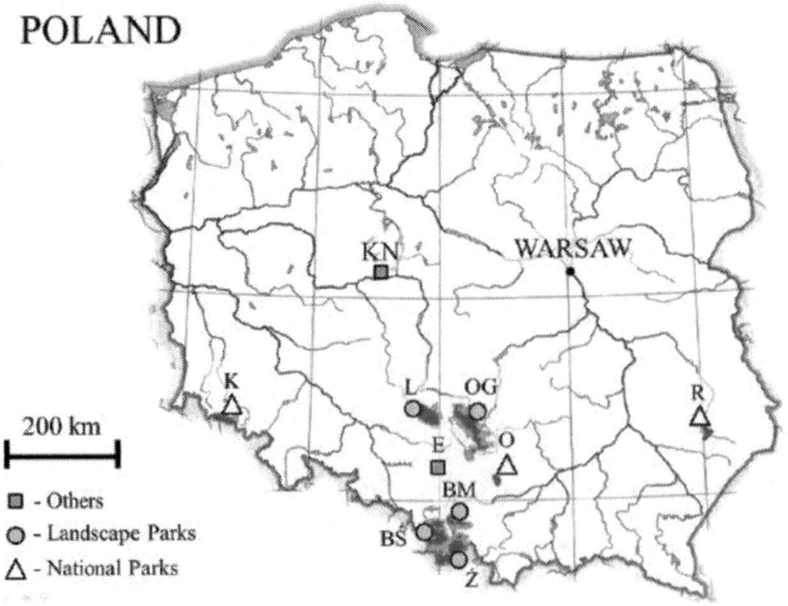

Fig. 1. Location of the study sites (abbreviations: K – Karkonoski NP, O – Ojcowski NP, RZ – Roztoczański NP, BM – Beskid Mały LP, BŚ – Beskid Śląski LP, OG – Orlich Gniazd LP, Ż – Żywiecki LP, L – Lasy nad Górną Liswartą LP, KN – Konin, E – Expresway).

The amount of PAH accumulation in needles was recalculated on the basis of dry needle weight [ng/g d. w.]. Characterisation was based on 2 parameters: the total amount of PAHs deposition within needles and carcinogenic potential (CP) of the PAHs deposit. Index CP was introduced with the following formula, on the basis of Table 1:

$$CP=[\sum(\textit{amount of carcinogenic PAH} \times \textit{\%its carcinogenic properties vs BaP})]\cdot 100^{-1} \quad (1)$$

Table 1. Characteristics of carcinogenic PAHs

Compound (abbreviation, number of rings)	Carcinogenic hazard vs BaP [%], EPA Tasmania (2012)
Benzo[a]antracene (BaA, 4)	10
Chrysene (Chr, 4)	1
Dibenzo[ah]anthracene (DahA, 5)	109.09
Benzo[b]fluoranthene (BpF, 5)	10
Benzo[k]fluoranthene (BkF, 5)	10
Benzo[a]pyrene (BaP, 5)	100
Indeno[123-cd]pyrene (IcdP, 6)	10

2.2 Chemical analysis

10 g of needles from each sample was dried at 30°C and grounded for the analysis. Organic compounds were extracted from the plant material using dichloromethane. PAHs were separated on florosil and analysed by HPLC (Hewlett Packard HPLC 1050) using a UV (naphtalene, acenaphthene) and fluorescence detector (the rest of PAHs). Identification of the separated compounds was carried out by the comparison of their retention times with

the values obtained from a mixture of the Supelco PAH – Mixture 610 – M [4]. All chemicals used were of analytical great.

The following polycyclic aromatic hydrocarbons (PAHs) listed by the United States Environmental Protection Agency (USEPA) and the European Community as priority pollutants were analysed: naphthalene (Naph), acenaphthene (Ace), fluorene (Fluo), phenanthrene (Phen), anthracene (Ant), fluoranthene (Flt), pyrene (Pyr),benzo[ah]-anthracene (BaA), chrysene (Chry), benzo[b]fluoranthene (BbF), benzo[k]fluoranthene (BkF), benzo[a]pyrene (BaP), dibenzo[ah]anthracene (DahA), benzo[ghi]perylene (BghiP), indeno[123-cd]pyrene (IcdP).

Detections limits [ng/g] were as follows: Naph (1.06), Ace (0.76), Fluo (0.76), Phen (2.67), Ant (0.28), Flt (2.76), Pyr (2.14), BaA (0.18), Chry (2.83), BbF (0.13), BkF (0.11), BaP (0.18), DahA (0.11), BghiP (0.31), IcdP (0.1). The limit of quantification (LOQ) were from 2 (BkF) to 7 (Flt) ng/g. Recovery results varied between 81 and 96%.

2.2 Statistical analysis

A dendrogram was used to determine the similarity in the content of the selected PAHs. Before the analysis was performed, data had been standardized. Dendrogram (distance measure - Euclidean distance, agglomeration method - single bond) was made by STATISTICA (StatSoft, lnc. 2017, ver.13).

3 Results and Discussion

The results for PAH biomonitoring using current year spruce needles are presented in Tab. 1. In general, as it was expected, the sites located in urban environment (Konin site) and near expressway had higher contamination levels when compared to the remote sites. The maximum concentration of 15 PAHs was found in Konin followed by the expressway (1097 and 687 ng/g (d. w.), respectively). In the national and landscape parks, the total concentration of 15 PAHs in needles ranged from 145 ng/g (d. w.) to 339 ng/g (d. w.) In all sites, with the exception of the Orlich Gniazd LP and Konin where 4 ring PAHs dominated, 2 and 3 rings compound prevailed making above 60% of the total PAHs (Tab. 1, 2, Fig. 2.). Konin and Orlich Gniazd LP also showed higher content of 5- and 6- rings PAHs than the others. Higher proportions of low molecular weight PAHs (LPAHs – 2 and 3 rings) especially Naph in the needles were due to their vapor phase which enables their transportation at wide distances [22]. There are large differences in the percentage of PAHs in the examined areas (Tab. 2). The highest variability of all the analyzed classes was found for 6-ring hydrocarbons (92% relative standard deviation). In the investigated sites all 15 PAHs were detected with the exception of DahA in some sites. Naph, Phen,Flt, Pyr, and Fluo were the most concentrated pollutants. These findings point towards the predominantly traffic-related sources for the semi-volatile PAHs [20, 22, 23].

Table 1. Content of PAHs [ng/g (d. w.)] in spruce (*Picea abies* (L.) Karst) needles.

PAHs	**Karkonoski NP**	**Ojcowski NP**	**Roztoczański NP**	**Beskid Mały LP**	**Beskid Śląski LP**	**Orlich Gniazd LP**	**Żywiecki LP**	**Lasy nad Górną Liswartą LP**	**Konin**	**Expressway**
Naph	86	92	120	33	104	45	92	21	213	148
Ace	15	4	2	12	4	11	5	2	2	29
Fluo	16	14	12	25	14	28	5	5	67	41
Phen	62	60	12	59	47	31	34	57	99	154
Ant	2	3	4	7	7	19	6	6	6	5
Flt	70	57	39	31	28	63	23	24	178	114
Pyr	70	58	11	13	13	33	6	11	119	148
BaA	4	5	7	4	4	16	3	3	36	7
Chry	4	11	8	6	6	18	4	4	99	16
BbF	3	7	3	8	7	13	3	3	73	9
BkF	2	3	2	4	3	8	2	2	32	5
BaP	2	1	2	4	3	11	2	1	52	3
DahA	0.0	0.0	0.0	1	0.0	2	2	0.0	3	6
BghiP	4	2	4	5	6	12	3	3	49	2
IcdP	0.2	0.1	0.1	0.1	6	11	0.1	4	69	0.3
$\sum_{\text{2-ring PAHs}}$	86	92	120	33	104	45	92	21	213	148
$\sum_{\text{3-ring PAHs}}$	95	80	29	102	73	89	49	71	174	224
$\sum_{\text{4-ring PAHs}}$	147	132	64	54	51	130	35	42	432	286
$\sum_{\text{5-ring PAHs}}$	6	11	8	17	13	33	8	5	160	23
$\sum_{\text{6-ring PAHs}}$	4	2	4	5	11	23	3	6	118	3
$\sum_{\text{PAHs}}$	339	317	224	211	252	319	187	145	1097	687
CP	3.0	2.6	3.3	6.8	5.1	18.2	5.0	2.2	77.3	11.8

The highest value of the CP index calculated on the basis of equation (1) was found for Konin site (77.3). High values of the index were also found for Orlich Gniazd LP and the Expressway - 18.2 and 11.8, respectively. The average value for CP for all other national and landscape parks amounted to 5.8 (Tab. 1). These values are close to the ones found in some plant species growing in remote sites in Poland (peat bogs). In that research noticeable increases in CP values in autumn and winter months in relation to spring - summer time were observed [25].

Table 2. Mean contents of PAHs [ng/g (d. w.)] in spruce needles collected from protected area (SD – standard deviation, RSD – relative standard deviation [%]).

PAHs	**2- rings**	**3- rings**	**4- rings**	**5- rings**	**6- rings**
Mean	70	71	77	13	7
SD	36.0	23.8	45.6	8.7	6.6
RSD [%]	51	33	59	68	92

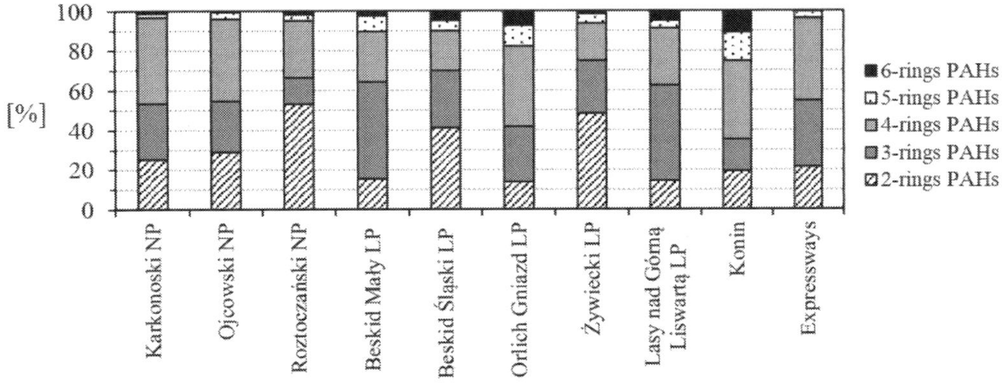

Figure 2. Contribution [%] of 2-, 3-, 4-, 5- and 6- ring PAHs to total PAH sum in spruce (*Picea abies* (L.) Karst) needles.

A dendrogram was used to isolate similar areas in terms of 15 measured PAHs. The greatest similarity was found for the two mountain sites - Żywiec LP and Beskid Śląski LP and Karkonoski NP and Ojcowski NP which are close to the Expressway site. The outlying areas are Konin and Orlich Gniazd Landscape Park.

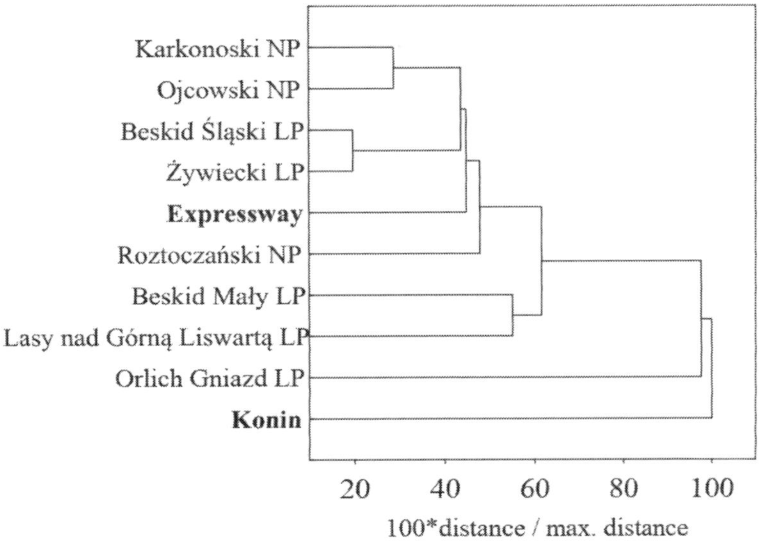

Figure 3. The results of the agglomeration of the examined objects.

In order to determine the presumed origin of PAHs, two diagnostics of PAH distribution ratios were plotted, as presented in Fig. 4. The Fluo/(Fluo + Pyr) ratio was selected for source discrimination because it covers the gaseous as well as the particle transported PAH [26]. These two compounds have comparable physical–chemical properties. It is supposed that variations in the ratio will thus be primarily source associated. Unmodified transfer of source-inherited Fluo/Pyr ratios into different matrices had previously been reported [27]. Ratios between 0.40 and 0.50 are more characteristic of liquid fossil fuel (vehicle and crude oil) combustion, whereas ratios >0.5 are characteristic of grass, wood or coal combustion

[27]. In our research this ratio suggests that in Karkonoski NP, Ojcowski NP and Expressway sites the emissions of PAHs from car traffic are predominant (Fig. 4).

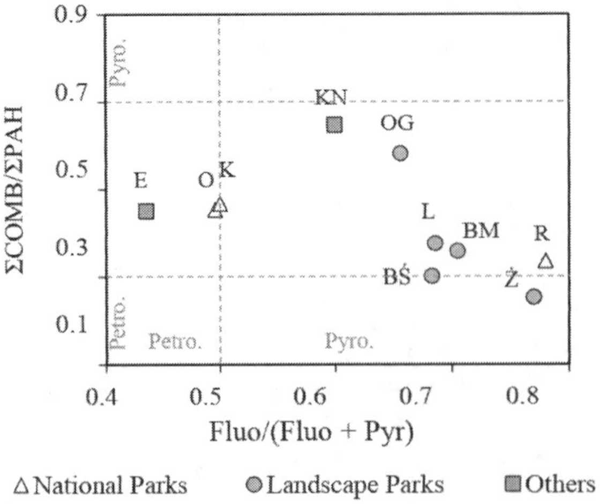

Figure 4. Bivariate plots of ΣCOMB/ΣPAH versus Fluo/(Fluo + Pyr) in spruce needles

Another ratio we used was ΣCOMB/ΣPAH. The ΣCOMB is the sum of concentrations of major non-alkylated PAHs`(Fluo, Pyr, BaA, Chr, BbF, BkF, BaP, IcdP and BghiP) and serves as a PAH indicator of combustion. ΣCOMB/ΣPAH <0.3 indicates petroleum combustion, when at ΣCOMB/ΣPAH >0.7 a pyrogenic origin is likely [19,23].

Generally, the needles sampled from our sites were characterized as low ΣCOMB/ΣPAH showing the mixed sources of PAHs. Higher ΣCOMB/ΣPAH ratio at Konin and Orlich Gniazd LP suggest, that these sites were probably under higher influence of the industrial emissions than the others. Our findings confirm the opinion about the limitations of the use of different ratios of PAHs for its source apportionment which perform well only at sites located near major sources [28].

3.3 Conclusions

The use of current year needles exposed for several months in non-heating time to PAHs deposition appears a good tool to assess their inflow and to show a potential hazard due to air pollution, for humans spending time in recreation areas. Carcinogenic potential of PAHs in the air (determined as their deposit on needles) found in the selected natural protected areas in the southern Poland may be regarded as low with the exception of Orlich Gniazd LP. The latter is located in the vicinity of Huta Katowice and industrial works of Olkusz. Nevertheless, the CP index in this site is several times lower than the value found in the industrialized urban area of Konin. Despite the seasonal lack of intensive emissions from domestic heating, PAHs profiles found in the majority of natural protective areas point to the pyrogenic sources. Profiles of PAHs found in Karkonoski and Ojcowski NPs suggest their petrogenic origin probably due to the intensive tourism (emissions from car engines) in those areas.

References

1. IARC, *Monograph on the Evaluation of the Carcinogenic Risk of Chemicals to Humans*. **32** (1983)
2. Thakker, D.R., Yagi, H., Levin, W., Wood, A.W., Conney, A.H., Jerina, D.M., *Polycyclic aromatic hydrocarbons: metabolic activation to ultimate carcinogens. In: Bioactivation of Foreign Compounds* (New York: Academic Press, 177, 1985)
3. K.C. Jones, G. Sanders, S.R. Wild, V. Burnett, A.E. Johnston, Nature **356**, 137 (1992)
4. S.R. Wild, M.L. Berrow, S.P. McGrath, K.C. Jones, Environ. Pollut. **76,** 25 (1992)
5. M. Riederer, Environ. Sci. Technol. **24**, 829 (1990)
6. Brorström-Lundén, C. Löfgren, Environ. Pollut. **102**, 139 (1998)
7. J. Niu, J. Chen, D. Martens, X. Quan, F. Yang, A. Kettrup, K.-W. Schramm, Environ. Pollut. **123**, 39 (2003)
8. T.S. Howe, S. Billings, R.J. Stolzberg, Environ. Sci. Technol. **38**, 3294 (2004)
9. J.B. Korosi, G. Irvine, E.K. Skierszkan, J.R. Doyle, L.E. Kimpe, J. Janvier, J.M. Blais, Environ. Pollut. **182**, 307 (2013)
10. H. Kylin, E. Grimvall, C. Oestman, Environ. Sci. Technol. **28**, 1320 (1994).
11. J.J. Slaski, D.J. Archambault, X. Li, J.J. Slaski, *Evaluation of Polycyclic Aromatic Hydrocarbon (PAH) Accumulation in Plants: The Potential Use of PAH Accumulation as a Marker of Exposure to Air Emissions from Oil and Gas Flares* (Alberta Environment, Edmonton, 2000)
12. M. Howsam, K.C. Jones, P. Ineson, Environ. Pollut. **108**, 413 (2000)
13. M. Howsam, K.C. Jones, P. Ineson, Chemosphere **44**, 155 (2001)
14. F. De Nicola, G. Maisto, M.V. Prati, A. Alfani, Chemosphere **61**, 432 (2005)
15. Official Journal of Laws No. 47 item 281 of 2008
16. Chief Inspectorate of Environmental Protection, *Zanieczyszczenie powietrza wielopierścieniowymi węglowodorami aromatycznymi na stacjach tła miejskiego w Polsce w 2013 r.* (Warsaw, 2014).
17. Voivodship Inspectorate for Environmental Protection, *Annual report* (Katowice, 2014)
18. M.B. Yunker, R.W. Macdonald, L.R. Snowdon, B.R. Fowler, Org. Geochem. **33**, 489 (2011)
19. M. Tobiszewski, J. Namieśnik, Environ. Pollut. **162**, 110 (2012)
20. Y. Liu, S. Wang, R. Lohmann, N. Yu, C. Zhang, Y. Gao, J. Zhao, L. Ma, Atmos. Environ. **107**, 129 (2015)
21. E. Stogiannidis, R. Laane, in:, D.M. Whitacre (Ed.), *Rev. Environ. Contam. Toxicol.*, (Springer International Publishing, Cham, 2015)
22. S.Y.N. Yang, D.W. Connell, D.W. Hawker, S.I. Kayal, Sci. Total Environ. **102**, 229 (1991)
23. H.-M. Hwang, T.L. Wade, J. Environ. Sci. Health Part A **43**, 1243 (2008)
24. N. Ratola, A. Alves, E. Psillakis, Water. Air. Soil Pollut. **215**, 189 (2011)
25. M. Mętrak, A. Ekonomiuk, B. Wiłkomirski, T. Staszewski, M. Suska-Malawska, Environ. Monit. Assess. 188, **456** (2016)
26. S. Orecchio, Atmos. Environ. **41**, 8669 (2007)
27. M.B. Yunker, R.W. Macdonald, R. Vingarzan, R.H. Mitchell, D. Goyette, S. Sylvestre, Org. Geochem. **33**, 489 (2002)
28. A. Dvorská , G. Lammel, J. Klánová, Atmos Environ **45**, 420 (2011)

Preliminary analysis of variability in concentration of fine particulate matter - $PM_{1.0}$, $PM_{2.5}$ and PM_{10} in area of Poznań city

Izabela Sówka[1], *Anna* Chlebowska-Styś[1,2,*], *Barbara* Mathews [3]

[1] Wroclaw University of Science and Technology Faculty of Environmental Engineering, Unit of Engineering and Atmospheric Protection, Wyb. Wyspianskiego 27, 50-370 Wroclaw, Poland
[2] Voivodship Inspectorate of Environmental Protection in Poznan, 4 Czarna Rola Str., 61-625 Poznan, Poland
[3] Institute of Environmental Engineering Polish Academy of Sciences in Zabrze, 34 Marii Skłodowskiej-Curie Str., 41-819 Zabrze, Poland

Abstract. It is commonly known, that suspended particulate matter pose a threat to human life and health, negatively influence the flora, climate and also materials. Especially dangerous is the presence of high concentration of particulate matter in the area of cities, where density of population is high. The research aimed at determining the variability of suspended particulate matter concentration ($PM_{1.0}$, $PM_{2.5}$ and PM_{10}) in two different thermal seasons, in the area of Poznań city. As a part of carried out work we analyzed the variability of concentrations and also performed a preliminary analysis of their correlation. Measured concentrations of particulate matter were contained within following ranges: PM_{10} – 8.7-69.6 μg/m^3, $PM_{2.5}$ – 2.2-88.5 μg/m^3, $PM_{1.0}$ – 2.5-22.9 μg/m^3 in the winter season and 1.0-42.8 μg/m^3 (PM_{10}), 1.2-40.3 μg/m^3 ($PM_{2.5}$) and 2.7-10.4 ($PM_{1.0}$) in the summer season. Preliminary correlative analysis indicated interdependence between the temperature of air, the speed of wind and concentration of particulate matter in selected measurement points. The values of correlation coefficients between the air temperature, speed of wind and concentrations of particulate matter were respectively equal to: for PM_{10}: -0.59 and -0.55 (Jana Pawła II Street), -0.53 and -0.53 (Szymanowskiego Street), for $PM_{2.5}$: -0.60 and -0.53 (Jana Pawła II Street) and for $PM_{1.0}$ -0.40 and -0.59 (Jana Pawła II Street).

1 Introduction

The sources of the initial emission of particulate matter are mainly combustion processes in stationary sources, where the highest share in Poland (about 50%) is constituted by emission from combustion processes originating from outside of industry [7, 13]. Many studies indicate so-called low-stack-emission of particulate matter from communal and living sector as a main factor forming the quality of air in Poland in the winter season [11, 16]. An important source of particulate matter emission, especially in cities, is also road transport, which besides emitting particulate matter by combustion of fuel, also emits it by

* Corresponding author: anna.chlebowska@pwr.edu.pl

abrasion of vehicle elements (brake blocks, tires) and surface of roads. Another significant place, in the list of emission sources, is taken by secondary emission of particulate matter, this is uplifting particulate matter from the ground caused by moving vehicles [1, 9].

The level of contamination of the atmosphere depends highly on meteorological conditions, especially on the speed and direction of the wind, the temperature of air, relative humidity of air and solar irradiation [5]. Meteorological factors may as well increase as decrease the pollutants concentration, large wind speed favors 'rarefaction' of contamination in the air, and high humidity of air enables aggregation of particulate matter what increases the speed of its fall. Especially important kind of influence of meteorological conditions is observed in the presence of excessive concentrations of particulate matter contaminants. Depending on existing conditions they may magnify or minimize the smog effect [6, 13, 14]. Dense urban development may significantly influence processes of dispersion of pollutants in cities, it disrupts the process of air circulation, therefore, makes it more difficult to exchange air in an area of a city.

The aim of this work was analysis of variability of particulate matter concentrations: $PM_{1.0}$, $PM_{2.5}$ and PM_{10} and preliminary assessment of the influence of the main meteorological parameters on particulate matter concentrations in Poznań.

2 Materials and methods

The research included two measurement campaigns in different thermal seasons: in the autumn-winter season (25.10-22.11.2016) and in the summer season (05.06-02.07.2017), at three measurement stations (Fig. 1.)

The collection of particulate matter - PM_{10} on Szymanowskiego Street during the first measurement campaign was done with usage of DERENDA PNS 16T-3.1 Automatic Sampling System (low-flow sequential impactor). During the second measurement campaign on Szymanowskiego Street and during two measurement campaigns on Polanka Street, the collection of samples was done with usage of LVS low-volume sequential impactor of a Umwelttechnik MCZ GmbH brand, (type MicroPNS LVS 16) which is characterized by a flow rate equal to 2.3 m^3/h. All of the impactors had a filter storage containing maximum 14 units. In the first case, as a supply material, Whatman QM-A quartz filters with a diameter of 150 mm and in the second case Whatman QM-A quartz filters with a diameter of 47 mm were used. In order to collect the suspended particulate matter - $PM_{1.0}$, $PM_{2.5}$ and PM_{10}, in own research, impactors of a Harvard type (MS&T Area Samplers, Air Diagnostics and Engineering, Inc., Harrison, ME, USA) was used. The air flow was forced by ultra-quiet oil free vacuum pumps (Air Diagnostics and Engineering, air sampling pump, SP-280E model). As a supply material we used Whatman QM-A quartz filters with a diameter of 37 mm. The flow rate during the collection of samples was respectively equal to: for particles ≤ than 1.0 μm - 23 dm^3min^{-1}; for particles ≤ than 2.5 μm and for particles ≤ than 10 μm - 10 dm^3min^{-1}. The volume of pumped air was controlled with usage of Ataris type flow meter. The time of sample collection was equal to 24h.

Filters, as well before and after the exposure, were conditioned and weighed in a weighing room with constant humidity of 45 ± 5% and the temperature of air equal to 20 ± 2°C. We used analytical balances Xa 60/220 of the RADWAG company as the measurement equipment. The conditioning period was equal to a minimum 48 hours, and the second weighing took place after 24 hours.

The measurement of particulate matter was done with usage of gravimetric method in accordance PN-EN 12341 standard [15].

In order to examine interdependence between meteorological factors and concentration of suspended particulate matter ($PM_{1.0}$, $PM_{2.5}$ and PM_{10}) correlation analysis was applied.

Assumed statistical significance was equal to 0.05, and the number of measurement samples in both cases was equal to 28. Interdependence between concentrations of particulate matter and meteorological conditions was presented with usage of a correlation matrix.

Fig. 1. Locations of measurement points in Poznań: 1 – Szymanowskiego Street, 2 – Jana Pawła II Street, 3 – Polanka Street (map source: By Oskarro.© users of OpenStreetMap. Copyright https://commons.wikimedia.org/w/index.php?curid=16437602)

3 Results of the research and related discussion

Performed analysis of the data (Table 1) indicates that during the first measurement campaign were dominating the winds blowing from southwest direction with an average speed of 2 m/s. The average air temperature was equal to 5°C, and the average relative humidity of the air was considerably high and equal to 87%. During the second measurement campaign were dominating the winds blowing from the west direction with an average speed of 2 m/s. The average air temperature was equal to 19 °C and the relative humidity was 20% lower than in the autumn season and was equal to 70%.

Table 1. Descriptive statistics determined for main meteorological parameters during both measurement series (data source: WIOŚ Poznań [Provincial Inspector of Environmental Protection in Poznań]).

Statistics	Atmospheric pressure	Wind speed	Temperature	Humidity
	[hPa]	[m/s]	[°C]	[%]
Autumn-winter season n=28				
Minimum	992	0	-1	79
Maximum	1027	4	11	95
Average	1010	2	5	87
Median	1009	1	6	87

Standard deviation	9,46	0,91	3,79	4,35
Summer season, n=28				
Minimum	986	0	14	53
Maximum	1016	5	24	88
Average	1005	2	19	70
Median	1005,5	2	18	67,5
Standard deviation	6,56	1,07	2,23	9,52

Higher concentrations of particulate matter ($PM_{1.0}$, $PM_{2.5}$ and PM_{10}) in the air were determined for the autumn-winter season (Table 2) for both of the measuring stations. The average concentration of PM_{10} in the autumn-winter season was higher at the station on Szymanowskiego Street (32.9 μg/m^3) than at the station on Jana Pawła II Street (25.4 μg/m^3), however, the concentration during the measurement session was comparable at both of the stations. At the station on Jana Pawła II Street the average concentration of PM_{10} in the summer season was 35% lower than during the first measurement session, in case of the station on Szymanowskiego Street, the concentration in the second measurement session was lower than the average concentration in the first measurement session (Figure 2a) by 55%. The concentrations of suspended particulate matter - $PM_{2.5}$ during the first measurement session were contained within the range of 2.2 – 39.9 μg/m^3 at the station on Jana Pawła II Street and at the station on Polanka Street concentration was contained within the range of 7.2 – 88.5 μg/m^3. The average concentration was respectively equal to 15.2 μg/m^3 for the station on Jana Pawła II Street and 30.8 μg/m^3 for the station on Polanka Street. The concentrations determined during the second measurement series were lower (Figure 2b) and were ranging from 1.2 to 40.3 μg/m^3 at the station on Jana Pawła II Street and from 7.2 to 17.1 μg/m^3 at the station on Polanka Street. The average concentration of $PM_{2.5}$ on Jana Pawła II Street, in the summer season, was 32% lower than in the winter season, and at the station on Polanka Street it was lower by 64%. The level of concentration of PM_{10} and $PM_{2.5}$ at both stations were similar to the concentration of PM_{10} observed in the same periods in previous years of measurements conducted by WIOŚ in cities of Wielkopolska, excluding Kalisz (in the city there are special conditions for dispersion of contaminants) [3, 4]. However, it should be noted, that the levels of concentration of particulate matter in the winter season, at all of the stations, were significantly higher than concentrations determined by WHO as safe for health and life of humans (PM_{10} – 20 μg/m^3, $PM_{2.5}$ – 10 μg/m^3) (7). Concentrations of $PM_{1.0}$ in the winter season were ranging within limits of 2.47 to 22.87 μg/m^3 and in the summer season they ranged between 2.72 and 10.35 μg/m^3. The average concentration in the winter season was equal to 8.50 μg/m^3 and was lower by 33% in the summer season and equal to 5.72 μg/m^3 (Table 2). In the graph presenting dependence between the concentration and the temperature (Figure 1 a, b) we might observe, that in the winter season, the lower was the air temperature the higher were concentrations of particulate matter - PM_{10}, $PM_{2.5}$ and $PM_{1.0}$ (Figure 1a). The dependence presented above was more significant for concentrations of particulate matter at background stations (Szymanowskiego and Polanka Streets) than for the concentrations at traffic station (Jana Pawła II Street) The summer season presented no observable dependence between the concentrations of particulate matter and the temperature of air (Figure 2b).

Higher concentrations of particulate matter at the stations of the city background may be related to a phenomenon of low emission. The area of the station on Szymanowskiego Street mainly consists of single-family housing developments, and the area of the station on Polanka Street consists of multi-family housing developments which is heated by local boiler rooms. The methods of heating in the heating season, in this case, influences the quality of air. It should be noted that the average concentration of PM_{10} in the summer season at the station on Szymanowskiego Street (14.7 μg/m^3) was lower than at the station on Jana Pawła II Street (16.6 μg/m^3). This situation also confirms, that eliminating such factor as heating of buildings, influences quality of air in a specific area. The differences in concentrations of particulate matter $PM_{1.0}$ between seasons are not so significant as in the case of other two fractions (PM_{10}, $PM_{2.5}$). Measured concentrations of $PM_{1.0}$ in Poznań are comparable to concentrations measured in other cities of Poland, but at the same time are higher than in countries of Western Europe [8, 10, 17]. High concentrations of $PM_{1.0}$ in cities are related to the presence of many sources of emission in that areas, however, it should be noted, that high level of concentration of $PM_{1.0}$ is also identified at stations located in suburban and rural areas in Poland. It is probably related to the process of transporting $PM_{1.0}$ from other areas, which are sometimes very distant. The finest fraction of the particulate matter is characterized by a long time of presence in the atmosphere, therefore it is possible to be carried far away [12].

Table 2. Basic statistics for concentrations of particulate matter ($PM_{1.0}$, $PM_{2.5}$ and PM_{10}) in the autumn-winter season (25.10-22.11.2016) and in the summer season (05.06-02.07.2017)

Statistics	**Concentration [μg/m^3]**	
	Autumn-winter season	**Summer season**
PM_{10} Szymanowskiego Street, n= 28		
Minimum	8,7	1,0
Maximum	69,6	22,6
Average	32,9	14,7
Median	34,5	15,7
Standard deviation	15,5	5,2
PM_{10} Jana Pawła II Street, n= 28		
Minimum	9,4	5,2
Maximum	68,6	42,8
Average	25,4	16,6
Median	21,5	15,1
Standard deviation	14,6	8,5
$PM_{2.5}$ Polanka Street, n= 28		
Minimum	7,2	7,2
Maximum	88,5	17,1
Average	30,8	11,0
Median	27,2	10,9
Standard deviation	19,1	2,5
$PM_{2.5}$ Jana Pawła II Street, n= 28		
Minimum	2,2	1,2
Maximum	39,9	40,3
Average	15,2	10,4
Median	12,9	8,7
Standard deviation	10,0	9,1
$PM_{1.0}$ Jana Pawła II Street, n= 28		
Minimum	2,5	2,7
Maximum	22,9	10,4
Average	8,5	5,7
Median	7,7	5,4
Standard deviation	5,3	1,9

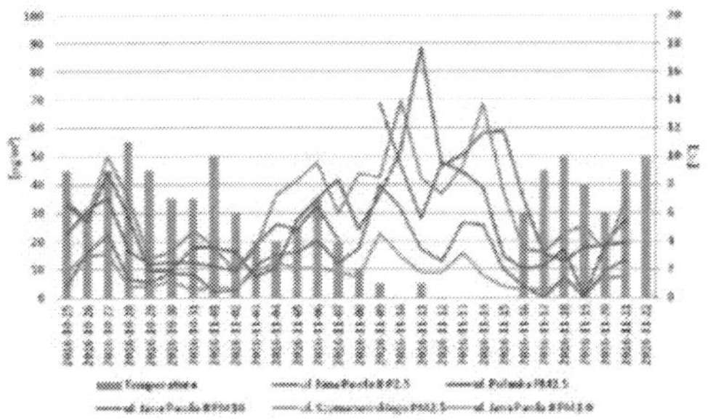
a)

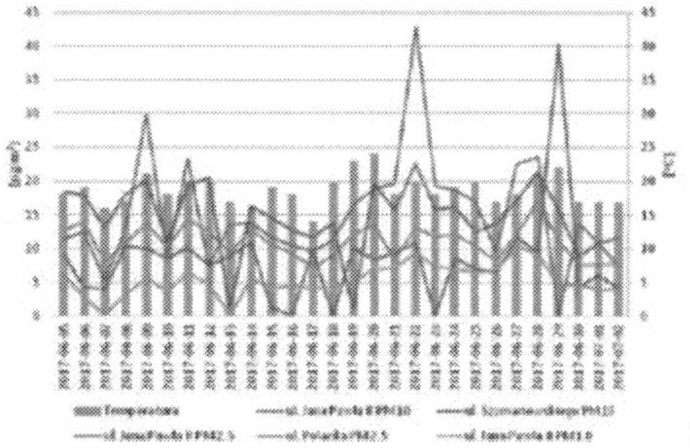
b)

Fig. 2. Changes of the variability of particulate matter concentrations: $PM_{1.0}$, $PM_{2.5}$ and PM_{10} and the temperature of air in Poznań, during two measurement series, in the autumn-winter (a) and summer (b) seasons

It is commonly known that suspended particulate matter is dependent on the temperature of air [1, 5, 9]. Basing on prepared correlation matrix, which elements were the coefficients of correlations for pairs of random variables described by a parameter of a line and a column of the matrix, in the winter season was determined statistically significant (with confidence level of $p<0.05$) negative correlations between the speed of wind, temperature and concentrations of particulate matter PM_{10} at the stations on Jana Pawła II Street (r=-0.55 and r=-0.59) and on Szymanowskiego Street (r=-0.53 and r=-0.53), $PM_{2.5}$ at the station on Jana Pawła II Street (r=-0.53 and r=-0.60) and $PM_{1.0}$ at the station on Jana Pawła II Street (r=-0.59 and r=-0.40) (Table 3). In case of concentrations of suspended particulate matter of $PM_{2.5}$ at the station on Polanka Street was also determined negative dependence between the concentration of particulate matter and the temperature of air (r=-0.64). In the summer season was determined reversed dependence in relation to the winter season, between the temperature of air and concentration of particulate matter. The positive correlation was established between the temperature of air and concentration of PM_{10} particulate matter at the station on Jana Pawła II Street (r=0.44) and particulate matter $PM_{2.5}$ (r=0.65) at the station on Polanka Street. In case of analysis of interdependence between concentrations of particulate matter and relative humidity of the air in summer season was determined negative dependence, for which the coefficients of correlation had following values: for PM_{10} particulate matter on Jana Pawła II Street r=-0.54, on Szymanowskiego Street r=-0.47, for $PM_{2.5}$ particulate matter at the station on Jana Pawła II Street r=-0.44, on Polanka Street r=-0.47.

Table 3. Acquired values of correlation coefficients between concentrations of particulate matter: PM_{10}, $PM_{2.5}$ and $PM_{1.0}$ and the speed of wind, temperature and relative humidity in the autumn-winter and summer seasons

Variable	Jana Pawła II Street PM_{10}	Szymanowskiego Street PM_{10}	Jana Pawła II Street $PM_{2.5}$	Polanka Street $PM_{2.5}$	Jana Pawła II Street $PM_{1.0}$
Autumn-winter season					
Wind Speed, m/s	**-0,55**	**-0,53**	**-0,53**	-0,30	**-0,59**
Temperature, C	**-0,59**	**-0,53**	**-0,60**	**-0,64**	**-0,40**
Humidity, %	0,09	0,31	0,05	**0,46**	-0,04
Summer season					
Wind Speed, m/s	-0,23	-0,33	0,17	**-0,53**	0,06
Temperature, C	**0,44**	0,17	0,10	**0,65**	0,32
Humidity, %	**-0,54**	**-0,47**	**-0,44**	**-0,47**	-0,29

4 Summary

Measurements of concentration of PM_{10}, $PM_{2.5}$ and $PM_{1.0}$ done as a part of the research, confirm previous results of analyses performed in the city of Poznań [4] and indicate that there is a problem of high concentrations of suspended particulate matter in the autumn-winter season. Measured values were limited within ranges: from 8.7 to 69.6 μg/m³ for PM_{10}, from 2.2 to 88.5 μg/m³ for $PM_{2.5}$ and 2.5 to 22.9 μg/m³ for $PM_{1.0}$. The analysis of correlation indicates that high concentrations of suspended particulate matter in the heating season may be dependent on the temperature of air (the lower the temperature, the higher concentrations of particulate matter) and stagnation conditions in the atmosphere (low values of the wind speed). The concentrations of particulate matter in Poznań, measured in the summer season, were lower and within ranges of 1.0 to 42.8 μg/m³ for PM_{10}, 1.2-40.3 μg/m³ for $PM_{2.5}$ and 2.7 to 10.4 μg/m³ for $PM_{1.0}$. A significant negative dependence between concentrations of particulate matter and the temperature of air was determined. It can be concluded that the quality of air in this season is formed by the secondary uplift of particulate matter and the emission from transport, including abrasion of elements of vehicles and the surface of roads. In the summer season it is the road transport which constitutes the significant source of air contamination in cities of the Eastern Europe [2]. High temperatures of the air cause, that the factor, which is heating of the buildings, is eliminated. At the same time low relative humidity of the air favors uplifting of the particulate matter from surfaces of the roads, mainly by moving vehicles. This was confirmed by the fact of higher concentrations of particulate matter in the summer season at communication measurement station (Jana Pawła II Street) than at the background stations (Polanka and Szymanowskiego Streets)

The investigations were co-financed within the framework of the order No. 0402/0088/16 with the specific subsidy granted for the Faculty of Environmental Engineering Wroclaw University of Science and Technology (W-7) by the Minister of Science and Higher Education to conduct research and development work and related tasks contributing to the development of young scientists and doctoral students in 2016/2017.

We would like to express our gratitude to the employees of Applications Division of Poznań Supercomputing and Networking Center affiliated to Institute of Bioorganic Chemistry of the Polish Academy of Sciences for providing a place and technical facilities necessary for carrying out the research in the field.

References

1. A. Badyda, G. Majewski, Scientific Review Engineering and Environmental Sciences, **33**, 146-157 (2006)
2. A. Chlebowska-Styś, I. Sówka, D. Kobus, Ł. Pachurka, EDP Sciences, 1-8 (2017)
3. A. Chlebowska-Styś, I. Sówka, Ł. Pachurka, Analisis of the particulate matter PM10 composition at the urban background measurement station in Pila. Oficyna Wydawnicza Politechniki Wrocławskiej, **8**, 36-53, (2016)
4. A. Chlebowska-styś, I. Sówka, Trends concentrations changes and compositions particulate matter (PM10 and PM2.5) on the example of selected cities in Wielkopolska Region, Oficyna Wydawnicza Politechniki Wrocławskiej, **6**, 40-53 (2015)
5. A. Gioda, L. Ventura, I. Lima A. Luna, EGU General Assembly Conference abstracts, **15**, 3256 (2013)

6. A. Nowicka, I. Rynkiewicz, E. Dragańska, Scientific Review Engineering and Environmental Sciences **13(28)**, 126-132 (2004)
7. Air quality in Europe — 2016 report, EEA Report No 28/2016
8. D. Dmochowski, A. Dmochowska, S. Biedugnis, Annual Set The Environment Protection, **1**7, 1403-1412 (2015)
9. G. Majewski, W. Przewoźniczuk,. Ann. Warsaw Agricult. Univ.-SGGW, Land Reclam, **37**, 55-67 (2006)
10. I. Sówka, M. Fortuna, A. Zwoździak, J. Rybak, M. Skrętowicz, K. Kwiecińska, *Analisis of concentrations of fine dust particles at selected points of Wroclaw,* Oficyna Wydawnicza Politechniki Wrocławskiej,**2**, 451-457 (2012)
11. J. S. Pastuszka, W. Rogula-Kozłowska, K. Klejnowski, P. Rogula-Kopiec, Atmosphere, **6**, 1521-1538 (2015)
12. J.O. Allen, P.R. Mayo, L.S. Hughes, L.G. Salmon, G.R Cass, Environ Sci Technol **35**, 4189–4197 (2001)
13. K. Juda-Rezler, M. Reizer, i J.P. Oudinet, Atmos Environ, **45**(36), 6557-6566 (2011)
14. K. L. Yang,. Atmos Environ **36**, 3403-3411 (2002)
15. PN-EN 12341:2014-07 Ambient air – A standard gravimetric measurement method for the determination of the PM10 or PM2.5 mass concentration of suspended particulate matter
16. W. Rogula-Kozłowska, Air Qual Atmos Health, 533–550 (2015)
17. W. Rogula-Kozłowska, K. Klejnowski, Bull Environ Contam Toxicol, , **90**,103–109 (2013)
18. Z. Chłopek, Proceedings of the Institute of Vehicles, **1**, 87 (2012)

Selected carbonyl compounds in the air of Silesia region

Marianna Czaplicka, and *Michał* Chrobok

Institute of Environmental Engineering Polish Academy of Science, M. Skłodowskiej-Curie 34 street, 41-819 Zabrze, Poland

Abstract. This study was carried out to characterize three aldehydes of health concern (formaldehyde, acetaldehyde, and acrolein) at a three sites in Silesian region (Poland) in January and June 2015. Aldehydes in polluted atmospheres comes from both primary and secondary sources, which limits the control strategies for these reactive compounds. Average aldehyde concentration in summer period lies in range from 3.13 μg/m³ to 10.43 μg/m³, in winter period in range from 29.0 μg/m³ to 32.2 μg/m³. Acetaldehyde was dominant compound in winter period, in summer formaldehyde concentration was highest of all determined aldehydes.

1 Introduction

Aldehydes and ketones are present in almost every part of enviroment. These compounds can be found in: rain, dew, mist, and in the air [1]. Their presence was observed in gases emitted to atmosphere from traffic and industrial sources, including combustion of biomass and solid fuels [2]. Mentioned compounds can be created in course of photooxidation of hydrocabons with help of radicals HO_2• and OH• and are part of photochemical smog [3]. Aldehydes play important role in tropospheric chemistry, since they are intermediate products of oxidation processes of hydrocarbons and subsequently generate photochemical products such as carbon monoxide and radical species that contribute to the production of ozone and peroxyacylnitrate in the troposphere. In order to understand the chemical cycles of ozone and its related species in the troposphere as well as the impacts of air pollutants on human health, it is important to investigate the behavior of aldehydes compounds in the ambient atmosphere [4]. Formaldehyde (FA) and acetaldehyde (AD) are dominant species in the aldehyde group in the atmosphere [5]. The most important man-made source of formaldehyde are exhaust gases from cars not fitted with catalytic converters. Most common acrolein (AC) sources are exhaust gases from cars powered by petrol, diesel or other petrochemical fuels. Removal of formaldehyde from the atmosphere can occur by: chemical transformations, rain and snow scavenging of vapors and particles, dry deposition of particles, and vapor exchange across the air–water interface [6]. The objective of this study was to measure the gas-phase concentration of aldehydes in 3 sites located in Silesian region along with establishing influence of local sources upon their concentration

https://doi.org/10.1051/e3sconf/20182801006

2 Experimental

2.1 Sample collection

Air samples were collected by passing through special sampler filed with silica gel and impregnated with solution of 2,4-DNPH (2,4-Dinitrophenylhydrazine) in hydrochloric acid. Potassium iodide was added to protect analytes from ozone influence. After collection samples were protected from sunlight and kept in temperature approximately equal to 4^0C during transportation to laboratory. Atmospheric conditions in course of whole campaign are presented in table 1. Samples were collected around communication nodes situated in centers of cities (Zabrze, Gliwice 1) of upper Silesia urban area and in proximity of residential area (Gliwice 2) in January and June. In every sampling point 10 samples were collected both in June and January. Sampling flow rate was equal to 200 ml/min. Amount of air collected varies from 0.1 m^3 to 0.2 m^3.

Table 1. Summary of meteorological condition and O_3 content

	Temperature [°C]	Wind speed [m/s]	Relative humidity [%]	Ozone, O_3 [µg/m^3]
January 2015				
Avarange	-1	1	89	27
Minimum	-8	0	65	4
Maximum	4	2	99	59
June 2015				
Avarange	15	0	70	68
Minimum	9	0	49	43
Maximum	23	1	94	113

2.2 Methods

Quantitative determination was conducted with help of Perkin Elmer Series 200 HPLC unit equipped with: UV-Vis diode array detector (DAD), Spheri - 10 RP-18 10µ (size 250 mm x 4,6 mm) column and 10 µl dispensing valve. Mobile phase composition changed from mixture of water:acetonitrile (40%:60%) to 100% of acetonitrile. Analysis was carried out at constant weavelenght equal 360 nm. Figure 1 presents chromatogram of mixture of standards.

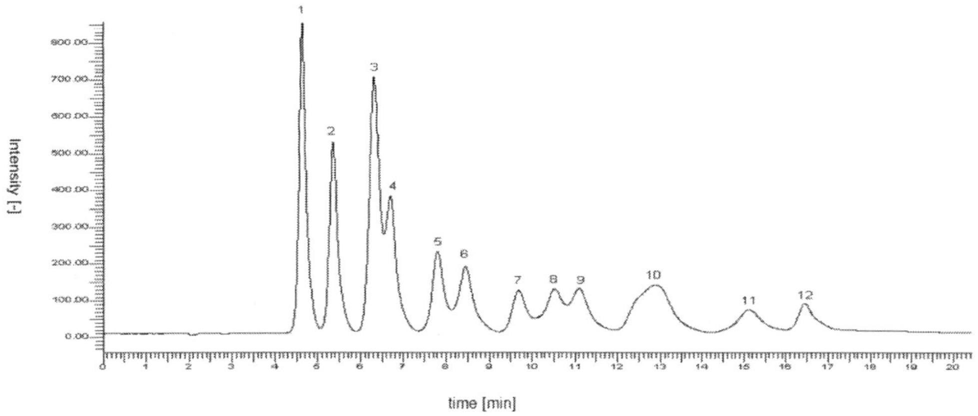

 https://doi.org/10.1051/e3sconf/20182801006

Fig.1. Chromatogram of standard solution 1; formaldehyde 2; acetaldehyde. 3; acetone 4; acrolein. 5; propionaldehyde. 6; Crotonic aldehyde. 7; butyl aldehyde 8; benzaldehyde. 9; *izo*valeric and valeric aldehyde . 10; methylbenzaldehydes 11; caproic aldehyde. 12; 2.5-dimetylbenzaldehyde

Quantitative determination was done by means of external standard method. DAD detector response was linear in concentration range from 0.15 µg/ml to 0.75 µg/ml. As limit of detection (LOD) three times value of signal to noise ratio was applied. Limit of quantification (LOQ) equals three times value of LOD. Tabel 2 presents statystical parameters of method.

Table 2. Values of LOD, LOQ and relative standard deviation for selected carbonyl compounds

Compound	Limit of detection [µg/m^3]	Limit of quantification [µg/m^3]	Relative standard deviation [%]
Formaldehyde	0.3	1.0	8.0
Acetone	0.2	0.5	6.0
Acrolein	0.2	0.6	9.5
Acetaldehyde	0.5	1.5	8.5
Benzaldehyde	0.5	1.5	14.6
Metyl-*iso*butylketone	0.3	1.0	14.0

3 Results and discussion

Conducted studies prove important influence of year period and sampling points characteristics upon aldehyde content in the air.

Table 3. The average total concentration of aldehydes (µg/m^3) during June and January at the sites Zabrze and Gliwice

Parameter	Mean (SD)	Max	Min
Zabrze			
June	10.43 (4.9)	16.36	4.19
January	29.07 (12.6)	51.00	12.33
Gliwice 1			
June	7.67(1.8)	9.58	4.69
January	32.20(12.7)	60.14	16.73
Gliwice 2			
June	3.17(1.5)	4.04	0.02
January	30.24(13.4)	53.99	14.92

In June, independently of location average total concentration of aldehyde was lower in comparison with data obtained from January sampling campaign. Lowest average aldehyde concentration – 3.17 µg/m^3 was recorded in June at point Gliwice 2 located at residential area, highest – 10.4 µg/m^3 in Zabrze at communication nod. In course of January sampling campaign minor differences occurs in concentration of aldehydes regardless of sampling points. Obtained results were similar to result of air quality monitoring of Milan surroundings obtained by Andreini et al. [7]. Andreini et al. indicated that during winter total concentration of aldehydes varies from 16.7 to 30.7 µg/m^3 at the urban locations and 11.7 µg/m^3 at the rural-industrial location.

 https://doi.org/10.1051/e3sconf/20182801006

3.1 Aldehyde profile

Formaldehyde, acetaldehyde and acrolein were present in all samples collected in January. Average concentration of formaldehyde in winter varies from 4.53 μg/m³ to 6.39 μg/m³ depending upon sampling point location and equals 14% to 21% of total aldehydes concentration (table 4). These concentrations were within the ranges previously measured at different sites around the world. By comparison, the rural site formaldehyde concentrations measured equals 4.17 ± 1.7 μg/m³ [8] in New Mexico USA and 0.05 – 9.1 μg/m³ at different sites in Canada [9].

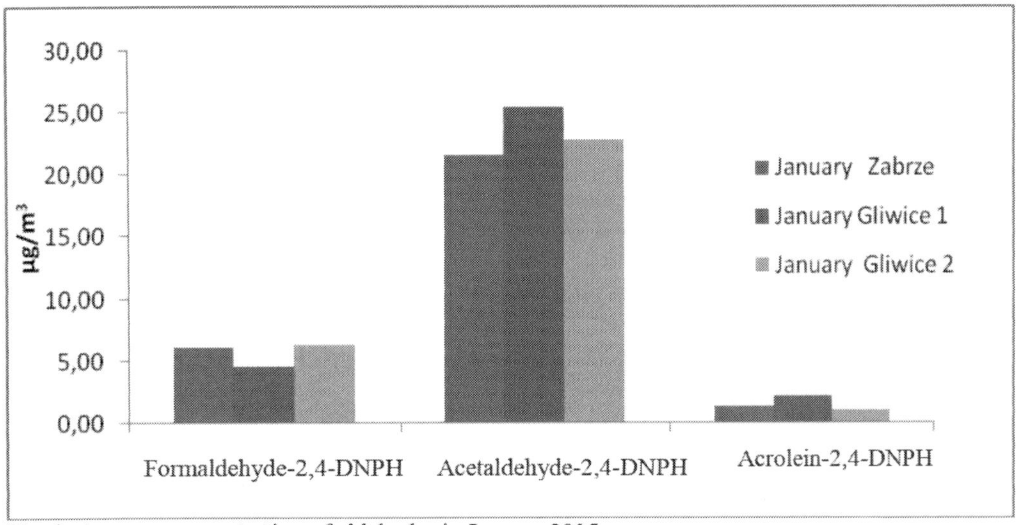

Fig. 2. Average concentration of aldehydes in January 2015

Table 4. Mass participation of FA, AD and AC in total concentration of aldehydes in January 2015

Compound	Zabrze [%]	Gliwice 1 [%]	Gliwice 2 [%]
Formaldehyde	21.0	14.1	21.1
Acetaldehyde	74.2	79.0	75.6
Acrolein	4.8	6.9	3.3

Acetaldehyde concentration exceeds 22 μg/m³ at every sampling point giving an maximal average value 25.5 μg/m³. Presented results indicate that dominant determined aldehyde was acetaldehyde, it's concentration equals 74% to 79% of total aldehyde concentration. Such high mass participation in total concentration may be result of influence of local combustion sources upon air quality.

Acetaldehyde is formed and released by the combustion of organic materials, a variety of natural processes and human activities. In addition, secondary formation of acetaldehyde occurs in the atmosphere through oxidation of natural and anthropogenic volatile organic compounds. Important mater is to consider half-life of determined compounds. It is widely known that under atmospheric conditions, formaldehyde is readily photo-oxidized in to carbon dioxide. Reaction with trace substances and pollutants in the air is relatively fast so that its half-life in urban air, under the influence of sunlight, is short. In the absence of nitrogen dioxide, the half-life of formaldehyde is approximately 50 minutes during the daytime. In the presence of nitrogen dioxide, this drops to 35 minutes. On the other hand the atmospheric half-life of acetaldehyde, calculated based on hydroxyl radical reaction rate, is less than six hours. The night time decomposition of acetaldehyde is expected to occur by the gas-phase reaction with nitrate radicals, this tends to be more significant in urban areas, where the concentration of the nitrate radical is higher than in rural areas. A half-life of 35 days was calculated based on an average atmospheric concentration of nitrate radicals in typical mildly polluted urban center.

The major sources of acrolein emissions were mobile and unspecified stationary combustion sources. Acrolein is relatively unstable in the atmosphere, therefore transport within the air is expected to be limited. On the other hand the relatively high vapor pressure of acrolein (274 mm Hg at 25 °C) suggests that this compound will not partition from the vapor phase to particles in the atmosphere. Acrolein concentration in January varies from 1 μg/m^3 to 2.2 μg/m^3 and its mass participation in total aldehyde content ranges from 3.3% to 6.9%.

Profile of determined aldehydes in summer period undergo significant qualitative and quantitative changes in same sampling points. In June samples, collected at point Gliwice 2, acrolein presence was not detected. Lack of this compound is most likely a result of sampling point location situated in low traffic residential area. In sampling points located near communication routes (Gliwice 1 and Zabrze) acrolein concentration equals to approximately 2.3 μg/m^3. Formaldehyde concentration ranges from 0.83 μg/m^3 to 4.81 μg/m^3 depending upon sampling point (Fig. 3). Highest concentration was recorded in Zabrze.

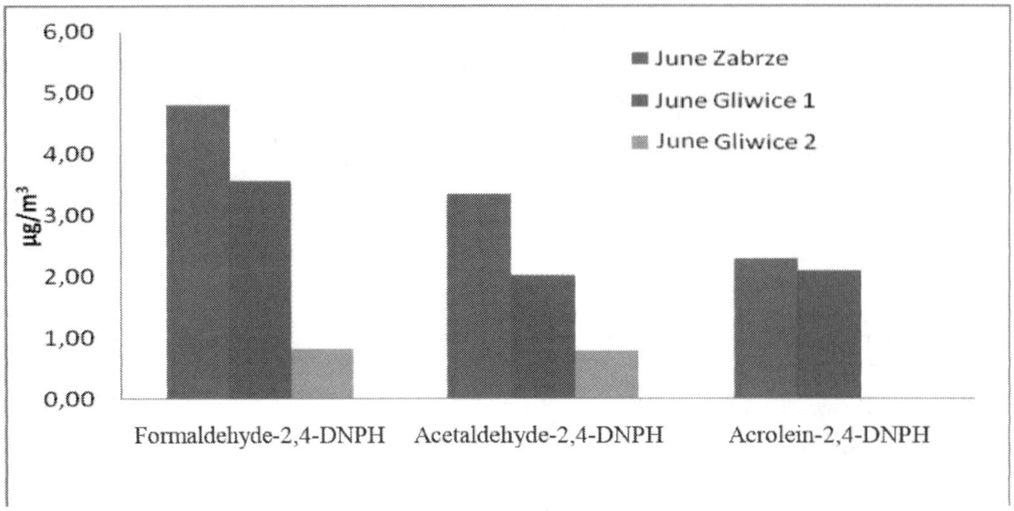

Fig. 3. Average concentration of aldehydes in June 2015

Table 3. Mass participation of FA, AD and AC in total concentration of aldehydes in June 2015

Compound	Zabrze [%]	Gliwice 1 [%]	Gliwice 2 [%]
Formaldehyde	46.1	46.5	51.5
Acetaldehyde	32.0	26.2	48.5
Acrolein	21.9	27.3	no

Formaldehyde content in total aldehyde concentration equals on average: 46% for Zabrze and Gliwice 1, 51% for Gliwice 2 (Table 3.). Acetaldehyde concentration in June was significantly lower in comparison to January and ranges from 0.8 μg/m^3 to 3.3 μg/m^3. Furthermore, its average share in determined compounds ranges from 26% to 48%.

Diversified aldehyde profile in summer and winter indicate important influence of local aldehyde emission sources upon air quality. Atmospheric conditions like: precipitation, solar radiation and ozone content can influence concentration of determined aldehydes.

4. Conclusion

Conducted studies indicate significant differences in aldehyde gas phase concentration in respect to year period. In summer aldehyde concentration was from three to ten times lower in respect to winter period, depending upon localization of sampling point. This tendency is most likely result of influence of local emission sources on air quality.

Differences were observed in profile of aldehydes gas phase concentration. In winter period dominant species present in the air was acetaldehyde in summer however highest concentration of formaldehyde was observed. In summer period acrolein was present only in places being communication nodes.

References

1. M. Czaplicka, K. Jaworek, and A.Wochnik, Archives of environmental protection. **41** 21-31 (2014)
2. W. Chongming, W Lixia, C Zhanjun, X. Hongming, D. Ritchie, and S. Shijin Combustion Science and Technology. **188**, 329-345 (2016)
3. R. Sommarival, J. A. de Gouw, M. Trainer, E. Atlas, P. D. Goldan, W. C. Kuster C. Warneke, and F. C. Fehsenfeld, Chem. Phys., **11**, 7081–7096 (2011)
4. R. M. LoPachin, and T. Gavin, Chem Res Toxicol, **27** 1081–1091 (2014)
5. C. Marchand, S. Bulliot, L Calvé, and P. Mirabel, Atmospheric Environment, **40,** 1336-1345 (2006)
6. M. Odabasi, R. Seyfioglu Atmospheric Environment **39,** 5149–5156 (2005)
7. B. P. Andreini, R. Baroni, E. Galimberti, U. G. Sesana, Microchemical Journal **67**, 11-19 (2000).
8. I.V. Fierro, C.J. Popp, R.S. Martin, Atmospheric Environment **38**, 249–260 (2004).
9. R. Che´nier, Human and Ecological Risk Assessment **9**, 483–509 (2003)

Spatial distribution of pollutants in the area of the former CHP plant

Robert Cichowicz[1,*]

[1] Lodz University of Technology, Faculty of Architecture, Civil and Environmental Engineering, Al. Politechniki 6, 90-924 Łódź, Poland

Abstract. The quality of atmospheric air and level of its pollution are now one of the most important issues connected with life on Earth. The frequent nuisance and exceedance of pollution standards often described in the media are generated by both low emission sources and mobile sources. Also local organized energy emission sources such as local boiler houses or CHP plants have impact on air pollution. At the same time it is important to remember that the role of local power stations in shaping air pollution immission fields depends on the height of emitters and functioning of waste gas treatment installations. Analysis of air pollution distribution was carried out in 2 series/dates, i.e. 2 and 10 weeks after closure of the CHP plant. In the analysis as a reference point the largest intersection of streets located in the immediate vicinity of the plant was selected, from which virtual circles were drawn every 50 meters, where 31 measuring points were located. As a result, the impact of carbon dioxide, hydrogen sulfide and ammonia levels could be observed and analyzed, depending on the distance from the street intersection.

1 Introduction

Formation of air, water and soil pollutants has always been inseparably linked with the history of human development. The problem of environmental pollution is to be considered both globally [1-6] and locally [7-12], since individual elements can affect the environment at different scales and at different times. This is a result of the simultaneous increase of industrial production, and consequently the increase of demand for heat and electricity, as well as the systematic increase of population and of its density in a given area and the creation or development of the so-called urban and rural tissue [13-15].

Therefore, in order to ensure that present and future generations can live in a unpolluted environment, pollution from anthropogenic sources, including in particular those related to energy industry and transportation should be controlled and reduced [16-19].

At the same time, the question arises: what to do with objects or entire areas that were previously heavily exploited, e.g. as a result of the operation of CHP plants, and have become degraded. The solution may be to carry out the reclamation and/or revitalization of such sites. This is especially true in cities where old boiler houses, heating plants and CHP

* Corresponding author: robert.cichowicz@p.lodz.pl

plants are located in the centers, and now their look and pollution of the neighborhood arouse repulsive sensations [20-21].

For this reason, it seems interesting to analyze such air pollutants as carbon dioxide (CO_2), hydrogen sulfide (H_2S) and ammonia (NH_3), especially when it refers to the area of nearly 20 hectares of a former CHP plant located in the vicinity of residential and commercial buildings and in the immediate vicinity of a big city center. In the world such objects/sites have now a second youth because of problems with the so-called urban sprawl which generates more and more costs and introduces additional planning and urban constraints. Therefore, more and more often a so-called 'return to the center' is observed. This, however, must be safe for future users [22-24].

The three selected pollutants are very common during the operation and closing of such industrial plants. Their excessive concentrations are or may be hazardous to human health (Table 1).

Table 1. Maximum allowable concentrations (MAC) of carbon dioxide (CO_2), hydrogen sulfide (H_2S) and ammonia (NH_3) in selected countries [25-28].

Country	**Carbon dioxide (CO_2)**		**Hydrogen sulfide (H_2S)**		**Ammonia (NH_3)**	
	MAC, mg/m³ (ppm)	**Instantaneous MAC, mg/m³ (ppm)**	**MAC, mg/m³ (ppm)**	**Instantaneous MAC, mg/m³ (ppm)**	**MAC, mg/m³ (ppm)**	**Instantaneous MAC, mg/m³ (ppm)**
Belgium	9131 (5000)	54784 (30000)	7 (5)	14 (10)	14 (20)	36 (50)
France	9000 (5000)	–	7 (5)	14 (10)	7 (10)	14 (20)
Germany	9100 (5000)	18200 (10000)	7.1 (5)	14.2 (10)	14 (20)	28 (40)
Poland	**9000**	**27000**	**7**	**14**	**14**	**28**
Sweden	9000 (5000)	18000 (10000)	7 (5)	14 (10)	14 (20)	36 (50)
Great Britain	9150 (5000)	27400 (15000)	7 (5)	14 (10)	18 (25)	25 (35)
USA: – NIOSH	9000 (5000)	54000 (30000)	–	15 (10)	18 (25)	27 (35)
– OSHA	9000 (5000)	–	(4)	30 (20) 70 (50)	35 (50)	–
UE	**9000 (5000)**	–	**7 (5)**	**14 (10)**	**14**	**36**

2 Methodology of measurements

The EC-2 plant (Fig. 1) is located in Lodz (Poland) in the district of Górna at the intersection of Wróblewskiego Street and Aleja Politechniki. The analyzed area in the north-west and west is adjacent to industrial buildings, while in the north, east, south and south-west direction with residential development. Additionally, in the north and east 300 m from the site there are 3 and 11 storey buildings (these are mainly the buildings belonging to the Lodz University of Technology: 19th century palaces and revitalized post office buildings, as well as student houses inhabited for 10 months a year).

Therefore, it can be assumed that around the former EC-2 plant dominate buildings with a height exceeding 5 m, for which the average coefficient of aerodynamic roughness is 2 m.

At the same time, according to [29-30], on the area affected by the EC-2 plant operation there have been no national parks or forest complexes and spa areas or historical monuments included in the World Heritage List as well as the areas belonging to the European network Natura 2000.

Fig. 1. Location of the analyzed CHP plant, i.e. EC-2 plant in Łódź [31].

The largest intersection of streets in the immediate vicinity of the studied site was chosen in the analysis as a reference point from which every 50 m virtual circles were drawn (Fig. 2) with 31 measuring points located on them (Fig. 3). For measurements, a VEGA-GC micro-chromatograph [32] was used, which allowed for analysis in field conditions. The device was equipped with 2 batteries, a tank with carrier gas (helium), sample pump and computer module, as well as 2 parallel columns to measure CO_2, H_2S and NH_3. The micro-chromatograph had a thermal conductivity detector (TCD) capable of analyzing samples with a minimum concentration of 500 ppb (0.005 ppm) in 6 to 300 seconds depending on the type of gas being tested.

A method for measurement of gaseous pollutants developed by Cichowicz [33-34] was used. Location of the measuring points on the virtual circles was dependent on the possibility of safe and repeatable adjustment of the device (at each measuring point three samples of air were taken). The measurements were carried out in 2 series/dates, i.e. 2 and 10 weeks after closing the CHP plant (17 April 2015 and 11 June 2015). During each measurement series, the concentration of pollutants was analyzed in 31 selected measuring points located on the circle lines with a common central point located at the crossing of the Wróblewskiego Street and Aleja Poltechniki, every 50 m (according to Fig. 2), collecting and analyzing three times the air sample at a 90 second interval and for further analysis taking the average value. At each measuring point, CO_2, H_2S, NH_3 and air temperature, as well as the wind direction and velocity were measured. The information on pressure and relative humidity was read from the meteorological station located about 5 km in a straight line from the Reymont airport in Łódź [35]. The measuring device was always positioned in the direction free from buildings and additionally an external device equipped with a GPS receiver was used. In the moment when a sample was dosed, the device recorded date, time and coordinates from the GPS system on the memory card (Table 2).

Fig. 2. Location of measuring points at EC-2 plant area.

Fig. 3. Location of the measuring point.

Table 2. Parameters necessary for numerical analyses (boundary conditions) [35].

Date of measurement/ Parameter	17.04.2015	11.06.2015
Temperature [°C]	-3 ÷ -1	5 ÷ 7
Cloudiness	light overcast	sunny
Wind flow velocity [km/h]	13 ÷ 15	9 ÷ 13
Wind direction	W ÷ NW	SW
Pressure [hPa]	1011	1022
Relative humidity [%]	40 ÷ 58	31 ÷ 43

3 Results of measurement

The measurements together with the analysis of literature data [29-30] are aimed at verifying if the area of the former CHP plant affects the environment (does not exceed the permissible air quality standards - Table 1) and if it might be subjected to post-industrial space revitalization.

On the virtual circles there were measuring points at which the following atmospheric air pollutants were measured: CO_2 (Fig. 4), H_2S (Fig. 5) and NH_3 (Fig. 6). All results obtained are below the maximum limit concentrations of specific pollutants and therefore do not adversely affect air quality in the analyzed area. It would be advisable to carry out this analysis again after a longer period of time to check the trend line for different levels of pollution and to eliminate the impact of the decommissioning activities that took place on the area of the former CHP plant during the two measurement series. In addition, to complete the knowledge of this area it might be useful to analyze pollutants present in the soil.

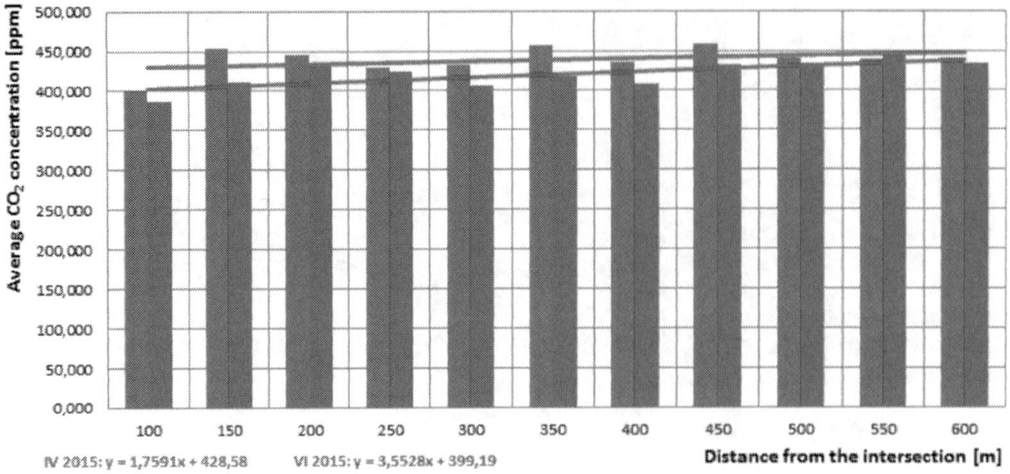

Fig. 4. Change in CO_2 concentration depending on the distance from the intersection (April and June 2015).

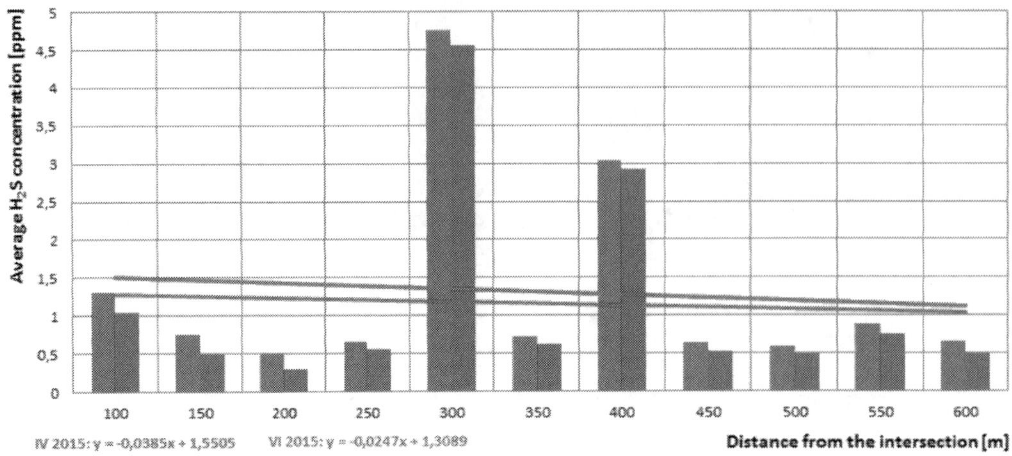

Fig. 5. Change in H_2S concentration depending on the distance from the intersection (April and June 2015).

On the basis of the measured data (Fig. 4), the average concentration of CO_2 was calculated for both series, which was 439.135 ppm and 420.503 ppm for April and June 2015, respectively. Thus, the concentration was reduced by 4.43%, but the level of pollutants in comparison to neighboring areas of public utility [32] was higher by about 13.51% on average.

In the analyzed period (Fig. 5), the average H_2S concentration was 1.319 ppm for April 2015 and 1.161 ppm for June 2015. So, there was a noticeable decrease in pollution amounting up to 13.65%. The presence of this type of gas can be hazardous to the environment and is also associated with the remainder of technical infrastructure because hydrogen sulfide occurs in both channels, waste tanks and over cesspool surfaces.

At the same time, despite the significantly different 2 average levels of H_2S (300 and 400 m from the reference point), the linear trend of the slightly decline in pollutant concentrations is generally visible, which varies with the distance from the streets crossing. However, 2 very high concentrations of pollutant are most probably associated with point sources of pollutant emissions, i.e. residues of chemicals and technical infrastructure occurring in this area.

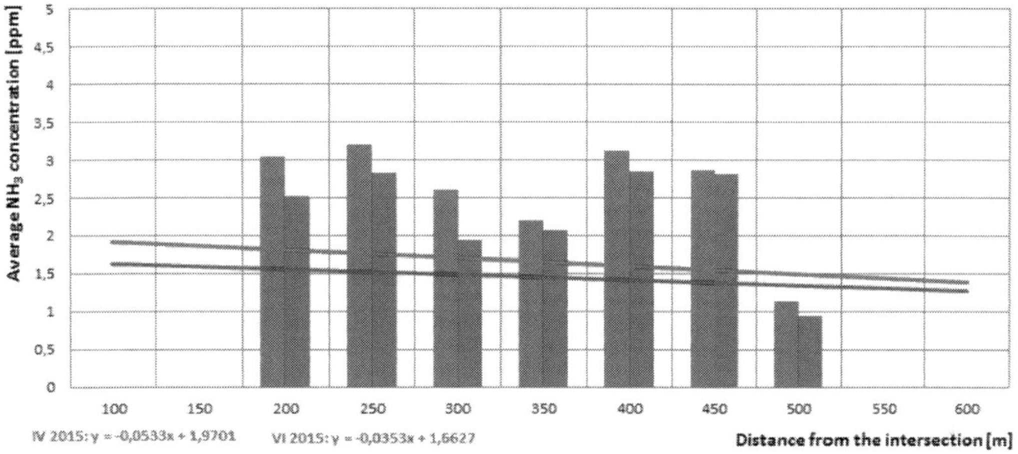

Fig. 6. Change in NH_3 concentration depending on the distance from the intersection (April and June 2015).

The situation was similar for NH_3 whose average level was 1.650 ppm for April 2015 and 1.451 ppm for June 2015, i.e. a 13.74% pollution reduction occurred.

In addition, literature data [36] have been analyzed, which shows that in 2015 all CHP plants in Lodz were responsible for 5.2% of the major pollutant emissions from point sources in the entire Lodz Province. All emissions from the CHP plants decreased by 17.2% year on year (probably due to the closure of EC-2 plant on March 31, 2015).

At the same time, based on [36] referring to air pollutants in 2015, the average annual concentration of SO_2 ranged from 4 to 9 $\mu g/m^3$, of NO_2 exceeded 24 $\mu g/m^3$ in the areas not directly affected by traffic emissions and was equal to 43.5 $\mu g/m^3$ near the roadway. CO concentrations were 600 $\mu g/m^3$ in city centers, while at main transport routes they ranged from 700 to 800 $\mu g/m^3$. The last of the selected pollutants were PM10 and PM2.5, which in city centers ranged from 25.6 $\mu g/m^3$ to 42.9 $\mu g/m^3$ and from 21.1 $\mu g/m^3$ to 22.1 $\mu g/m^3$, respectively.

4 Summary

The measurements and analysis of literature data [29] showed that the area of the former CHP plant had only slight environmental impact and did not exceed the permissible air quality standards.

The use of gas micro-chromatograph allowed accurate measurements of concentrations of carbon dioxide, hydrogen sulfide and ammonia as well as helped to investigate the influence of the communication path on the levels of particular pollutants.

The results are consistent with the trend observed in the city of Lodz where in recent years there has been a noticeable improvement in the quality of air, but this is not related to increasing exceedance of pollutant emissions associated with the use of low energy sources and the increasing number of mobile sources (rapidly increasing number of diesel vehicles).

It should be remembered that the role of the EC-2 CHP plant (as well as other CHP plants in Lodz) in shaping the air pollution immission fields in this city was small or even negligible owing to the size of the emitters and waste gas treatment installations. This is confirmed by the 'Air Protection Program in Łódź' in which the CHP plants are not indicated as a source of threats to the aerosanitary conditions in Lodz [36].

Such analyzes may help answer the question: whether and in what direction should reclamation and/or revitalization in a given post-industrial site be carried out and whether it is safe to use such degraded sites.

The author thanks Prof. Grzegorz Wielgosiński for help, valuable suggestions and opinions.

References

1. R. P. Schwarzenbach, T. Egli, T. B. Hofstetter, U. von Gunten, B. Wehrli, Annu. Rev. Env. Resour. **35**, 109–136 (2010)
2. T. Asami, Contaminated Soil '88, 261–263 (1988)
3. D. Bucke, Symposia of the British Society for Parasitology **106**, 1, S25–S37 (1993)
4. G. A. Folberth, T. M. Butler, W. J. Collins, S. T. Rumbold, Environ. Pollut. **203**, 235–242 (2015)
5. R. Bossi, K. Vorkamp, H. Skov, Environ. Pollut. **217**, 4–10 (2016)
6. A. de la Torre, P. Sanz, I. Navarro, M. Á. Martínez, Pollut. **217**, 26–32 (2016)
7. D. T. Lobdell, V. Isakov, L. Baxter, J. S. Touma, M. B. Smuts, H. Özkaynak, Environ. Health Persp. **119**, 487–493 (2011)
8. R. Cichowicz, G. Wielgosiński, W. Fetter, Environ. Monit. Assess. **189**, 605 (2017)
9. G. Wielgosiński, R. Cichowicz, A. Targaszewska, J. Wiśniewski, Ecol. Chem. Eng. S. **24**, 2, 263–275 (2017)
10. D. G. Atanassov, Int. J. Environ. Pollut. **14**, 1-6, 217–224 (2000)
11. D. J. Nowak, S. Hirabayashi, A. Bodine, E. Greenfield, Environ. Pollut. **193**, 119–129 (2014)
12. H. Flocas, A. Kelessis, C. Helmis, M. Petrakakis, M. Zoumakis, K. Pappas, Theor. Appl. Climatol. **95**, 3–4, 265–277 (2009)
13. X. Tang, Urbanization, Energy, and Air Pollution in China: The Challenges Ahead: Proceedings of a Symposium, National Academies Press, 47–54 (2004)
14. B. Prieto, P. Sanmartín, Urban Pollution and Changes to Materials and Building Surfaces **5**, 193–224 (2016)

15. H. Martins, A. Miranda, C. Borrego, Air Pollution - A Comprehensive Perspective **2**, 27–72 (2012)
16. G. Fernandez-Martinez, P. Lopez-Mahia, S. Muniategui-Lorenzo, D. Prada-Rodriguez, E. Fernandez-Fernandez, Atmos. Environ. **35**, 5823–5831 (2001)
17. D. Li, Y. Guo, Y. Li, P. G. Ding, Q. Wang, Z. Cao, Open Journal of Air Pollution **1**, 37–41 (2012)
18. Y. Qin, L. Y. Chan, Atmos. Environ. **27B**, 275–282 (1993)
19. L. Y. Chan, Y. M. Liu, Atmos. Environ. **35**, 2637–2646 (2001)
20. M. K. Ng, Soc. Indic. Res. **71**, 1–3, 441–465 (2005)
21. E. Ozus, V. Dokmeci, International Real Estate Review **8**, 1, 144–159 (2005)
22. M. Matysiak, Problemy Turystyki i Rekreacji (PTiR) **4**, 46–70 (2013)
23. M. P. Johnson, Environ. Plann. A **33**, 4, 717–735 (2001)
24. S. Angel, J. Parent, D. L. Civco, A. Blei, D. Potere, Prog. Plann. **75**, 2, 53–107 (2011)
25. http://limitvalue.ifa.dguv.de/.
26. J. Stetkiewicz, Podstawy i Metody Oceny Środowiska Pracy **4**, 70, 97–117 (2011)
27. RTECS, Registry of Toxic Effects of Chemical Substances (2007)
28. ACGIH, Guide to occupational exposure values, Cincinnati (2008)
29. Application for issuing an integrated permission (IPPC) for introduction of pollutants into the environment by the Zespół Elektrociepłowni in Łódź S.A. (ZEC) **5**, Lodz University of Technology (2005)
30. G. Wielgosiński, Atmosphere Protection Analysis (Operat Ochrony Atmosfery) for ZEC in Łódź S.A. Department of EC-2, Lodz University of Technology (2004)
31. Google Earth 7.1.2.2041 (freeware)
32. Catalogue card of micro-chromatograph VEGA GC (DI-EN0022-1 VegaGC)
33. R. Cichowicz, G. Wielgosiński, Ecol. Chem. Eng. S. **22**, 2, 189–200 (2015)
34. R. Cichowicz, G. Wielgosiński, Ecol. Chem. Eng. S. **22**, 4, 513–525 (2015)
35. http://freemeteo.com/
36. Lodz: Regional Environment Protection Inspectorate, Report on the state of the environment in the Lodz region in 2015 (2016)

Particle size distribution of fly ash from co-incineration of bituminous coal with municipal solid waste

Ewelina Cieślik[1*], *Tomasz* Konieczny[1] and *Bartłomiej* Bobik[1]

[1] Polish Academy of Sciences, Institute of Environmental Engineering, M. Skłodowskiej-Curie 34, 41-819 Zabrze, Poland

Abstract. One of the source of air pollutants is emission from local coal-fired boiler-houses and domestic heating boilers. The consequence of incineration of municipal waste is the introduction of additional pollutants into the atmosphere, including fly ash. The aim of this work was to evaluate the particle size distribution of fly ash emitted by coal combustion and co-incineration of coal with municipal waste in a domestic 18 kW central heating boiler equipped with an automatic fuel feeder. Mixtures of bituminous coal with different types of solid waste (5, 10 and 15% of mass fraction) were used. Solid waste types consisted of: printed, colored PE caps, fragmented cable trunking, fragmented car gaskets and shredded tires from trucks. During the incineration of a given mixture of municipal waste with bituminous coal, the velocity of exhaust gas was specified, the concentration and mass flow of fly ash were determined together with the physico-chemical parameters of the exhaust gas, the samples of emitted fly ash were taken as the test material. Particle size analysis of fly ash was performed using laser particle sizer Fritch Analysette 22. The PM10 share from all fly ashes from incineration of mixtures was about 100%. Differences were noted between PM2.5 and PM1.

1 Introduction

One source of air pollutants is the emission from local coal-fired boiler-houses and domestic heating boilers. About 37% of households in Poland are heated with single-function or combi solid-fuel boilers [1].

The use of feeders in boilers with permanent grates provides better fuel combustion conditions, but does not prevent the emission of harmful substances. The construction of such boilers allows them to incinerate solid waste, combustion of which is prohibited in Poland in boilers unsuitable for this purpose. During combustion in these boilers, the resulting gases are introduced into the atmosphere without purification, providing a major source of harmful emissions (including PM10 and PM2.5) [2].

* Corresponding author: ewelina.cieslik@ipis.zabrze.pl

Solid particles are considered hazardous pollutants because they can be small enough to deposit deep in the lungs and cause respiratory and cardiovascular problems. The most dangerous particulate matter is dust with dimensions not exceeding 2.5 μm [2].

The tests were carried out in a boiler actually used in central heating systems in small buildings, which, unlike physical or mathematical models, allows for an accurate assessment of the effects of fuel combustion in such boilers.

Research on the emission of gaseous and particulate pollutants by waste incineration in open spaces [3-5] and in laboratory conditions [6-9] has been carried out. Emissions from coal combustion [10], waste paper [11-12], polyethylene (PE) [13-15], polyvinyl chloride (PCV) [13-14], tires [3, 12 and 16] were determined. The studies were conducted for incineration of particular solid wastes. The impact of their co-incineration with coal has not been studied so far, as well as their incineration in domestic boilers equipped with an automatic dispenser and a fuel feeder.

The aim of this work was to compare the fly ash emission and to analyze the particle size distribution of fly ash emitted by coal combustion and co-combustion of coal with municipal waste in a domestic central heating boiler of 18 kW equipped with an automatic fuel feeder.

2 Materials and methods

The following fuels were used in the comparative tests of co-incineration of bituminous coal with individual solid wastes (ethylene propylene diene monomer (EPDM), tires, waste paper, polyvinyl chloride, polyethylene):

- bituminous coal (grain size: 5-25 mm), from the Kazimierz Juliusz mine in Sosnowiec (coal);
- shredded, printed xerographic paper (waste paper);
- fragmented car gaskets (EPDM rubber);
- shredded tires from trucks (tires);
- colored caps from juice, water bottles etc. (PE);
- fragmented cable trunking (PVC).

The examined fuels were subjected to chemical and technical analysis, which included determination of ash and volatile substances. In the case of solid waste co-incineration, 3 mixtures were used which differed by the mass fraction of waste in the fuel. The mixtures were prepared by mixing the previously weighed coal and proper types of municipal solid waste in the appropriate weight ratios. The mixtures containing 5, 10 and 15% of waste were analyzed. Table 1 shows the calculated ash and volatile substances values for each mix used. Coal combustion and co-incineration took place in a 18 kW central heating boiler equipped with an automatic fuel feeder.

Research material - fly ash, was collected during eight-hour combustion cycles of a given mixture of municipal waste with bituminous coal using a measuring probe, from the flue gas directed to the chimney, under the operating conditions of the boiler with nominal load, with full control of fuel combustion parameters. During each combustion cycle, the qualitative and quantitative composition of the main gaseous pollutants (NO_X, CO, SO_2) was continuously analyzed. The mass flow of fly ash in the exhaust gas was determined using the gravimetric method.

Samples of fly ash were subjected to granulometric analysis on PM10, PM2.5 and PM1 content. The results are shown in Table 2.

Particle size analysis of fly ash was performed using laser particle sizer Fritch Analysette 22. The analyzer is used for rapid and automatic particle size analysis of solids in suspension by laser diffraction. The range of operation of the analyzer varies from 0.16

to 1160 μm. Analysette 22 uses the physical principle of electromagnetic wave scattering during particle size analysis after its delivery to the analyzer.

3 Results and discussion

The boiler with the fuel feeder works periodically with two alternating modes - operation (with air and fuel supply to the combustion chamber) and supervision (blower and fuel feeder switched off). This results in a gradual increase and decrease of pollutant concentrations, depending on the mode of operation of the device. Sampling of fly ash to determine the concentration of ash in the exhaust gas was carried out in one-hour periods throughout the measurement period. The results obtained were averaged.

The highest proportion of ash in the fuel was obtained from PVC incineration (26.27%) and lowest from PE incineration (0.61%). All municipal waste showed a higher proportion of volatile substances compared to coal. Waste paper, EPDM rubber, PVC and tires had a comparable level of volatile substances, ranging from 65 to 70%, while PE had a higher share, up to 99%. Table 1 shows the calculated ash and volatile substances values for each mix used.

Table 1. Content of ash and volatile substances in the used mixtures (%m/m).

fuel/ mixture	waste paper			EPDM			PVC		
	5%	10%	15%	5%	10%	15%	5%	10%	15%
ash	4.75	5.50	6.25	4.24	4.36	4.61	5.11	6,23	7,34
volatile substances	32.64	34.61	36.59	34.1	35.82	39.26	32.38	34,1	35,81
fuel/ mixture	PE			tires			coal		
	5%	10%	15%	5%	10%	15%	100%		
ash	3.83	3.66	3.49	4.12	4.24	4.35	4.00		
volatile substances	34.09	37.53	40.96	32.44	34.21	35.99	30.66		

Table 2 shows the calculated dust emission rates per unit of fuel and the estimated share of PM1, PM2.5 and PM10 in the fly ashes collected.

Table 2. Fly ash emission factor (g/kg_{fuel}) and PM1, PM2.5 and PM10 share in fly ashes (%).

fuel/ mixture	waste paper			EPDM			PVC		
	5%	10%	15%	5%	10%	15%	5%	10%	15%
EF_{ash}	2.58	1.50	1.86	3.89	5.01	7.38	4.05	5,93	7,44
PM1	50.8	50.4	50.9	43.5	37.8	44.1	33.4	36,8	46,1
PM2.5	87.1	86.1	87.1	74.8	62.1	68.6	63.9	68,0	78,7
PM10	100	100	100	100	99.3	100	99.9	99,9	99,9
fuel/ mixture	PE			tires			coal		
	5%	10%	15%	5%	10%	15%	100%		
EF_{ash}	3.43	4.25	4.51	1.65	1.96	3.57	3.61		
PM1	40.1	42.1	37.7	42.3	48.4	26.9	42.9		
PM2.5	83.0	81.5	76.7	67.0	72.1	63.4	75.3		
PM10	100	100	100	100	100	98.7	100		

The fly ash emission was lowest in the co-incineration of coal with waste paper and coal with tires (mean values of 2.0 and 2.4 g/kg$_{fuel}$ respectively). It was lower than the corresponding rate in the case of coal combustion, which amounted to 3.6 g/kg$_{fuel}$. During EPDM rubber, PE and PVC co-incineration, the fly ash emission figures were higher (mean values of 5.4, 4.1 and 5.8 g/kg respectively). The fly ash emissions obtained from the co-incineration of the waste are on average seven times lower than in the case of incineration of clean waste under laboratory and open burning conditions [8-9]. The dust emission values obtained from co-incineration of PVC and PE mixtures are not compatible with the literature data. Incineration of these wastes in laboratory conditions caused lower emission in the case of PVC and higher in the case of PE [6].

The particle size distribution of fly ash from coal combustion, PVC and tires mixtures is shown in Fig. 1 - 3. According to literature data, the bulk equivalence ratio has an influence on particle size distribution of fly ash [6].

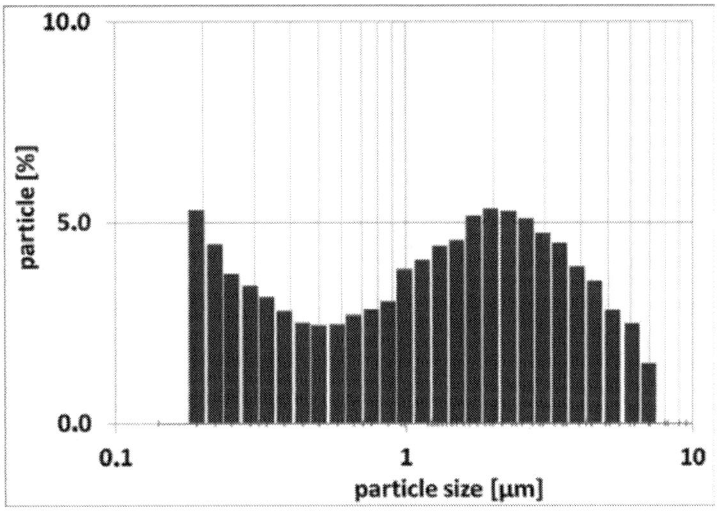

Fig. 1. Particle size distribution of fly ashes from coal combustion.

PM10's share in all fly ash from mixtures' incineration was about 100%. Differences in the PM2.5 fraction were noted. Fly ash from co-incineration of waste paper and PE was characterized by the highest content of PM2.5 (86.8 and 80.4%, respectively). In the case of co-inicineration of EPDM rubber, tires and PVC, PM2.5 fraction share was lower than from coal combustion (75.3%) and equaled to 68.5, 67.5 and 70.2%, respectively. According to literature, PVC combustion is associated with the highest share of particles above 2 μm [6].

In the case of PVC co-incineration, the share of PM2.5 in fly ash increased with the increase of the mass fraction of waste in the mix, whereas in the case of PE co-incineration an opposite was observed - the increase of the waste content in the mix reduced the PM2.5 share in fly ash.

The granulometric composition of fly ash obtained from mixtures containing 5 and 10% of tires was similar. Large differences were noted in the combustion of the mixture containing 15% of tires, as shown in Fig. 3. The PM1 content in fly ash from this mixture was the lowest and was about 27%. Differences have also been observed in fly ash from the combustion of a mixture containing 15% of PVC in relation to mixtures containing 5 and 10% of this waste (Fig. 2). This ash was characterized by a higher proportion of PM1 and PM2.5 than fly ash from combustion of other mixtures of this waste.

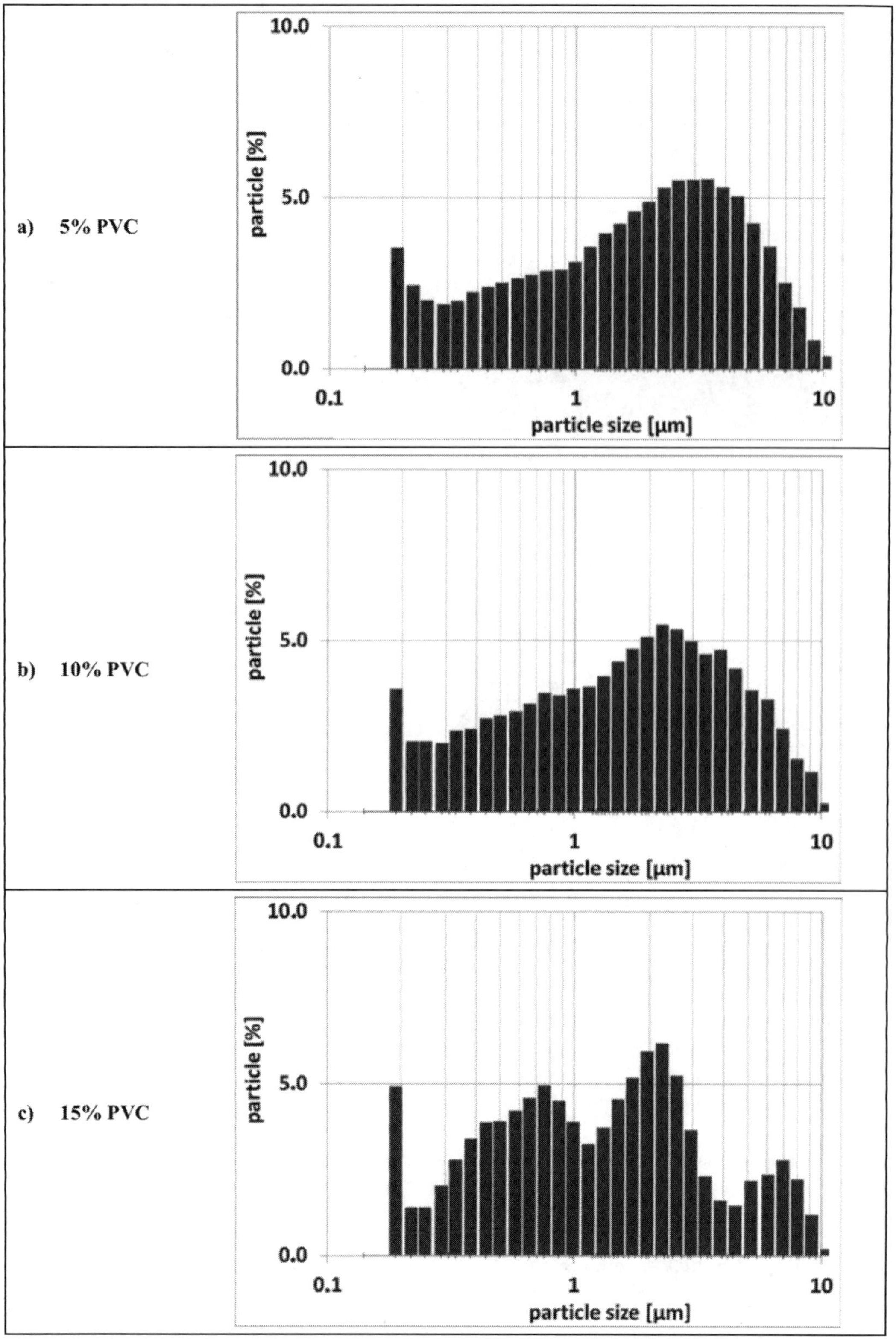

Fig. 2. Particle size distribution of fly ashes from co-incineration of coal with PVC.

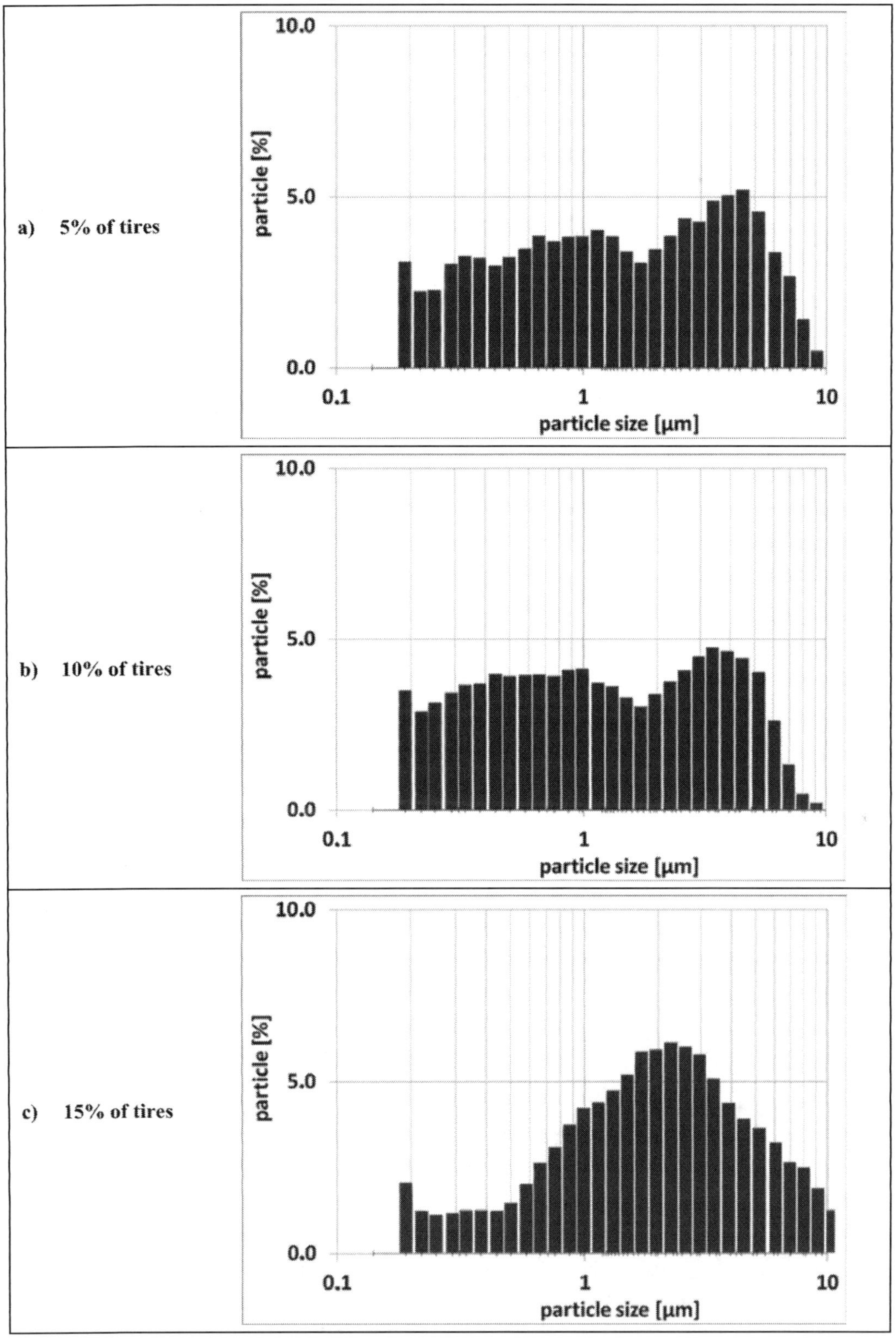

Fig. 3. Particle size distribution of fly ashes from co-incineration of coal with tires.

During the co-incineration of waste paper, the mass increase of waste did not contribute to the change of PM2.5 share in the samples tested. Analysis of the measurement results for EPDM rubber and tire co-incineration showed no correlation.

The PM2.5 and PM1 share in fly ash from coal combustion was lower than the data available in the literature for domestic heating boilers, which is related to the construction of the heating device used and the removal of larger particles while flue gas passes through the baffles in the boiler.

4 Summary

Small solid-fuel boilers emit fly ash with a high share of fine particles. Emission of fly ash from coal combustion equaled 3.6 and from coal co-incineration with municipal waste it ranged from 1.5 to 5.9 g/kg_{fuel} and was lower than in the case of waste incineration under laboratory conditions.

The PM10 share in the fly ash from the incineration of the mixtures was about 100%. Differences were noted between PM2.5 and PM1.

The PM10 share in fly ash was over 98% of emitted dust. The share of PM2.5 ranged between 62-82%, while PM1 varied between 27 and 51%.

The fly ash from the co-incineration of waste paper was characterized by the highest content of PM1 fraction (50.4%) and together with fly ash from PE co-incineration was characterized by the highest content of PM2.5 fraction (86.7% for waste paper and 80.4% for PE).

In the case of PVC co-incineration, the share of PM2.5 and PM1 in fly ash increased with the increase of the mass fraction of waste in the mix, whereas in the case of PE co-incineration an opposite was observed - the increase of the waste content in the mix reduced the PM2.5 and PM1 share in fly ash.

The project has been financed by National Science Centre in Poland under the project no. DEC-2013/11/N/ST10/00995.

References

1. CSO *Energy consumption in households in 2015.* Warszawa 2017
2. K. Juda-Rezler and B. Toczko, Eds., *Pyły drobne w atmosferze. Kompendium wiedzy o zanieczyszczeniu powietrza pyłem zawieszonym w Polsce.* Warszawa: Główny Inspektorat Ochrony Środowiska (2016)
3. A. Lönnermark, P. Blomqvist, S. Marklund, Chemosphere, **70**, 626 (2008)
4. S. Sidhu, B. Gullett, R. Striebich, J. Klosterman, J. Contreras, M. DeVito, Atmos. Environ., **39**, 801 (2005)
5. T.J. Christian, R.J. Yokelson, B. C′ardenas, L.T. Molina, G. Engling, S.-C. Hsu, Atmos. Chem. Phys., **10**, 565 (2010)
6. B.E. Shemwell, Y.A. Levendis, J Air Waste Manag Assoc., **50,** 94 (2000)
7. T. Nakao, O. Aozasa, S. Ohta, H. Miyata, Chemosphere, **62**, 459 (2006)
8. A. Lönnermark, P. Blomqvist, *Emissions from Tyre Fires,* SP Fire Technology, SP REPORT, **43** (2005)
9. F. Reisen, *Inventory of major materials present in and around houses and their combustion emission products*, Report to Bushfire CRC, (2011)

10. A.B. Ross, J.M. Jones, S. Chaiklangmuang, M. Pourkashanian, A. Williams, K. Kubica, J.T. Andersson, M. Kerst, P. Danihelka, K.D. Bartle, Fuel, **81,** 571 (2002)
11. J.S. Choi, D.K. Lee, K.H. Kim, Y.M. Jo, *Evaluation of Emission Factor of Harmful Elements from Open Burning*, (ISWA/APESB World Congress 2009, Lisbon 2009)
12. J.A. Conesa, R. Font, A. Fullana, I. Martín-Gullón, I. Aracil, A. Gálvez, J. Moltó, M.F. Gómez-Rico, J. Anal. Appl. Pyrolysis, **84,** 95 (2009)
13. A. Valavanidis, N. Iliopoulos, G. Gotsis, K. Fiotakis, J. Hazard. Mater., **156**, 277 (2008)
14. T. Panagiotou, Y.A. Levendis, J. Carlson, P. Vouros, Symp. (Int.) Combust., **26,** 2421 (1996)
15. B.R.T. Simoneit, P.M. Medeiros, B.M. Didyk, Environ. Sci. Technol., **39**, 6961 (2005)
16. P.M. Lemieux, Ch.C. Lutes, D.A. Santoianni, Prog. Energy. Combust. Sci., **30**, 1 (2004)

Eco-driving: behavioural pattern change in Polish passenger vehicle drivers

Piotr Oskar Czechowski[1]*, *Aneta* Oniszczuk-Jastrząbek[2], and *Tomasz* Czuba[2]

[1] Gdynia Maritime University, Faculty of Entrepreneurship and Quality Science, Gdynia, Poland
[2] Gdańsk University, Faculty of Economics, Sopot, Poland

Abstract. In Poland, as in the rest of Europe, air quality depends primarily on emissions from municipal, domestic and road transport sources. The problems of appropriate air quality are especially important within urban areas due to numerous sources of emissions being concentrated in relatively small spaces in both large cities and small/medium-sized towns. Due to the steadily increasing share of urban population in the overall number of population, the issue of providing clean air will over the years become a more significant problem for human health, and therefore a stronger incentive to intensify research. The key challenge faced by a modern society is, therefore, to limit harmful substance emissions in order to minimise the contribution of transport to pollution and health hazards. Increasingly stringent emission standards are being imposed on car manufacturers; on the other hand, scant regard is paid to the issue of drivers, i.e. how they can help reduce emissions and protect their life and health by applying eco-driving rules.

1 Introduction

The health effects of the impact of pollution on the humans are sometimes difficult to estimate, because they depend on the substance and its concentration, as well as the time of exposure. According to WHO, 1.3 million city-dwellers die prematurely around the world each year as a result of air pollution [1]. In Europe (according to EEA data [2]), the death toll includes 350 thousand premature deaths due to exposure to $PM_{2,5}$ as well as 20 thousand premature deaths due to exposure to high concentrations of O_3 [3]. WHO data [4] also shows that exposure to particulate matter is a worldwide cause of 8% deaths due to lung cancer, 5% deaths due to cardiovascular disease, 3% deaths due to respiratory infections. WHO also demonstrates [5] that an increased morbidity / mortality risk due to respiratory diseases is associated with exposure to NO_2, including at concentrations below the limit values.

At a time when human activity is having more of a negative environmental impact, contributing to increased carbon dioxide emissions into the atmosphere, and fuel prices are rising, the knowledge and practice of eco-driving techniques are soon bound to become an important life skill.

This article surveys primary research based on interviews with Polish drivers, providing the motivation for the author's own statistical models and reflections on the subject. The research was undertaken with the basic aim of examining eco-driving awareness and

* Corresponding author: e-mail: oskar@am.gdynia.pl

behaviour. To achieve the aim of stepping up measures to promote eco-driving, however, it is necessary to explain what eco-driving involves. The research has provided a way to evaluate the awareness level of eco-driving and identify measures which further responsible driving behaviour in Polish drivers, while also recognizing factors which contribute to fuel savings in city driving. The research involved seeking information on average fuel consumption in city and out-of-city conditions when obeying and ignoring eco-driving rules. The models proposed compare fuel consumption in eco-driving and non-eco-driving situations.

The issues which the models seek to address include:

1. Empirical distributions of selected variables.
2. Factors contributing to lower fuel consumption in city driving.
3. Eco-driving awareness, including driver profiles of those who recognize and apply eco-driving techniques.

2 Review of the literature

Air pollution and noise levels are becoming more and more of a problem every year. Air pollution contributes significantly to health issues, leading to numerous respiratory and circulatory diseases. Having regard to health and environment protection as a whole, it is especially important to combat emissions. Industrial operations and transportation are responsible for dust and toxic gas emissions which combined with humid and windless conditions result in the formation of an unnatural weather phenomenon – smog. The effects of smog are most visible in large city areas, causing urban decay and health problems, as growing numbers of people are struggling with allergic reactions, asthma and respiratory failure [6].

According to WHO, inhaling harmful particles from polluted air is the cause of 2 million deaths every year. Scientists agree that implementing solutions designed to limit poisonous gas emissions would reduce the number of premature deaths by up to 2/3. PM10 particles can make their way into the lungs and bloodstream, resulting in cardiac failure, lung cancer, asthma and acute respiratory infections. These are the findings of a WHO report based on research conducted by that organization. Pollution comes mainly from coal-burning power plants and vehicle engines [7] .

A review of CO_2 emissions figures by sector and projections of changes in emissions expected up until 2030 confirms that the decarbonisation of the transport industry requires the implementation of a series of initiatives [8] based on innovative technologies, organizational improvements and behavioural adjustments. Despite technological advancement, there is a continuing (and growing) source of climate-changing CO_2 emissions in the form of high traffic flow in city areas and urban driving style characterized by frequent stopping and moving off.

One of the initiatives to limit CO_2 emissions is eco-driving. Eco-driving is basically a smooth and controlled method of driving a car. The following essential driving tips are offered: plan journeys ahead, anticipate traffic flow, move into the highest possible gear while keeping a constant speed, brake gently by gearing down, turn off the engine when stopped for longer than a minute, avoid engine pre-heating (including in winter), maintain correct tyre pressure, limit air conditioning usage to a reasonable minimum [9, 10]. The relations between individual items on the list are widely discussed by Andoa, Nishihori [11], Andrieu, Saint Pierre [12]. To achieve its aim, the article examines driving awareness and behaviour.

Research from pilot studies shows that eco-driving reduces average fuel consumption by 10% per kilometre [13]. Also, this manner of driving reduces CO_2 emissions by 6 – 10% [13] and fuel consumption in buses by 2% in the 12 months following training [14] 4.35%

in the two-month follow-up period [15]. The mean change in fuel consumption for all drivers in the study after their eco-driving course was a reduction of 5.8%, but with large differences between individuals [16].

Various factors may contribute to improving driver behaviour and eco-driving awareness. Research conducted by Boriboonsomsin, et.al. shows that 40% of respondents have changed their driving style in reaction to a rise in fuel prices. Other factors may include, e.g. vehicle weight, road quality, weather conditions, road congestion or the need for reducing carbon dioxide emissions [10].

According to Jack N. Barkenhaus, eco-driving brings substantial benefits to individuals (lower costs and better personal safety) and society (reduced CO_2 emission, lower oil imports, less conventional emissions and lower death rate) [17, 18, 19, 20]. The authors of the article decided to investigate which factor – economic or ecological – contributes more to Polish drivers following eco-driving rules.

3 Methodology

The nationwide study (using the direct interview method (PAPI)) examined how drivers / passenger vehicle owners employ the rules of eco-drive and how this affects their health. The research was carried out in the period between 12 November 2015 and 30 November 2015 on a random sample of 1000 drivers / passenger vehicle owners throughout the country. The attitude to eco-driving was one of the points of this study that focused on driver behaviour and preference in terms of patterns of petrol station visits.

The aim of the study was to assess driver awareness of eco-driving techniques and verify their use in practice. The separate point was to identify factors contributing to reducing fuel consumption. The bulk of research work was aimed at proving that transport pollution levels could be reduced dramatically through individual action, leading to significant reductions of emissions globally.

In order to verify the hypotheses, the selected models were used, depending on the measurement scales of researched empirical stochastic and exploratory variables and assumptions. Among the key models, the following should be mentioned: simple models of the ANOVA family, multidimensional principal component (PCA) and correspondence analysis (CA) models, as well as advanced generalised regression models (GRM). The obtained results fully addressed all of the research goals, while the cognitive value gained in the process proved to be useful in a practical dimension. It shows how well the rules of eco-driving are known and applied in practice. The awareness of principles of eco-driving is at a low level in Poland and undertaking educational action is the only way to bring the desired (including health-related) effects for the general public. It is especially important in the context of constantly growing number of cars and risks related to air pollution. High awareness of eco-driving among drivers / passenger vehicle owner will have an advantageous impact on health protection and natural environment.

A basic research problem is to answer questions in which both dependent and independent variables are measured on weak, i.e. nominal and ordinal, scales. This is due to the fact that a classic regressive model cannot be adopted here.

An alternative solution is to use linear log models, and specifically, a logistic regression model. It assumes that a dependent variable is a dichotomous variable, while independent variables can be measured on any measurement scale.

Further on, a model of factors contributing to eco-driving awareness will be used as an example to explain the model estimation results. The knowledge of eco-driving is a dichotomous dependent variable, in which a negative answer is coded as "0" and a positive one as "1." Independent variables relating to the knowledge of eco-driving are measured on various measurement scales, such as "age" (ordinal scale), "gender" (nominal scale).

A multiple regression model is the most appropriate quantitative analysis method to determine which variables under consideration are factors with a relevant impact on the knowledge of eco-driving. A standard solution, such as, for example, multi-dimensional regression would be insufficient to examine this type of cause-and-effect relationship, as different scales have been used to evaluate variables. It has been assumed that a logistic regression model will allow identifying relevant factors. The identification in this particular case is limited to seeking to establish a regression-like connection between the occurrence probability of a given variant (contact / no contact) with a group of independent variables, such as respondent answers and other factors, e.g. gender.

4 Factors contributing to eco-driving awareness

4.1 Analysis of variance (ANOVA)

The aim of analysis of variance (ANOVA) is to analyse differences between group means, both within specific groups and across groups in populations from which samples were drawn.

Generally speaking, it is a technique of analysing variables (experiment results, observations) which depend on one or several co-occurring factors. Here, unlike in exploratory methods, these factors are known. The most frequently considered factors in this research include gender, university, place of origin, place of residence and others (these are sometimes referred to as grouping, classifying or manipulating factors). Each factor has a number of variants, such as male / female gender, six tertiary education centres in the Tricity area, three variants for place of origin (places with population ranges from up 5 thousand, from 5 to 50 thousand, and over 50 thousand) and four variants for place of residence while in college (dormitory, rented quarters, family home, home-owner). ANOVA allows verifying whether the factors under analysis have an impact on the observed variables.

A variable analysed in this way is known as a dependent variable and must be at least interval-scaled, e.g. respondent age (in years). A single-factor analysis of variance test examines factor impact on the outcomes of a relevant feature. The analysis of variance test is presented in Table 1.

Table 1. Table of analysis of variance (ANOVA)

Source of variability	Sum of squares	Degrees of freedom	Variance	F-Test
Across populations (groups)	$\sum_{i=1}^{k}(\bar{x}_i - \bar{x})^2 n_i$	k – 1	$\hat{S}_1^2$	$F = \frac{\hat{S}_1^2}{\hat{S}_2^2}$
Within groups (random effect)	$\sum_{i=1}^{k}\sum_{j=1}^{n_i}(x_{ij} - \bar{x}_i)^2$	n – k	$\hat{S}_2^2$	

Source: Author's own research

The *F*-statistic value is compared with critical value $F\alpha_{,k-1,n-k}$ taken from F-Snedecor distribution tables for specific relevance level α and specific number of degrees of freedom k-1 i n-k. If $F \geq F\alpha_{,k-1,n-k}$, then H_0 hypothesis on the equality of means in the population should be rejected. On the other hand, for $F < F\alpha_{,k-1,n-k,}$ there is no basis for rejecting hypothesis H_0. Rejecting a zero hypothesis constitutes proof of a relevant impact on the populations under survey. Otherwise, all the populations could be considered equal in terms of feature values obtained, which means that the dependent variable is not impacted by the factor.

4.2 Kruskal-Wallis Test

The classic ANOVA is not applicable to situations where the variance equality condition is not satisfied. An alternative to one-way ANOVA is provided by non-parametric Kruskal-Wallis test (KW ANOVA).

This test assumes that a variable under consideration is a continuous variable which has at least been interval-scaled. The hypothesis tested here is that samples have been drawn from a population following the same distribution or from distributions having the same median. The method of interpretation for this test is the same as in a single-factor parametric ANOVA except here it is based on ranks, not on means.

$$H = \frac{12}{n(n+1)} \sum_{i=1}^{k} \frac{T_i^2}{n_i} - 3(n+1) \tag{1}$$

where: n – total size $= \sum_{i=1}^{k} n_i$, n_i – group size, T_i – sum of ranks for a given group, k – number of groups / n – *is the total number of observations across all groups* $= \sum_{i=1}^{k} n_i$, n_i – *is the number of observations in group i,* T_i – *is the sum of rank (among all observations) of observation from group i, k – number of group*

5 Factors contributing to reducing fuel consumption in city driving – research results

5.1 Relationship between eco-driving awareness and reduced fuel consumption in city driving

The research question was to establish the relationship of eco-driving awareness and reduced fuel consumption in city driving (Table 2).
Because of failure to satisfy the condition of variance homogeneity in a classic ANOVA model, an alternative KW ANOVA (Kruskal-Wallis) model has been applied, as presented in Table 2.

Table 2. Relationship between eco-driving awareness and reduced fuel consumption in city driving

Eco-driving – Have you heard…?	**DifferencesM (4_3-4_1)**	**DifferencesM (4_3-4_1)**	**DifferencesM (4_3-4_1)**
	Mean	Valid	Standard deviation
I have never met with this concept	-0.66319	47	1.445497
I have met with this concept, but I do not know exactly what it means	-0.86642	134	1.301066
I have met with this concept, and I know exactly what it means	-0.87823	271	0.888573
I do not know / hard to say	-2.25000	6	3.342903
Total	-0.87068	458	1.149414

Source: Author's own research

Both models point to a significant influence of eco-driving awareness on fuel consumption, but excessive dispersion in the distribution of respondents declaring uncertainty prevents a clear-cut answer. It is necessary to develop more detailed models taking account of more factors (GRM).

5.2 Identification of factors contributing to reducing fuel consumption

https://doi.org/10.1051/e3sconf/20182801009

A GRM model has been applied to identify factors contributing to reducing fuel consumption. Five relevant variables have been identified. The strongest factors turned out to be age, eco-driving awareness and gender, as presented synthetically in the Pareto graph (Fig. 1.) using t-statistic value modules for each factor.

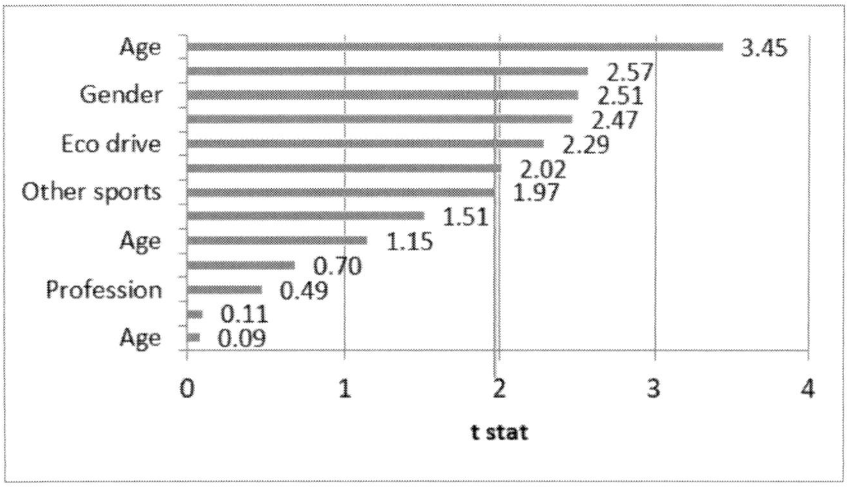

Fig. 1. Pareto-model of fuel consumption factors
Source: Author's own research

Individual factor values have varying impacts on fuel consumption savings. Largest fuel savings have been observed in the youngest age group under 25 but with the highest degree of statistical dispersion, which points to a strong heterogeneity of that group. This may also suggest that this age group is need of saving money. The asymmetry is strongly left-skewed (alpha3= -2.13; Table 3), suggesting a majority of respondents, whose fuel consumption saving is above the average value in the distribution, and who therefore do not drive economically. The age group over 60 is the only one to display right-skewed asymmetry, suggesting a majority of economical drivers who closely follow eco-driving rules (Table 3).

Table 3. Average decrease fuel consumption

Differences (without and with eco-driving) in fuel consumption in the city								
Age	**N**	**average**	**ME**	**Min**	**Max**	**S**	**alfa3**	**alfa4**
Age over 60	36	-0.64	-0.60	-3.00	4.00	1.14	1.63	7.14
Age 51 - 60	41	-0.73	-0.20	-10.00	0.50	1.65	-4.73	26.12
Age 31 – 40	170	-0.93	-1.00	-9.00	3.00	1.02	-3.27	25.07
Age 26 - 30	74	-0.77	-0.70	-6.87	4.00	1.34	-0.97	7.22
Age 41 - 50	108	-0.84	-1.00	-3.50	1.00	0.80	-0.46	0.42
Age up to 25	29	-1.41	-1.00	-7.00	0.00	1.49	-2.13	6.22

Source: Author's own research

From an analysis of the eco-driving awareness factor, it can be seen that there is considerable variety in factor values, which makes mean-based interpretation rather difficult. As in the case of age, however, value distribution should be taken into account. The group which is the most homogenous in terms of dispersion (V(S)=101%) and least asymmetrical is represented by those who declare eco-driving awareness and practice (alfa3=-0.22; Table 4).

Table 4. Contribution of eco-driving awareness to fuel consumption savings in city driving

Eco-drive perception	N	average	ME	Min	Max	S	V(S)	alfa3	alfa4
I have never met with this concept	47	-0.66	-0.50	-6.87	4.00	1.45	-217.96	-1.26	8.45
I have met with this concept, but I do not know exactly what it means	134	-0.87	-1.00	-10.00	3.00	1.30	-150.17	-3.46	21.37
I have met with this concept, and I know exactly what it means	271	-0.88	-1.00	-5.00	4.00	0.89	-101.18	-0.22	5.38
I do not know / hard to say	6	-2.25	-1.00	-9.00	0.00	3.34	-148.57	-2.33	5.57

Source: Author's own research

From an analysis of the occupation factor's value distribution, it seems that the best eco-drivers are self-employed respondents. Although the saving on fuel consumption (-0.83) is not the lowest in this group, self-employed respondents exhibit the smallest dispersion and a right-skewed asymmetry, suggesting that a majority of these people apply eco-driving techniques correctly (fuel consumption below average) (Table 5).

Table 5. Contribution of occupation to fuel consumption savings in city driving

Occupation	N	average	ME	Min	Max	S	V(S)	alfa3	alfa4
pensioner	28	34.57	-1.01	-0.50	-10.00	1.95	-191.96	-3.88	17.78
employee	296	47.90	-0.88	-1.00	-9.00	1.16	-132.12	-2.15	13.73
entrepreneur	103	46.19	-0.83	-1.00	-4.00	0.87	-105.10	1.05	9.46
unemployed	31	39.74	-0.81	-1.00	-3.00	0.93	-115.08	-0.75	0.48

Source: Author's own research

It remains an open question if, besides ecological considerations, the decisive contributor to eco-driving is the economic motivation (saving money).

6 Conclusions

Eco-driving is a technique which contributes to reducing fuel consumption, demand for resources and, as a result, limiting atmosphere degradation thanks to lower carbon dioxide emissions. It also leads to savings on transport costs (e.g. road infrastructure), lower street noise levels and improving living standards as a result of reduced dust content in the air and alleviating the danger of smog.

Poland is seeing more measures being taken in that direction. There are driving schools instructing students in eco-driving techniques. Various educational centres provide courses, training and lessons. Drivers are becoming more aware of and knowledgeable about eco-driving. It is hoped that eco-driving will gain widespread approval as a way of environmental care, a response to the need for savings and for greater social awareness.

The research suggests that the essential factors contributing to eco-driving awareness and practice, and in turn to fuel savings are gender, age and occupation. The models confirm the hypotheses and allow a precise quantitative description of the trends under survey.

References

1. http:// www.who.int /mediacentre /factsheets / fs313/en/, (2008)
2. http://www.eea.europa.eu/data-and-maps/indicators/air-pollution-by-ozone-1/assessment
3. *WHO Regional Office for Europe.* Copenhagen (2006)
4. World Health Organization. WHO Library Cataloguing-in-Publication Data, (2009)
5. Technical report. WHO, (2013)
6. EEA: Towards clean and smart mobility. Transport and environment in Europe (2016)
7. I. Sówka, Czysta Energia, **1-2** (2017)
8. Tracking Progress Towards Europe's Climate and Energy Targets for 2020, EEA, **45** (2014)
9. M. Lempart, P. Malik, Organizacja i Zarządzanie, Zeszyty Naukowe Politechniki Poznańskiej, **60** (2013)
10. K. Boriboonsomsin, A. Vu, M. Barth, University of California Transportation Center, (2010)
11. R. Andoa, Y. Nishihori, , Procedia Social and Behavioral Sciences **20** (2011)
12. C. Andrieu, G. Saint Pierre, Procedia - Social and Behavioral Sciences **54** (2012)
13. A. Schroten, Domain Report Transport, CE Delft (2012)
14. A. E. Wahlberg, International Journal of Industrial Ergonomics 37 (2007)
15. M. Zarkadoula, G. Zoidis, E. Tritopoulou, Transportation Research Part D **12**, (2007)
16. B. Beusen, S. Broekx, T. Denys, C. Beckx, B. Degraeuwe, M. Gijsbers, K. Scheepers, L. Govaerts, R. Torfs, L. L. Panis, Transportation Research Part D **14** (2009)
17. J. N. Barkenbus, Energy Policy **38** (2010)
18. G. Majewski, W. Rogula-Kozłowska, P.O. Czechowski, A. Badyda, A. Brandyk, Atmosphere **6** (2015)
19. W. Rogula-Kozłowska, G. Majewski, P.O. Czechowski, P. Rogula-Kopiec, EPE **43**, 1 (2017)
20. A. Badyda, P. Dąbrowiecki, P.O. Czechowski, G. Majewski, Respiratory Physiology & Neurobiol. **209** (2014)

E3S Web of Conferences **28**, 01010 (2018) https://doi.org/10.1051/e3sconf/20182801010
Air Protection in Theory and Practice

Preliminary comparative assessment of PM_{10} hourly measurement results from new monitoring stations type using stochastic and exploratory methodology and models

Piotr Oskar Czechowski[1]*, *Tomasz* Owczarek[1], *Artur* Badyda[2], *Grzegorz* Majewski[3] , *Mariusz* Rogulski[2] and *Paweł* Ogrodnik[4]

[1]Faculty of Entrepreneurship and Quality Science, Gdynia Maritime University, Gdynia, Poland
[2]Faculty of Building Services, Hydro- and Environmental Engineering, Warsaw University of Technology, 20 Nowowiejska St., PL00-653 Warsaw, Poland
[3]Warsaw University of Life Sciences, Faculty of Civil and Environmental Engineering, Nowoursynowska 166, 02-776 Warsaw, Poland
[4]The Main School of Fire Sevice, Faculty of Fire Safety Engineering, Slowackiego 52/54, 01-629 Warsaw, Poland

Abstract. The paper presents selected preliminary stage key issues proposed extended equivalence measurement results assessment for new portable devices - the comparability PM_{10} concentration results hourly series with reference station measurement results with statistical methods. In article presented new portable meters technical aspects. The emphasis was placed on the comparability the results using the stochastic and exploratory methods methodology concept. The concept is based on notice that results series simple comparability in the time domain is insufficient. The comparison of regularity should be done in three complementary fields of statistical modeling: time, frequency and space. The proposal is based on model's results of five annual series measurement results new mobile devices and WIOS (Provincial Environmental Protection Inspectorate) reference station located in Nowy Sacz city. The obtained results indicate both the comparison methodology completeness and the high correspondence obtained new measurements results devices with reference.

1. Introduction

The direction in which follow modern society is the development and growing importance of urban industry-agglomerations – smart cities. It is essential that process proceeded in a way that ensures balances economic development as a consequence of providing a high quality of life for residents. In classical terms, the model concept of "smart city" the environment is one of six main elements [12].

* Corresponding author: e-mail: oskar@am.gdynia.pl

Automatic air pollutants monitoring (APM) systems are a key part of the "nervous" smart city [12]. The functions it must ensure, among others, such as reporting on the state of the air, weather warnings or ultimately control the emission determine the adopted solutions. Such system must provide above all the credibility of measurements results. Obtained data allow to use them in key models: impact air pollution on health [1, 3]. It is indispensable in the statistical methodology [4].

The use of quantitative methods, including stochastic and exploratory techniques in environmental studies does not seem to be sufficient in practical aspects. There is no comprehensive analytical system dedicated to this issue, as well as research regarding this subject.

Automatic air monitoring systems in the agglomerations are built on the basis of reference devices (WIOŚ, ARMAAG). Currently, they provide the highest credibility of measurement results, and thus – mathematical modeling, environmental assessments including health impacts and forecasts. The disadvantages are the high cost of purchasing equipment and their maintenance over a long period of time, to ensure the highest quality of measurement results. This is a barrier to the development of monitoring networks even for large industrial urban areas. The purchase of reference equipment is sometimes costing several hundred thousand dollars, depending on the equipment and measured impurities.

This is one of the reasons why there is a need for cheaper measuring devices. Such devices, apart from significantly lower purchase and maintenance costs, have a number of other advantages, including:

- allow making measurements in places not yet available for large and expensive stations such as public transport, inaccessible hilly and marine areas,
- the network of such stations, through their large number, allows for more accurate modeling and forecasting, as errors are eliminated. with mathematical spatial interpolation necessary for reference point measurements,
- allow making measurements in the immediate vicinity of the impact of pollutants on humans so that a precise assessment of the impact of pollutants on health is possible.

These three examples show how many potential changes in air pollution measuring can introduce new devices.

In addition to the advantages, these measures, however, have a major drawback, namely the risk of erroneous quality of measured results. The quality of the measurement itself determines the further use of the acquired data.

In this case, the term "quality" does not mean rating the quality of measured data and the cause of its origin [1]. In the first stage of work on new methods of balancing differences of measurement results, it is necessary to evaluate the conformity with the reference measurement standard - this is a problem of equivalence of measurement results. In the case of PM_{10}, the gravimetric method (manual weighting) [2] is used as the reference method according to PN-EN 12341: 2014 which results in daily results (24 hours). For continuous measurements, it is necessary to average the hourly data.

Particulate matter in one of the major air pollutants. It is a complex mixture of particles of different chemical composition and size that strongly interacts with human health [3, 4, 5]. It can be found both from natural sources (marine aerosol, rock erosion) as well as anthropogenic (transport emissions, municipal and other emissions). It is considered as derivative pollution [6, 7] due to on the complexity of processes that shape the number of particles of a certain size at a given location.

The European Union has set the requirements for permissible concentrations of atmospheric dusts [8] and obliges Member States to take effective action in the event of exceeding the limit values. The Polish legal system is obliged to carry out annual assessments of air quality. Quality assessment is prepared by the Provincial Inspector for

Environmental Protection, which is the result of the obligation imposed by Art. 89 and 90 of the Environmental Protection Law [9].

2. Materials and methods

2.1 Construction of new portable devices

The prototype air quality measurement device was constructed by scientists from the Warsaw University of Technology. The devices consist of microcontroller, temperature, humidity, particular matter sensors and communication modules.

Optical dust sensors are used in the devices. In order to select a suitable sensor, several models were tested. The DFRobots particular matter sensor in the preliminary tests showed the best correlation with the DustTrak portable device, so it was chosen as the primary particular matter sensor for further testing.

According to the producer, the sensor detects three types of particular matters: PM_{10}, $PM_{2.5}$ and PM_1 ranging from 0 μm/m³ to 1000 μm/m³ and has a response time of less than 10 seconds. It allows continuous measurements.

The devices were also equipped with temperature and humidity sensors, a modem and additional components to provide the correct voltage specifications for each component. The central part of the device is based on the Arduino Mega microcontroller. Communication with the particulate matter sensor and modem is via serial ports. The temperature and humidity sensors communicate via a digital interface. The devices operate 24 hours a day. They send measurements to the server with the database [10, 11].

2.2 Data source

The preliminary analysis of the comparability of PM_{10}measurements was based on the results of continuous measurements from five mobile stations u1 to u5 compared to the reference station PL0550A (Fig. 1.) in Nowy Sacz for an annual period from 27.09.2016 to 30.09.2017.

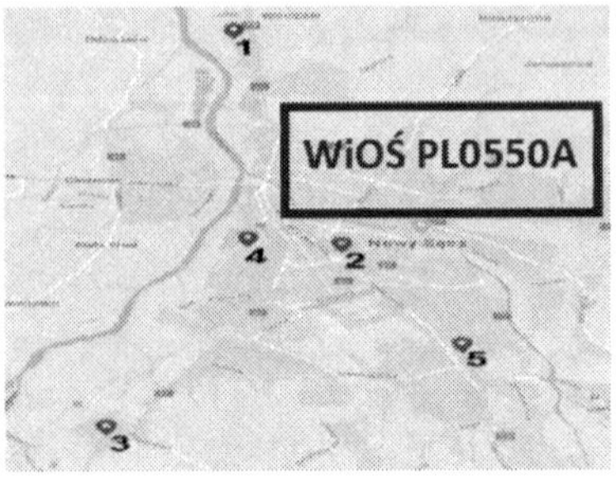

Fig. 1. Measurement units localization in Nowy Sacz city
Source: Author's own work based on research

Table 1 shows average concentrations of PM_{10} and average values of temperatures and relative humidity in the analyzed months at all stations.

Designed measuring devices have been used to build a measuring network operating under real conditions. As their location was chosen the area of the Nowy Sacz. Nowy Sacz is a city of over 80 thousand inhabitants, it covers area of 57 km² and is located in the Malopolska province. The city is characterized by diverse terrain – the lowest point of the city is situated at 272 m AMSL, while the highest – 475 m AMSL. In the city there are typically urban areas with town houses, parks and green areas and single-family housing. Some areas are powered by district heating (therefore, theoretically low emissions should be there reduced), while others, especially single-family housing, have their own fireplaces, which, as may be supposed, can be a source of low emissions. There is one air quality monitoring station owned by polish General Inspectorate of Environmental Protection. The closest measurement station is located in Tarnow (about 45 km in straight line).

Table 1 Monthly averaged values of PM_{10} dust concentration on five mobile (u) stations and reference (Ref), temperature, relative humidity, where Ref – reference station, u1 to u5 measuring devices.

month	PM_{10} [μg/m³]						Temp [°C]	Humidity [%]
	Ref	u1	u2	u3	u4	u5		
1	113.21	97.93	156.74	108.72	121.42	20.17	-2.49	94.99
2	75.05	78.63	138.64	98.25	100.17	70.93	4.13	94.33
3	46.53	44.38	68.56	50.50	62.94	41.31	9.74	89.74
4	28.11	26.33	34.52	27.62	33.93	16.34	11.59	83.48
5	22.86	24.97	25.00	20.60	24.51	5.62	19.12	73.57
6	18.68	13.54	12.10	9.74	14.04	8.02	24.76	58.78
7	16.85	10.15	12.00	9.53	17.23	9.09	25.44	64.42
8	22.91	18.14	17.32	13.84	23.35	19.56	25.48	71.13
9	20.30	20.12	21.34	17.17	28.42	18.39	18.06	80.78
10	26.78	33.08	44.67	35.95	42.43	2.37	11.42	83.42
11	50.88	52.01	72.09	61.29	65.92	6.23	6.85	95.30
12	79.49	67.24	112.47	72.49	83.03	13.89	2.45	98.07

Source: Author's own work based on research

Under the agreement between the Nowy Sacz, Faculty of Building Services, Hydro and Environmental Engineering in Warsaw University of Technology and Gdynia Maritime Academy, the first prototype devices were installed in September in 2016 and started operating in five locations in Nowy Sacz. The location, areas and method of assembly were determined by the employees of the Environmental Protection Department of the Nowy Sacz. The data from the individual devices are available to the residents at: http://www.nowysacz.pl/pomiary-powietrza, so that it is possible to check the current air quality from individual measuring devices.

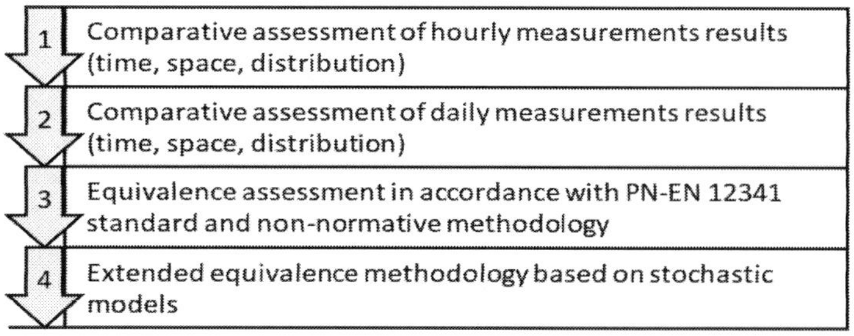

Fig. 2. Proposed measurement results equivalence methodology based on stochastic models
Source: Author's own work based on research

3. Methodology

The present equivalence methods have been described in detail in [3, 4]. These methods allow for the correction of measurement results of various types of new measuring devices based on correction factors and models of linear and orthogonal regression. The main objective of the study is to propose a new, extended methodology of equivalence based on statistical functions and models, including nonlinear models, to improve the quality of the measurement results from new equipment, ie. to reduce the difference in measurement results to reference devices. New correction models are designed to operate in real time, ie. the correction mechanisms will be built directly into devices. The proposed methodology (Fig. 2.) comprises four stages, the first of which is the subject of this study.

The first two steps aim is to identifying the differences in the statistical properties of measurement results comparable sets of new portable devices in relation to reference device in three domains of construction stochastic and exploratory models (time, frequency and space). A comparison of the correctness in the internal structure of the measurement results allows us to identify potential causes for differences, which, according to as the Jenkins Box models [1], will allow to build more accurate correction models in the final stage. Phase three includes the equivalence models currently in use.

The last stage of the extended equivalence methodology includes proposals for correcting the measurement results based on the identified differences in earlier stages.

3.1 Idea

The essence of the first stage of the proposed stochastic correction models is a detailed comparison of internal correctness in the results of measurements, presented on the example of PM_{10} concentrations.

The alignment of the series of measurement results of the examined devices and reference devices is not sufficient. It only allows the final assessment of the differences and the extent to which the measurement results differ from the reference.

The comparison of the causes of these differences, ie. the accuracy of the results of the measurements, allows, on the one hand, for accurate and reliable comparative assessment, and on the second hand, in case of strong differences, to propose correction models based on the causes of differences and not on the results alone (Fig. 3.).

In the time domain, aside from the results of measurements, the key comparisons of correctness are, among others:

- process constancy measured by the basic function of the total autocorrelation and partial autocorrelation, in the form of:

$$r_\tau^* = \frac{\sum_{t=\tau+1}^{n} X_t^* X_{t-\tau}^*}{\sum_{t=\tau+1}^{n} (X_{t-\tau}^*)^2} \tag{1}$$

where: τ - *lag in hours*, X_t – time serie, μ - average, n – total size

- homoscedasticity, the estimation of which is the variability of the process variance over time,

- the stability of mutual relations in selected, previously identified, statistically significant periodic periods for a given concentration.

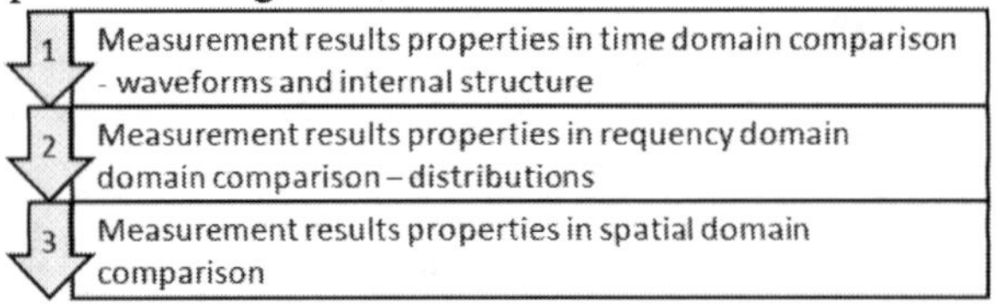

Fig. 3. Proposed measurement results comparison methodology
Source: Author's own work based on research

In the frequency domain, it is crucial to compare the spectral models (Fourier) with the results of periodograms and the form of distributions.

Comparison of the spatial domain is not obligatory - it is required in the case where the measuring devices are located, for independent reasons, in different spatial locations. In this case, the probable causes of the differences should be recognized.

In the final phase, this stage must be eliminated. Monitoring devices should be located in the immediate vicinity of the reference device to minimize the impact of additional interference factors on the measurement differences (eg. meteorological factors, additional emission sources).

This is not always possible in practice, as in this example, for reasons independent of researchers. It is therefore necessary to take this into account.

The property comparison in the results of hourly and continuous measurements averaged daily must lead to common conclusions. If this is not the case, or the deviations in the regularities are very strong, this may indicate instability in the measurements and consequently the impossibility of building a stable correction model for the measuring device.

4. Results

Before comparing, PCA (Principal Component Analysis) model was used to analyze the relationships between mobile station measurements and differences in relation to reference results (Fig. 4.).

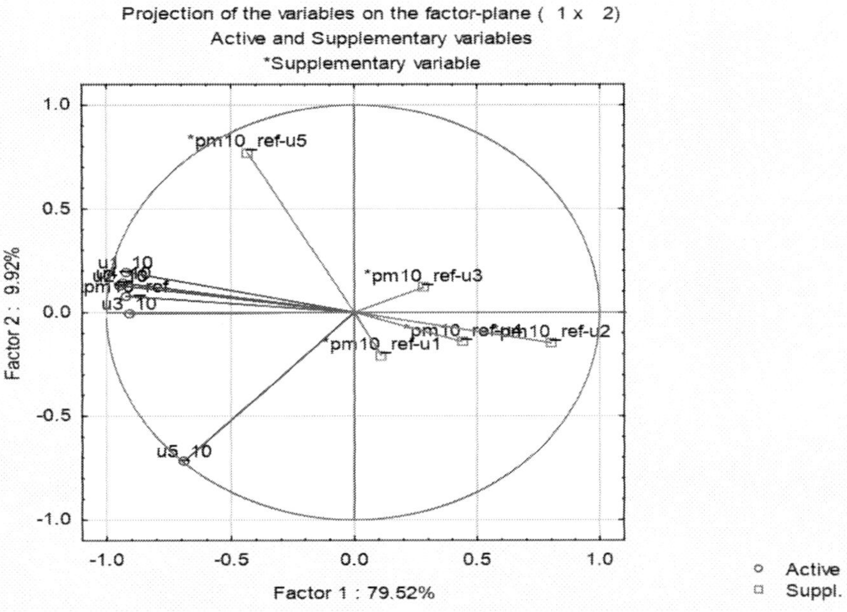

Fig. 4. PM_{10} concentrations results projection and differences from reference results on the two-dimensional plane of the factors in the PCA model.
Source: Author's own work based on research

The obtained results show strong spatial similarities between PM_{10} measurements at stations u1 to u4. Station u5 deviates from the rest due to the location in which there is no impact of traffic and municipal traffic.

A slightly different situation occurs in the case of differences in the results of measurements in relation to reference device. In addition to the PM_{10} measurements at station u5, u3 is also characterized by greater variation. The differences are related to location and confirm the need to build different correction models for each station in the future.

Fig. 5 shows the time series of 1 hour PM_{10} measurements of all analyzed mobile stations (marked as "u") and reference station.

The waveforms indicate a higher variation in the measurement results in the first half of the year, with a high degree of compliance with the reference results. Confirmation is the R2 determination coefficients of linear regression models of mobile stations performance with reference device in the daily cycle (Fig. 6).

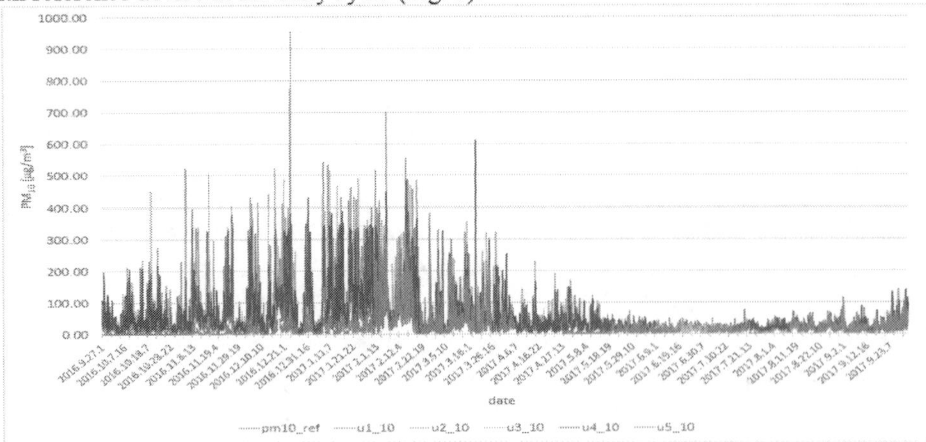

Fig. 5. PM_{10} 1-hourly concentrations results time series; 27.09.2016 to 30.09.2017 in Nowy Sacz. Source: Author's own work based on research

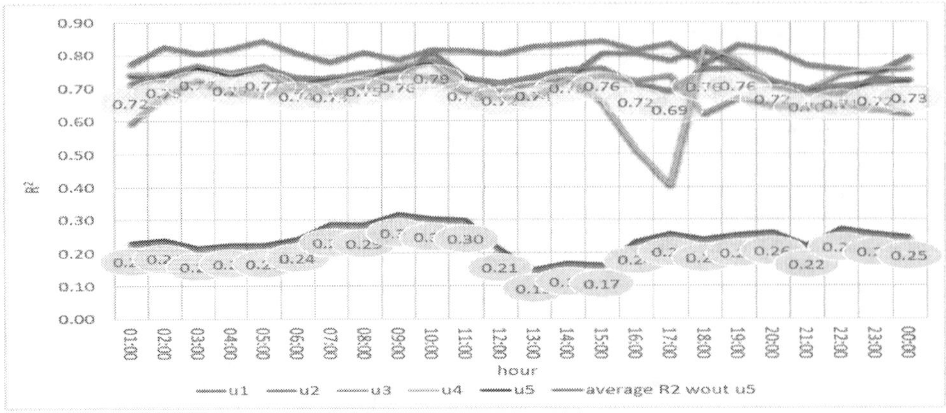

Fig. 6. Determinations coefficients R2 PM_{10} regression models of reference measurements with test stations (u) for each hour in daily cycle.
Source: Author's own work based on research.

Fitting of linear regression models for each station, except u5, is high (average from 0.69 to 0.79), which makes it possible to assert that the introduction of a more accurate correction function can significantly improve the quality of measurement results obtained with new meters.

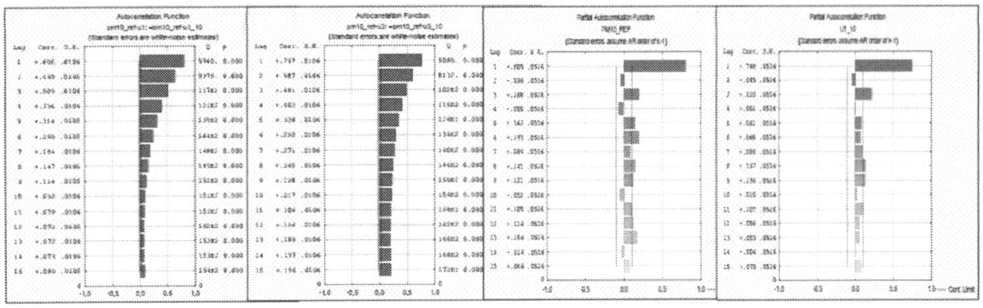

Fig. 7. Selected differences PM_{10} autocorrelation function (a, b) and partial autocorrelation functions of selected PM_{10} results (c, d).
Source: Author's own work based on research.

Comparison of the total and partial autocorrelation of measurement differences from mobile devices to reference device and PM_{10} measurements indicates a high degree of similarity between the PM_{10} measurements of new stations and reference station. This observation leads to the conclusion that the probability of obtaining effective corrections with new correction models using the stochastic methodology is high.

5. Conclusions

Due to the limited volume of the paper, all research results for one hour data, including Fourier models, time correlation coefficients and detailed regression models of PM_{10} measurement results with PM_{10} reference results, are not presented. A separate study presents conclusions for daily averaged data and is therefore directly used for estimation of equivalence.

All obtained results, including these which are not presented, confirm the high compatibility of PM_{10} measurements from new mobile stations with the reference station. Lower differences in the results were observed for the warm season, higher for the cooler. Regression models are also highly adjustable regardless of the time.

A key synthetic conclusion is the high probability of proposing new effective models and correction functions, based on identified regularities, for continuous and daily PM_{10} measurements of new measurement devices. It will also be possible to implement them as the measurement device software, so that the correction will be done in real time.

References

1. P. O. Czechowski, A. Badyda, G. Majewski, Arch. of Environ. Protect., **39**, 4 (2013)
2. J. Gębicki, K. Szymańska, Atmos. Environ. **54** (2012)
3. G. Majewski, W. Rogula-Kozłowska, P. O. Czechowski, A. Badyda, A. Brandyk, ISSN 2073-4433, Atmosphere, **6** 2015
4. A. Badyda, J. Grellier, P. Dąbrowiecki, Adv. in Exper. Med. and Biol., **44** (2016)
5. A. Badyda, P. Dąbrowiecki, P. O. Czechowski, G. Majewski, Resp. Phys. & Neur., **209** (2014)
6. J. Gębicki, K. Szymańska, W. Wardencki, Inst. Inż. Ochr. Środ. Poli. Wroc. (2012)
7. J. Gębicki, K. Szymańska, Pol. J. Environ. Sci. **20**, 6 (2011)
8. Directive of the European Parliament and of the Council 2008/50 / EC, 21 may (2008)
9. Official Gazette Republic of Poland **25**, 15 (2008)
10. M. Rogulski, B. Dziadak, Appl. Manag. (2017)
11. M. Rogulski, **128**, Energy Proc. (2017)
12. BSI Standards Publication, ISBN 978 0 580 84320 4, (2014)

Application of satellite images analysis to assess the variability of the surface thermal heat island distribution in urban areas

Janina Fudała[1] *, *Ádám* Nádudvari[1], *Joachim* Bronder[1] *and Marta* Fudała[1]

[1] Institute for Ecology of Industrial Areas, 6 Kossutha St., 40 – 844 Katowice, Poland

Abstract. One of the elements of the urban plans for adapting to climate change is to identify the range the urban heat island (UHI). To a relatively rare ground station network air temperature, one of the possible methods to identify this phenomenon in cities is the analysis of satellite images, and in particular the thermal images surface cities in conjunction with the land-use structure. In the publication is presented the application of indirect methods of determining surface characteristics of heat island in the cities of Upper Silesia Agglomeration on the basis of the analysis of the thermal images from the satellite Landsat for the period 1986-2016. It presents ways to interpret these images depending on the needs of determination the areas sensitive to the impact of the (UHI) and define the areas where adaptation actions to the climate change should be undertaken.

1 Introduction

The effects of changing climate, particularly increases in temperature, frequency and severity of extreme phenomena occurring in the last few decades, arose. They represent the same threat to the social and economic development of many countries in the world, including Poland. It is therefore necessary to take measures to adapt (adaptation) to the projected impacts of climate change, which should be implemented simultaneously with the activities of limiting greenhouse gas emissions (adaptation).

Observed trends in changes of climatic factors for the period 1981 – 2015 for Katowice synoptic station show:

- increase of the maximum temperature,
- increase of the number of days with temperature > 30°C,
- increase of the number of heat waves during the year,
- increase of the length of heat waves,
- increase of the number of days with temperature > 25°C without precipitation.

* Corresponding author: j.fudala@ietu.pl

The below Figures 1, 2 and 3 document some of the trends of climatic factors for Katowice synoptic meteorological station, representative for Upper Silesian Agglomeration.

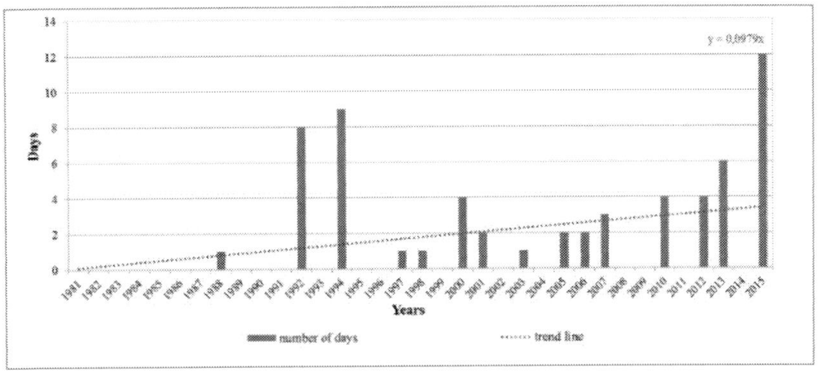

Fig. 1. The number of days with T_{max} > 98th percentile (32.5 °C), Katowice (Poland) synoptic station, 1981-2015; Source: IMGW

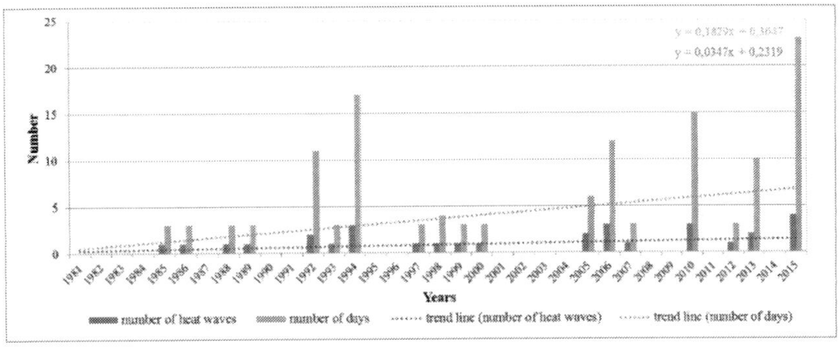

Fig. 2. Heat waves (T_{max} > 30°C), Katowice (Poland) synoptic station, 1981-2015; Source: IMGW

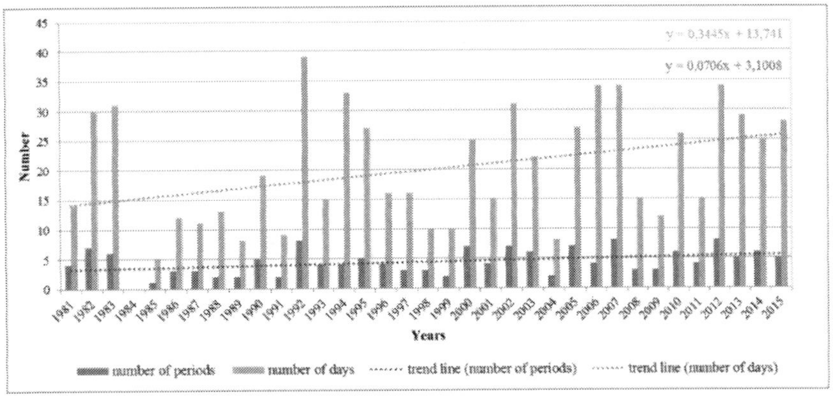

Fig. 3 Precipitation – free days with T_{max} > 25°C, Katowice (Poland) synoptic station, 1981-2015; Source: IMGW

It reflects on the severity of the phenomenon of the UHI and the deterioration of the urban quality of life.

The UHI is microclimatological phenomenon, commonly occurring in urban areas, which is characterising the significant increase of the temperature in the town area compared to surrounding areas. It can be compared to the heat island (or sometimes archipelago), relative coolness. In the literature and the internet we can find the picture of the UHI as on below Figure 4.

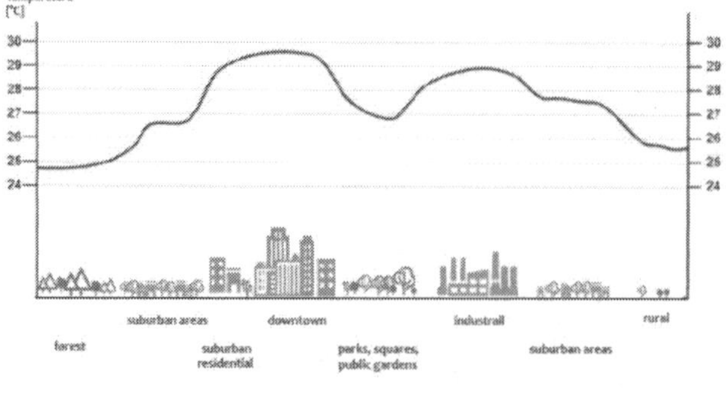

Fig.4. Urban Heat Island (UHI). Source: IETU

The rise of the urban heat island is due to [3]:

- reduced albedo variety of artificial surfaces in the city and the geometry of the city,
- the revised structure of long-term radiation (heat) in built-up areas,
- a small share of natural surface plant,
- human activity,
- the greenhouse effect accompanying the city.

In urban areas are generally not available detailed meteorological data for the determination of the spatial distribution of urban heat island (UHI). Most measurements are performed on representative synoptic stations, located in open areas, in some distance from the city centre and without data from the representative station in the surrounding areas. Therefore the satellite images of surface temperature of the difference land use in the town can be an alternative method for determining of the spatial distribution of the UHI. This method have been used in many urban areas in the world [1, 2, 7] and in Poland [3, 4, 5, 6].

The UHI is most effected on human quality of life during summer time, our investigations were concerned intensity UHI during summer period.

2 Methodology

The current methods for determination of UHI are used both with the standard station meteorological, as well as data from a special observation network, mobile measurement or thermal images of air from satellite.

UHI based on the ground measurements is called the atmospheric heat islands, and this evaluated on the basis of satellite images in literature is called as the surface heat island.

At the heart of the designation of a range of urban heat island (UHI) for Upper Silesia Agglomeration cities lies a few conditions. First, due to the vast range of contemporary surround towns, diversity in land use within the UHI that have been adopted will be set on the basis of the temperature field in the city. The point is that modern cities include areas with highly different thermal properties, reflecting the variation in the land use and also the humidity of the substrate. Secondly, the spatial coverage of UHI was established on the basis of the statistical parameters of temperature distribution within the administrative unit. In this case, the point is that within the city, it can be isolated areas in terms of statistics differ significantly from the surrounding areas. In addition, due to the diversity of land use structure of cities, as well as the nature of the satellite data, for each city, you can designate a different limit absorption isotherms by defining the range of the UHI.

Third, the urban heat island affect generally residential area. Not every residential (e.g., disperse build - up) creates the conditions for the emergence of the UHI. However, it is in the form of residential building compact or dense, with little land unspoiled area creates conditions for the formation of the urban heat island. The following methodology to designate urban heat islands for the cities of Upper Silesia Agglomeration has been used.

The first step in extracting the heat island in the city is mapping the ground thermal systems based on a series of satellite images. On this basis, the average temperature of the map was obtained of land obtained based on the dozens of satellite images. For localisation of urban heat islands the Landsat (4-5TM, 7 ETM+, 8 OLI) and ASTER satellite images (using the thermal infrared bands – TIRS, resolution of 4-5TM: 120m, 7 ETM+: 60m, 100m, ASTER: 90m) have been used. These images were cloud free, summer-time from 2006-2016. According to: https://landsat.usgs.gov/using-usgs-landsat-8-product [8], firstly the radiance had to be calculated, then the brightness temperature:

$$L_\lambda = MLQcal + AL \tag{1}$$

where:
L_λ = TOA spectral radiance (Watts/(m^2 * srad * μm))
ML = Band-specific multiplicative rescaling factor from the metadata file
AL = Band-specific additive rescaling factor from the metadata file
Qcal = Quantized and calibrated standard product pixel values (DN)

Then, TIRS band data can be converted from spectral radiance to brightness temperature using the thermal constants provided in the metadata file:

$$T = \frac{K_2}{ln\left(\frac{K_1}{L_\lambda} + 1\right)} \tag{2}$$

where:
T = At-satellite brightness temperature (K)
L_λ = TOA spectral radiance (Watts/(m^2 * srad * μm)
K_1 and K_2 = Calibration constants for thermal bands of Landsat.

After calculations, the prepared images were layer stacked. Generally these urban heat islands are constant thus layer stacking and averaging can eliminate several atmospheric or surface distortions. Important to mention is that, the calculated images are not surface temperatures, however in case of urban heat island localisation for longer period, these precise surface temperatures are not necessary.

After layer stacking and averaging there were visible the constant heat emitters such as city centres and residential areas with non - or low vegetation, industrial areas, coal wastes dumps without vegetation, roads and relative cold surfaces e.g. forests, vegetated areas, lakes. Also such layer stacking has prepared for Landsat 4-5TM images from approx. 30 year ago and it could be compared with at present ones. It was also noticed that, better to prepare such layer stacking for max. 2-3 years to have more updated data, than 10 years.

The second step was the designation of the isotherms within the city which is the boundary of the maximum range the urban heat island. Above the absorption isotherms are designated according to the following formula:

 https://doi.org/10.1051/e3sconf/20182801011

$$T_{UHI} = \left(\frac{T_{MAXi}+T_{MINi}}{2}\right) - (k * STD_i) \quad (3)$$

where:
T_{UHI} – temperature of the urban heat island, the minimum kinetic temperature of UHI constituting the maximum extent of the UHI
T_{MAXi} - maximum kinetic temperature of surface in town $_i$
T_{MINi} - minimum kinetic temperature of surface in town $_i$
STD_i - standard deviation of mean kinetic temperature in town $_i$
k - derivative of land use pattern in a $_i$ town

Parameter *k* constitutes the percentage of area of all artificial surfaces in a town. This data can be obtained from Corine Land Cower, Urban Atlas or national land use surveys like for instance in Poland there is a database of topographic objects in scale 1:10 000. Parameter *k* may also be calculated on the basis of normalized difference vegetation index (NDVI) value.

Parameters T_{MAXi}, T_{MINi} and STD_i are determined on the basis of raster map of the mean surface kinetic temperature.

The first part of the above equation represents middle surface kinetic temperature in the town. Due to the fact that statistical distribution of the surface kinetic temperature in a given town is not symmetric or even bimodal like in case of analysed example (Figure 5) it turn out that mean or median of the surface kinetic temperature overestimate the area of the UHI while middle temperature underestimate the area. It is important to obtain appropriate temperature corresponding to lower temperature boundary of UHI. To low temperature results in encompassing in the UHI green areas, to high – omitting built up areas. This is the reason why the second part of the equation was added.

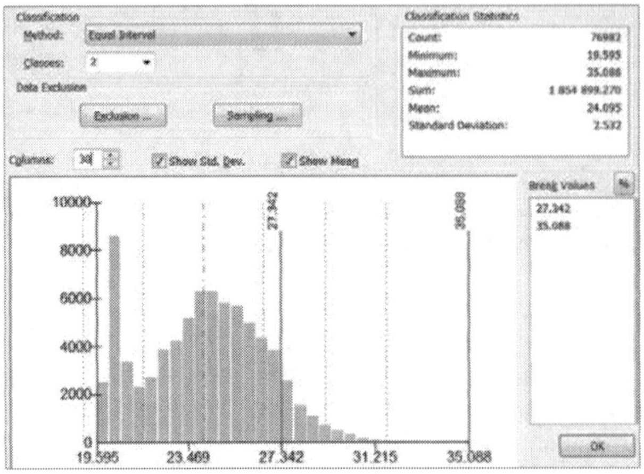

Fig. 5. Histogram of the surface kinetic temperature of the Bytom town

3 Results

At the first step the surface heat kinetic temperature had been determined for Upper Silesian Agglomeration as the mean values from the period 1986 – 2016 which is presented on the Figure 6. This resolution was not proper for detail recognizing the differences of the surface temperatures inside the cities. Only the analyses for particular towns allowed on the determination of changes in surface temperatures during the summer period of different land use areas inside the cities in the analysed period 1986 - 2016. There were chosen three representative areas for introducing different heat emitters (Figure 6). In the case of Bytom, where the city center has been densely built up may indicate the occurrence of UHI. Furthermore the shape of these problematic areas has not changed since 1987-1988, rather increased e.g. on the southern part of the city where new coal waste dumps (with bare surfaces) were deposited. Another problematic areas are introduced by Katowice where a new heat emitter has constructed (shopping centres) in comparison to 1987-1988 (see Figure 6, Katowice, middle part of the image). The applied roof material for such newer buildings should be reconsidered as they serve as potential heat source. Similar problematic areas can be described by Dąbrowa Górnicza, where the industrial facilities occurring as a large heat source together with newer buildings with same problems as previously mentioned by Katowice (roofs). As Figure 6 indicating the extended deep lakes (Dąbrowa Górnicza) or construction new parks, increased number of trees in city center can serve as a cooling feature.

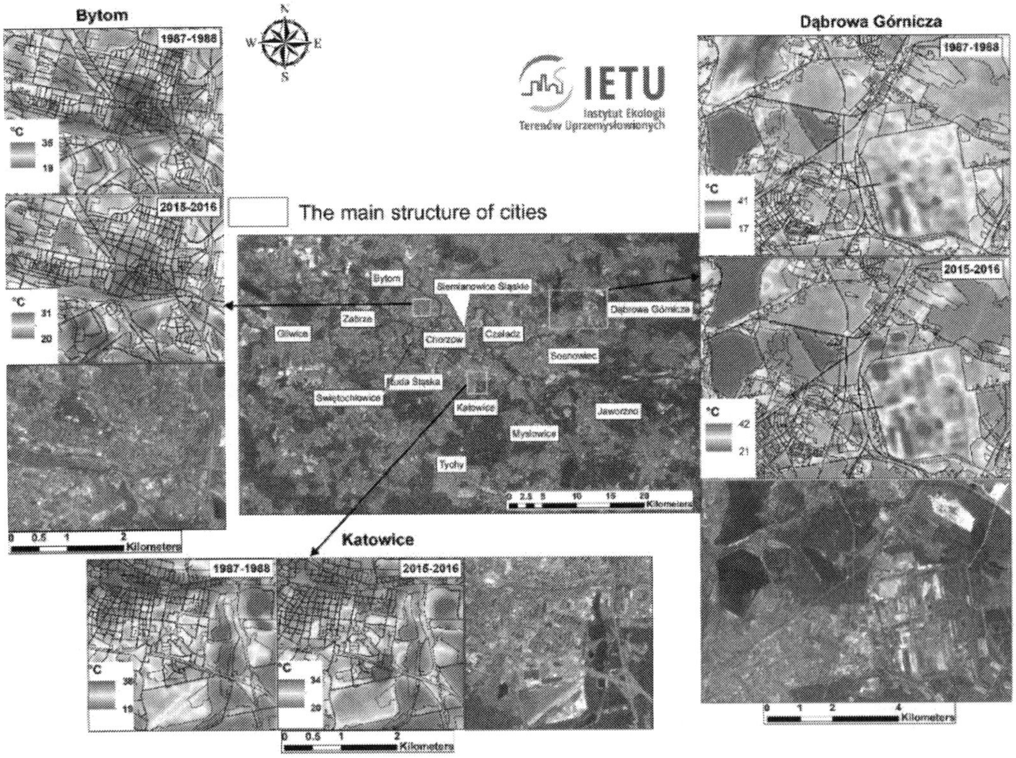

Fig. 6. Chosen three representative problematic areas in the Upper Silesian Agglomeration.

The examples outlined above, indicate the ability to monitor changes to the spatial distribution of surface temperatures within the city due to its development and the identification of the major emitters of the heat into the environment, affecting the inhabitants quality of life.

The boundary of the heat island inside the town due to equation 2 omit the areas with the surface temperature below the mean value. The value of this isotherm depends on the value of k factor used as it is presented on the example of Bytom city.

Extent of the urban heat island presented on Figure 7. is the case applied in Municipal Plan for Adaptation (MPA) project. The lowest temperature of the UHI was determined as middle kinetic temperature minus standard deviation of the kinetic temperature observed in a given city. In case of the Bytom city thus obtained T_{UHI} equals 24.9°C. The area defined by isoline of T_{UHI} is 3145.41ha, what constitutes 45.4% of the whole city area. The residential area covers area of 1042.07ha what constitutes 15.03% of the city area.

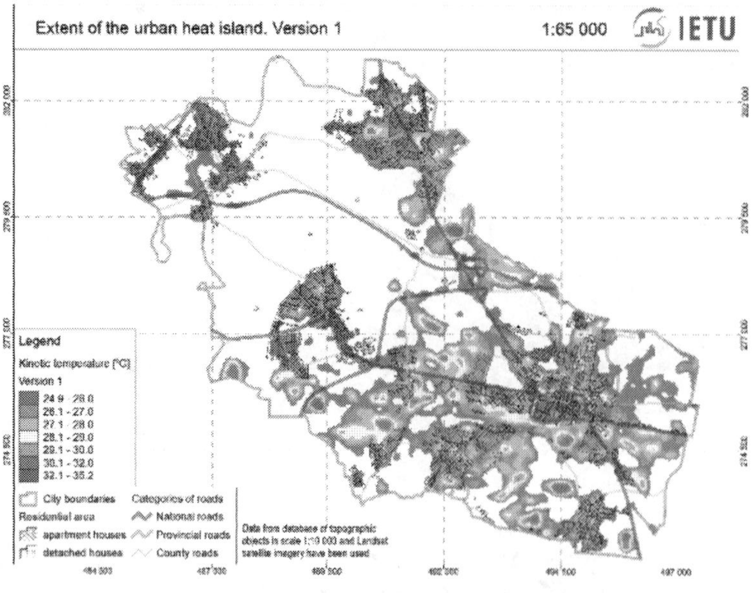

Fig. 7. Extent of the urban heat island of Bytom determined by use of $k = 1$

Extent of the urban heat island presented on Figure 8. is the case when the standard deviation was multiplied by half of percentege of artifical areas (see equation 2). In case of the Bytom city thus obtained T_{UHI} equals 26.86°C. The area defined by isoline of T_{UHI} is 1237.23ha, what constitutes 17.8% of the whole city area.

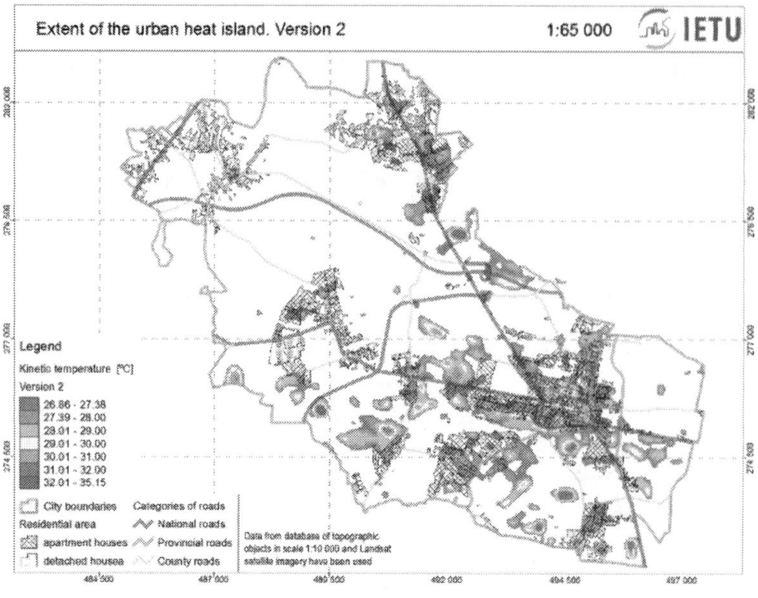

Fig. 8. Extent of the urban heat island of Bytom determined by use of $k = ½$ percentage of artificial area in town

The correctness of the application of the k factor value requires further research, however, it should be noted that the use of $k = ½$ percentage of artificial area in town allows to pinpoint most vulnerable areas to the adverse effects of UHI, where adoption activities should be undertaken immediately.

4 Conclusions

Examples of the use of satellite images technique described above for determine the kinetic surface temperature distribution in the cities area, with different way of development, seems to be the only possible method in the analysis of the urban heat island (UHI) at the stage of preparation of urban plans for adapting to climate change (MPA) in the different size of cities. This method allows to monitor changes in the occurrence of critical areas in the city over the past 30 years, resulting from the development of the city. These critical areas emit the heat into the atmosphere and affecting the temperature of the ambient air, as well affecting the quality of human life. In particular, vulnerable groups (children, persons age 65 and > cardiovascular diseases and respiratory diseases). This method also allows to determine the isotherm range the urban heat island, limited to the most vulnerable to the adverse effects of surface heat emitters on the comfort of life. Although the using $k = ½$ percentage of artificial area in town allows to pinpoint most vulnerable areas to the adverse effects of UHI, where adoption activities should be undertaken immediately, however the correctness of the application of the k factor value requires further. The use of satellite imagery for this type of analysis is much cheaper than ground-based measurements of the air temperature in a network of dozens of measurement points in the city and in surrounding areas for at least a year.

References

1. B. Dousset, F. Gourmelon, E. Mauri. Urban Remote Sensing Joint Event, *Application of satellite remote sensing for urban risk analysis: a case study of the 2003 extreme heat wave in Paris*, (Telecom Paris, 2007)
2. L. Liu , Y. Zhang, Remote Sens., *Urban Heat Island Analysis Using the Landsat TM Data and ASTER Data: A Case Study in Hong Kong*, **3**, 1535-1552 (2011)
3. K. Błażejczyk, M. Kuchcik, P. Milewski, W. Dudek, B. Kręcisz, A. Błażejczyk, J. Szmyd, B. Degórska, C. Pałczyński, Miejska wyspa ciepła w Warszawie. (Publishing house Sedno, 2014)
4. J. Walawender, M. Hajto, P. Iwaniuk, Rocz. Geomat., *Zastosowanie algorytmu „pojedynczego okna" do opracowania map temperatury powierzchni ziemi na podstawie danych satelitarnych LANDSAT*, **9**, 4, 139–150 (2011)
5. J. Walawender, M. Hajto, P. Iwaniuk, Proc. IEEE IGARSS, *A new ArcGIS toolset for automated mapping of land surface temperature with the use of LANDSAT satellite data*, 4371–4374, (2012)
6. J. Walawender, M. Szymanowski, M. J. Hajto, A. Bokwa, Pure and Applied Geophysics, *Land Surface Temperature Patterns in the Urban Agglomeration of Krakow (Poland) Derived from Landsat-7/ETM+ Data*, 913–940 (2014)
7. EPA, Reducing Urban Heat Islands: Compendium of Strategies Urban Heat Island Basics, (2014)
8. https://landsat.usgs.gov/using-usgs-landsat-8-product

E3S Web of Conferences **28**, 01012 (2018) https://doi.org/10.1051/e3sconf/20182801012
Air Protection in Theory and Practice

Instrumental measurement of odour nuisance in city agglomeration using electronic nose

Bartosz Szulczyński[1,*], *Tomasz* Dymerski[2], *Jacek* Gębicki[1] and *Jacek* Namieśnik[2]

[1]Department of Chemical and Process Engineering, Faculty of Chemistry, Gdansk University of Technology, 11/12 G. Narutowicza Str., 80-233 Gdańsk, Poland
[2]Department of Analytical Chemistry, Faculty of Chemistry, Gdansk University of Technology, 11/12 G. Narutowicza Str., 80-233 Gdańsk, Poland

Abstract. The paper describes an operation principle of odour nuisance monitoring network in a city agglomeration. Moreover, it presents the results of investigation on ambient air quality with respect to odour obtained during six-month period. The investigation was carried out using a network comprised of six prototypes of electronic nose and Nasal Ranger field olfactometers employed as a reference method. The monitoring network consisted of two measurement stations localized in a vicinity of crude oil processing plant and four stations localized near the main emitters of volatile odorous compounds such as sewage treatment plant, municipal landfill, phosphatic fertilizer production plant. The electronic nose prototype was equipped with a set of six semiconductor sensors by FIGARO Co. and one PID-type sensor. The field olfactometers were utilized for determination of mean concentration of odorants and for calibration of the electronic nose prototypes in order to provide their proper operation. Mean monthly values of odour concentration depended on the site of measurement and on meteorological parameters. They were within 0 – 6.0 ou/m^3 range. Performed investigations revealed the possibility of electronic nose instrument application as a tool for monitoring of odour nuisance.

1 Introduction

Unpleasant odours are naturally associated with danger, feeling of discomfort and they can cause negative psychosomatic symptoms. Long-term exposure to odorous compounds with unpleasant odour can result in loss of pleasure in living in particular area or building, in extreme cases it can lead to persistent changes in human organism and to chronic diseases [1-2]. Nowadays unpleasant odours present in a vicinity of air pollution sources are one of the main reasons of the complaints on environment quality, including ambient air quality. The major reasons of the complaints include [3-5]:

- actual odour nuisance,
- too short distance between sources of emission of odorants with unpleasant odour and residential areas,

[*] Corresponding author: bartosz.szulczynski@pg.edu.pl

- health hazard,
- lack of hermetization or deodorization of odour nuisance sources,
- increasing environmental concern among society.

Reduction or almost total elimination of emission of odorants with unpleasant odour become a priority for these branches of industry, the activity of which contributes to introduction of those compounds to ambient air. This reduction can consist in:

- prevention of emission of odorants with unpleasant character,
- introduction of suitable unit operations (absorption, adsorption, catalytic combustion, non-thermal combustion, biofiltration) to deodorization for existing industrial units,
- suitable planning of building and localization of new-built industrial units.

Independently of applied purification technology resulting in a limitation of emission of odorants characterized by unpleasant odour it is necessary to monitor the effectiveness of deodorization process via measurement of odour concentration at inlet and outlet of an installation [6-8]. The most popular techniques dedicated to this type of applications are olfactory techniques, especially dynamic olfactometry normalized in accordance with the standard EN 13725:2003 "Air quality. Determination of odour concentration by dynamic olfactometry". Sense of smell is a sensor allowing odour detection and evaluation of ambient air quality with respect to odour. Popularity of this technique is exemplified by the fact that it is the most often proposed measurement technique in these countries, which undertake the attempts of solving the problems connected with odour nuisance [9-13]. Apart from unquestionable advantages of this technique there are also certain drawbacks and inconveniences including:

- lack of on-line monitoring and ability to record short episodes of high concentration of odorants,
- need for qualified personnel (team of assessors) with suitable, regularly controlled odour perception,
- need for a laboratory with precisely defined parameters such as temperature, relative humidity as well as noise level.

In some cases dynamic olfactometry technique does not provide satisfactory results, for example in case of low odour concentrations (at the level of 1-50 ou/m^3). It is also difficult to determine odour concentration upon short-lasting changes of emission over particular area. In these situations it seems more advantageous to apply field olfactory technique. This approach allows characterization of concentration of odorants with unpleasant odour present in ambient air via dilution of investigated air stream with odourless gas (air purified on carbon filters). Among many devices utilized for field olfactometry measurements the Nasal Ranger (St. Croix Sensory, USA) field olfactometer is the most frequently employed one.

Application of the instrumental techniques enabling on-line measurements would make it possible to conduct the measurements, which cannot be carried out with dynamic or field olfactometry techniques. For many years an electronic nose has been an instrumental tool, which fulfils this requirement. The electronic nose is defined as an analytical device intended for fast detection and discrimination of odorous substances mixtures due to imitation of human sense of smell. In case of the electronic nose an identification of gas mixture composition is possible thanks to excitation of all chemical sensors comprising the electronic nose. This excitation (sensor response in the form of analytically useful signal) occurs in every sensor with different extent. Such excitation yields characteristic set of signals from particular sensors called a response vector of sensor matrix or electronic nose [14-17].

The main cause of odour nuisance in city agglomerations is emission of volatile compounds characterized by unpleasant odour from such sectors of industrial activity as: utilization and disposal of municipal waste, sewage treatment plants, chemical plants, meat and crop processing plants [18]. In the Tri-city agglomeration under consideration apart from the aforementioned sources of emission of odorants with unpleasant odour there is also a crude oil processing plant. Even such superficial analysis of the emission sources over particular area reveals a need of creation of an aerial monitoring network employing electronic nose-type devices, which would provide reliable information to the society and industry about occurrence of odour nuisance hazard.

The authors of this paper would like to present the results of investigation on ambient air quality with respect to odour nuisance near municipal landfill, sewage treatment plant and crude oil processing plant, which were performed using designed and built odour nuisance monitoring network. The aim of the investigation was verification of variation in odour intensity depending on location of the measurement site and meteorological conditions during a six-month period. Moreover, usability of the electronic nose instrument for monitoring of ambient air with respect to odorants presence was examined.

2 Experimental

2.1 Structure of electronic nose

The measurement set-up consisted of: a Tecfluid flow meter, the prototype of electronic nose, a suction pump and a PC class computer. Volumetric flow rate of the ambient air sucked equalled 1 dm^3/min. The prototype design involved a set of six commercial semiconductor sensors by FIGARO Co. (TGS 823, TGS 826, TGS 832, TGS 2600, TGS 2602, TGS 2603) and one photoionization sensor of PID-type (PPB MiniPID). Fig. 1 presents a block diagram of the electronic nose prototype.

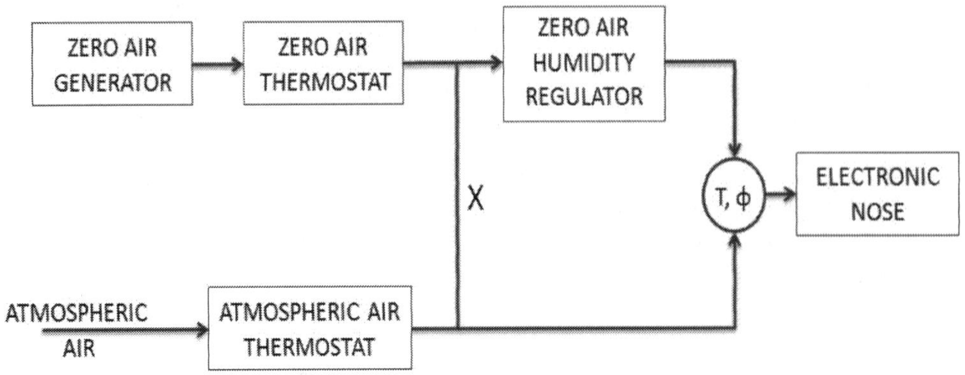

Fig. 1. Block diagram of electronic nose-type instrument. X-bypass, T-temperature sensor, ϕ-humidity sensor.

An analogue-to-digital converter (ADC) was used to process an output signal from the sensor set of the prototype. The output signal was converted into digital form in the range from 0 to 16 bits. The values of particular sensor signal taken for data analysis originated from the range, where the sensor signal attained steady value (steady state).
The operation mode of the electronic nose was as follows:

- suction of ambient air – 1 min.,
- stop-flow and measurement – interruption of ambient air flow in the sensor chamber – 0.5 min.,
- washing – zero air is passed through the entire pneumatic system in order to provide cleaning and stabilization of the sensors signals at the baseline signal level – 13.5 min.

Each cycle was followed by another repetition, so 4 measurements of odour concentration were performed during one hour.

2.2 Odour nuisance monitoring network

The monitoring network consisted of two measurement stations localized in a vicinity of the crude oil processing plant and four measurement stations localized near the main emitters of volatile odorous compounds such as sewage treatment plant, municipal landfill and phosphatic fertilizer production plant. Fig. 2 schematically presents distribution of six odour nuisance monitoring stations in the city agglomeration. Each of these stations was equipped with the e-nose type device and a system for data transmission to the central unit. Correct operation of the odour nuisance monitoring network required preliminary investigations of the electronic nose prototypes in field conditions supported with the field olfactometry tests treated as a reference method, with respect to which the electronic nose instrument was calibrated.

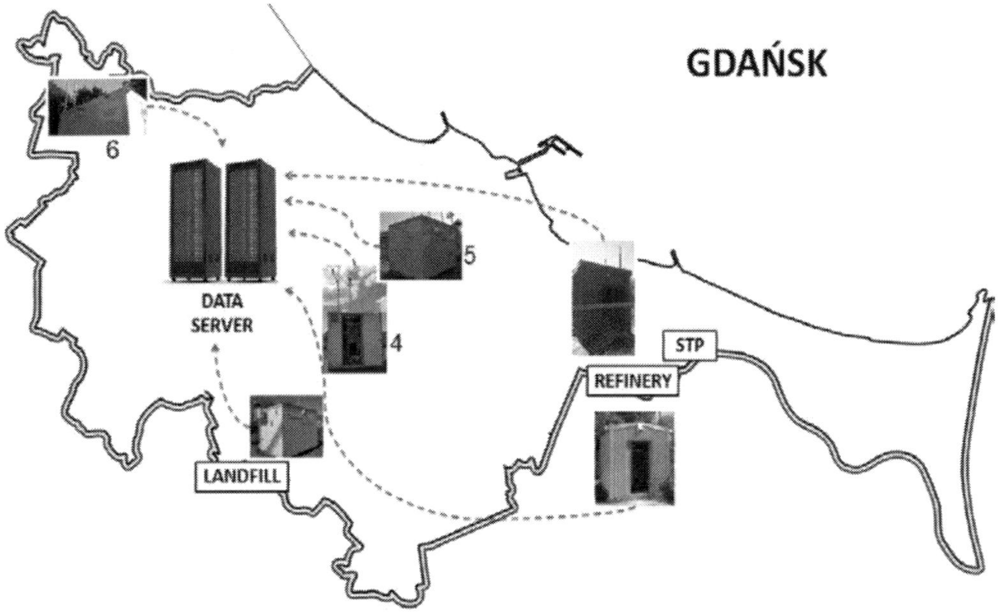

Fig. 2. Diagram of odour nuisance monitoring network in city agglomeration.STP-sewage treatment plant, 4,5,6 - Location of other measurement stations.

2.3 Investigations using Nasal Ranger field olfactometer

Four persons (a team of assessors) took part in the investigation carried out with the field olfactometers Nasal Ranger (St. Croix Sensory, USA). These persons were selected from a larger group following a standard procedure elaborated by the St. Croix Sensory, Inc. (St. Croix Sensory 2006). The aim of this procedure was selection of the assessors characterized by particular odour sensing threshold. Moreover, the assessors were trained with respect to sensory analysis using the field olfactometers. The measurements with field olfactometers allowed determination of D/T values (dilution to sensing threshold), at which the odour was sensed. Based on the D/T values it was possible to calculate individual odour concentration $c_{od,\,i}$ at given measurement point using the following formula (1):

$$c_{od,i} = \sqrt{Z_{no} \times Z_{yes}} \tag{1}$$

where: Z_{no}=D/T+1 – D/T value, at which the odour is imperceptible prior to the D/T value, at which odour is perceptible, Z_{yes}=D/T+1- D/T value, at which the odour is perceptible following the D/T value, at which the odour is imperceptible.

The value of odour concentration c_{od} [ou/m^3] was calculated as a geometrical mean of the n set of all individual odour concentrations $c_{od,i}$ for a given measurement point using the formula (2):

$$c_{od} = \sqrt[n]{c_{od,1} \times c_{od,2} \times c_{od,3} \times \ldots \times c_{od,n}} \tag{2}$$

where: $c_{od,n}$ is the individual odour concentration provided by a given panellist.

2.4 Calibration of electronic nose instrument with respect to field olfactometer

The signals obtained from particular sensors were subjected to statistical analysis in order to eliminate statistically insignificant signals. Then statistically significant signals were used to determine a ratio of sensor signal change ΔS to reference signal S_0 and the value of mean geometric signal was calculated using the formula (3):

$$Signal = \sqrt[n]{S_1 \times S_2 \times S_3 \times \ldots \times S_n} \tag{3}$$

In the next step the calculated value was compared with the value of mean odour concentration determined by the team of assessors using Nasal Ranger field olfactometers. A series of analysis enabled calibration of the electronic nose instrument in the range from 0 to 25 ou/m^3. Generally, a process of sensor signal conversion into odour concentration can be presented with the scheme illustrated in Fig. 3.

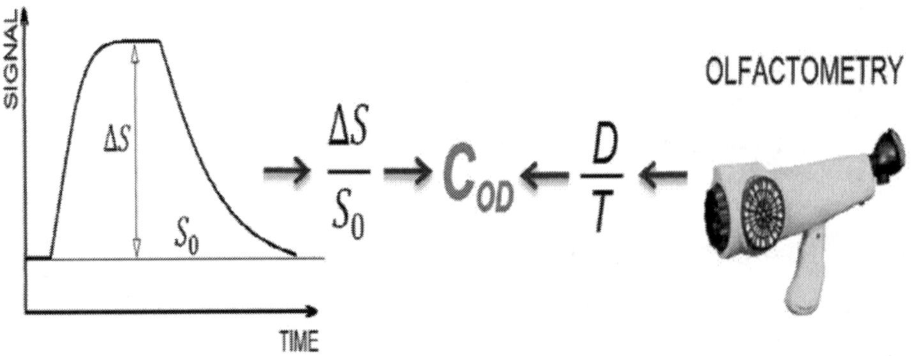

Fig. 3. Process of sensor signal conversion into odour concentration.

3 Results

Fig. 4 presents the changes of mean odour concentration versus time and localization of the measurement station. During the six-month time span it can be seen that higher odour concentrations occur in the summer season. As far as localization of the measurement station is concerned, the highest odour concentrations are noticed over the area neighbouring the municipal landfill, then next to the sewage treatment plant and finally in a vicinity of the crude oil processing plant. A range of mean odour concentration observed in a vicinity of the municipal landfill was from 2.5 ou/m^3 to 6.0 ou/m^3. Corresponding range in case of the sewage treatment plant was from 2.0 ou/m^3 to 4.3 ou/m^3. A range of mean odour concentration for the area next to the crude oil processing plant was from 1.9 ou/m^3 to 3.7 ou/m^3. In the remaining measurement points mean odour concentration varied between 0 and 2.3 ou/m^3.

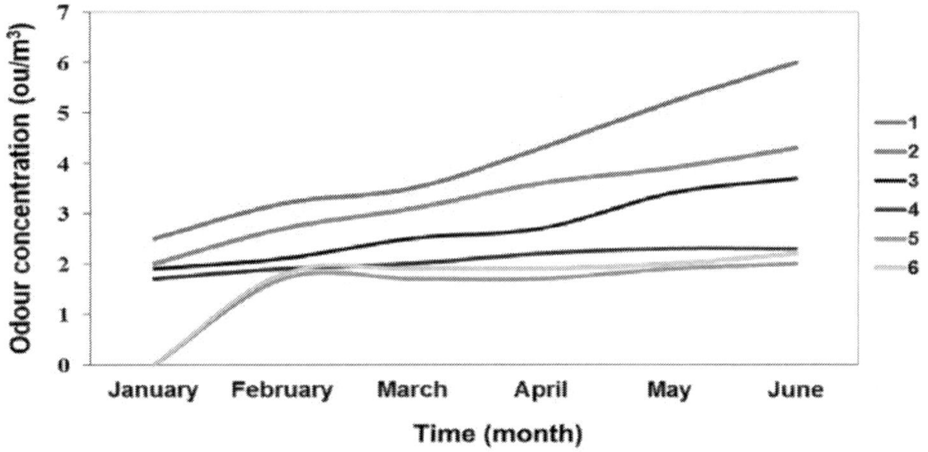

Fig. 4. Changes of mean odour concentration versus time and localization of measurement station. 1-landfill, 2-sewage treatment plant, 3-refinery, 4,5,6-other measurement stations.

Table 2 shows the dependences between the maximum odour concentration and time as well as localization of the measurement station. Also here it can noticed that the highest concentrations occurred in the summer season and the values of concentration were also dependent on localization of the measurement station. The highest odour concentration was observed in June at the station localized near the municipal landfill. Its value was equal 24.0 ou/m^3.

Table 1. Dependence between maximum odour concentration and time as well as localization of measurement station.

	maximum odour concentration					
measurement station	**January**	**February**	**March**	**April**	**May**	**June**
landfill	7.2	8.4	10.2	14.1	17.2	24.0
sewage treatment plant	4.5	5.8	8.0	11.2	13.5	17.4
refinery	3.0	4.5	5.6	5.8	6.7	7.4
4,5,6	1.7	2.5	3.5	4.2	5.5	5.9

This situation was probably caused by the chemical and biological processes occurring inside waste piles, which can be treated as a bioreactor. An increase in temperature contributed to acceleration of the microbiological and biochemical processes generating significant amount of volatile compounds, including those malodorous ones. These processes yield the products of anaerobic denitrification and desulphurization. In case of the sewage treatment plant there is intensification of digestion processes. In case of the crude oil processing plant it is probable that an increase in ambient air temperature contributes to enhanced sweating of tanks, leakages of valves, pipelines and to non-organized emission. These factors can influence on increased emission of malodorous compounds.

Fig. 5. illustrates the changes of mean odour concentration versus relative humidity of ambient air for three measurement stations characterized by the highest odour concentrations during a given measurement period (the stations localized in a vicinity of municipal landfill, sewage treatment plant and crude oil processing plant). It can be noticed that the values of mean odour concentration decrease with an increase in ambient air relative humidity. Most probably it is caused by a phenomenon of sorption and absorption of odorous compounds on and inside water vapour droplets (fog), hence a wet deposition occurs, which results in a decrease in odour concentration level in ambient air [19].

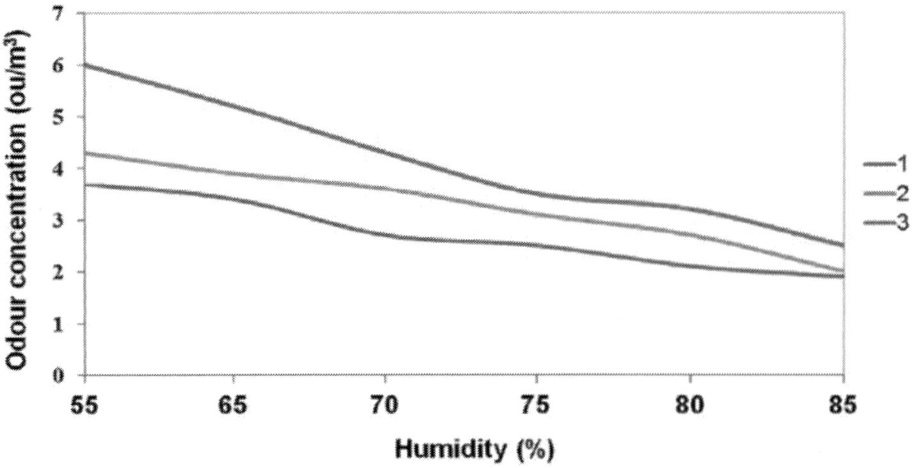

Fig. 5. Changes of mean odour concentration versus relative humidity of ambient air for three measurement stations characterized by the highest odour concentrations over given area. 1-landfill, 2-sewage treatment plant, 3-refinery.

4 Summary

The investigations of odour concentration level in a vicinity of the main sources of odorants emission in the city agglomeration, performed with the electronic nose-type instruments revealed variation of odour concentration depending on the time of year and the measurement site. The area near the municipal landfill was characterized by the highest mean as well as maximum odour concentrations. It was also observed that the value of odour concentration over given area depended on climatic conditions. An increase in temperature and a decrease in relative humidity of ambient air are associated with an increase in odour concentration. Reverse situation occurred for a decrease in temperature and an increase in relative humidity of ambient air. Proper operation of the electronic nose depended on its calibration with respect to the field olfactometry measurements. Obtained completeness of measurement results was at the level of 60%. The obtained mean odour concentration are in the range of 1 to 6 ou/m^3. This indicates that the European standard of some countries has been exceeded. Work on the proper functioning of the electronic nose used to monitor air quality for odor compounds should be continued and more and more sensitive chemical sensors should be added to these devices.

The investigations were financially supported by the Grant No. PBSII/B9/24/2013 from the National Centre for Research and Development and the Grant No. UMO-2015/19/B/ST4/02722 from the National Science Centre.

References

1. M. Kampa, E. Castanas, Environ. Pollut. **151**, 362–367 (2008)
2. V. Naddeo, T. Zarra, S. Giuliani, V. Belgiorno, Chem. Eng. Trans. **30**, 85-90 (2012)
3. I. Sówka, P. Sobczyński, U. Miller, Rocz. Ochr. Sr. **17**, 1339-1349 (2015)
4. J.A. Nicell, Atmos. Environ. **43**, 196-206 (2009)
5. M. Trincavelli, S. Coradeschi, A. Loutfi, Sens. Actuators B-Chem. **139**, 265–273 (2009)
6. L. Capelli, S. Sironi, R. Del Rosso, J.M. Guillot, Atmos. Environ**.** **79**, 731–743 (2013)
7. L. Capelli, S. Sironi, R. Del Rosso, P. Centola, Water Res. **43, 7**, 1977–1985 (2009)
8. C. Alfonsín, R. Lebrero, J.M. Estrada, R. Munoz, N.J.R. Kraakman, G. Feijoo, M.T. Moreira, J. Environ. Manage. **149**, 77–84 (2015)
9. A.H. Bokowa, 2010, Chem. Eng. Trans. **23***,* 31-36 (2010)
10. L. Capelli, S. Sironi, R. Barczak, M. Il Grande, R. Del Rosso, Water Sci. Technol. **66**, 1607–1613 (2012)
11. J.M. Guillot, Chem. Eng. Trans. **30**, 295-300 (2012)
12. I. Sówka, Environ. Prot. Eng. **36,** 2, 133-141 (2010)
13. L. Capelli, S. Sironi, R. Del Rosso, P. Centola, S. Bonati, Water Sci. Technol. **61**, 1267–1278 (2010)
14. A.D. Wilson, M. Baietto, Sensors **9**, 5099–5148 (2009)
15. J. Gębicki, T. Dymerski, J. Namieśnik, Environments **4,** 1, 7 (2017)
16. R. Munoz, E.C. Sivret, G. Parcsi, R. Lebrero, X. Wang, I.H. Suffet, R.M. Stuetz, Water Res. **44**, 5129–5149 (2010)
17. B. Szulczyński, J. Gębicki, Environments **4,** 1, 21 (2017)
18. J. Gębicki, T. Dymerski, J. Namieśnik, Environ. Prot. Eng. **42,** 2, 97-106 (2016)
19. H. Byliński, P. Kolasińska, T. Dymerski, J. Gębicki, J. Namieśnik, Monats. Chem. **148,** 9, 1651-1659 (2017)

Long-term trends of pollutant concentrations in selected sites in Silesian Voivodeship

Szymon Hoffman*

Department of Chemistry, Water and Wastewater Technology
Faculty of Infrastructure and Environment
Czestochowa University of Technology
69 Dabrowskiego St., 42-200 Czestochowa, Poland

Abstract. The assessment of the air pollution quality in selected sites of Silesian Voivodeship was presented in this paper. The evaluation based on the sets of long-term data, recorded by the state air monitoring network. Concentrations of main air pollutants such as PM10, O_3, CO, SO_2, NO_2, NO were considered. The basis for the calculations were 12-year time series of hourly concentrations. Using this data, the monthly averages of pollutant concentrations were calculated. Long-time trends of concentration changes were determined for each pollutant separately. Based on the analysis of trends, risks that may arise in the future were identified.

1 Introduction

Silesian Voivodeship has the opinion of the most polluted region of Poland. Cities like Zabrze and Rybnik are often mentioned as extremely contaminated places [1]. Long-term measurements of the air monitoring stations allow to answer the question how the air quality has changed in these towns. The aim of the study is to identify the trends of changes in the concentration level of main air pollutants.

Currently in Poland, the state network of air monitoring operates according to the new regulations, which are in line with the European Union legal acts. The present monitoring program started functioning shortly before Poland's accession to the EU [2, 3]. In previous years, air monitoring measurements were actually carried out, but on other principles. Therefore, in this examination, 2005 was adopted as the starting point in the analysed time series. 12-year observations should allow to determine the real trends. Designing trends allow to identify the air quality hazards that may emerge in the future.

2 Methods

Concentrations of basic air pollutants such as: O_3, NO, NO_2, CO, SO_2, PM10, registered in the years 2005-2016 at air monitoring stations in Zabrze and Rybnik, were used in the analysis. Both stations are located in big industrial centers and belong to the

* Corresponding author: szymon@is.pcz.czest.pl

 https://doi.org/10.1051/e3sconf/20182801013

type of urban background monitoring sites. The analysed data were received from the Voivodeship Inspectorate of Environmental Protection in Katowice. Using time series of 1-hour (instantaneous) concentrations, the monthly mean concentrations were calculated. For each pollutant, 12-year courses of monthly concentrations were determined. Linear trends were adjusted to the courses. Trend lines were determined by the least squares method. Based on the analysis of trends, risks that may arise in the future were identified.

3 Results

Results of the analysis were shown in figures 1-12. The figures picture long-term courses of monthly concentrations as well as trend lines, for each pollutant and each site separately.

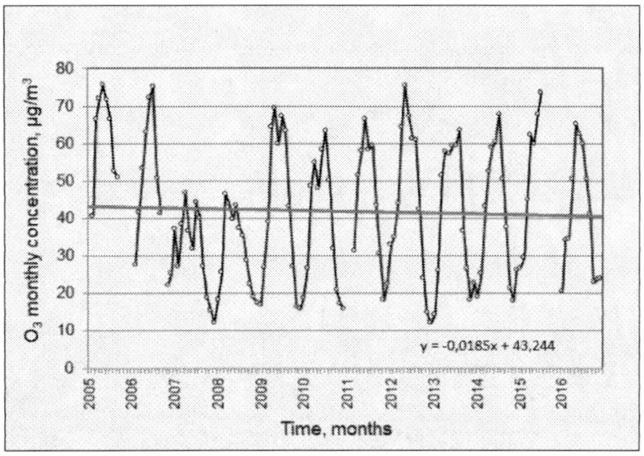

Fig. 1. The course of monthly average concentrations of O_3 at the Zabrze station in the period 2005-2016 with adjusted linear trend.

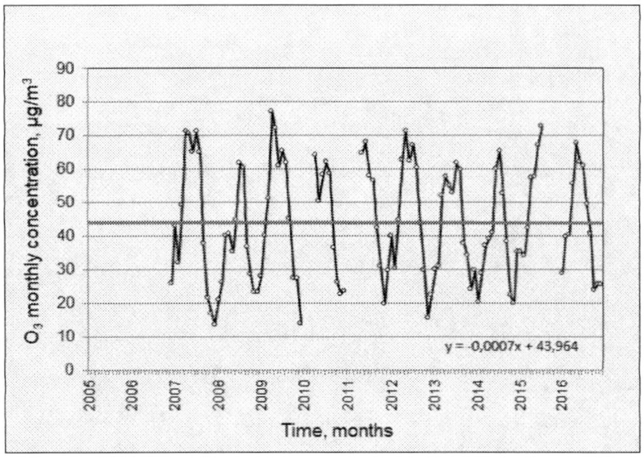

Fig. 2. The course of monthly average concentrations of O_3 at the Rybnik station in the period 2005-2016 with adjusted linear trend.

https://doi.org/10.1051/e3sconf/20182801013

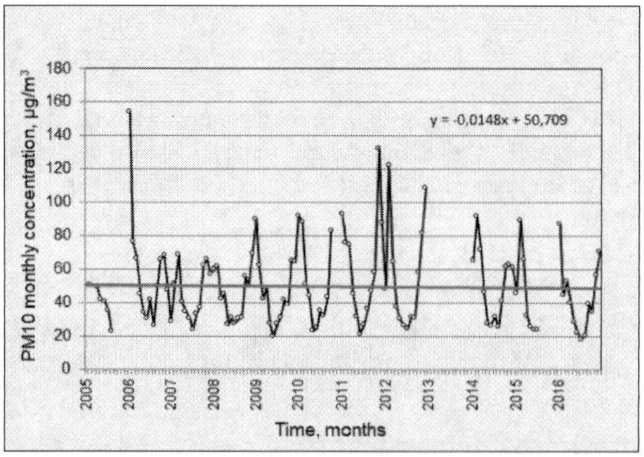

Fig. 3. The course of monthly average concentrations of PM10 at the Zabrze station in the period 2005-2016 with adjusted linear trend.

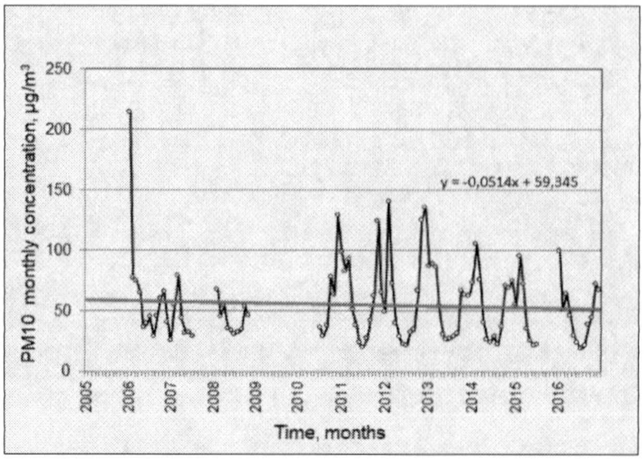

Fig. 4. The course of monthly average concentrations of PM10 at the Rybnik station in the period 2005-2016 with adjusted linear trend.

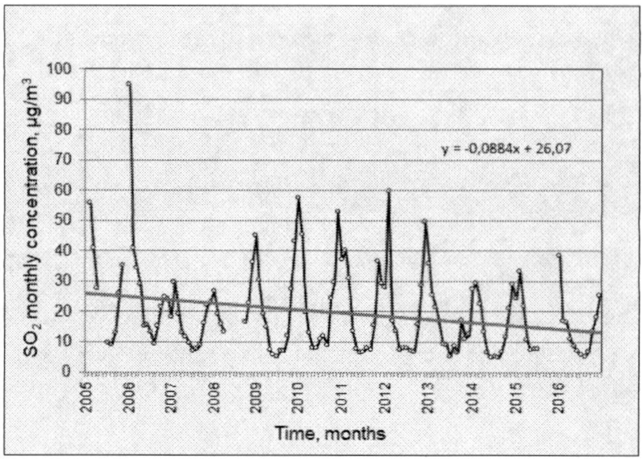

Fig. 5. The course of monthly average concentrations of SO_2 at the Zabrze station in the period 2005-2016 with adjusted linear trend.

 https://doi.org/10.1051/e3sconf/20182801013

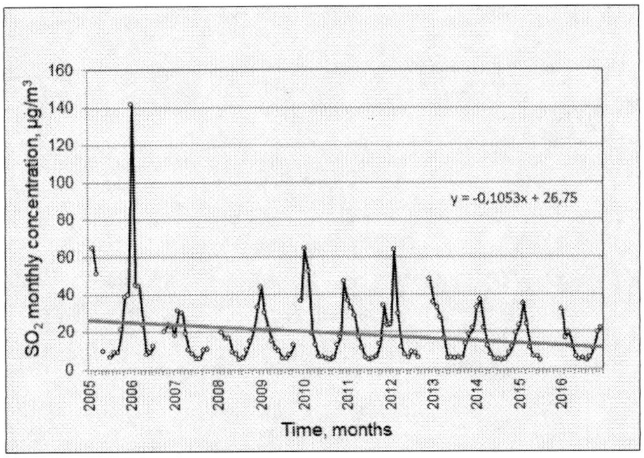

Fig. 6. The course of monthly average concentrations of SO_2 at the Rybnik station in the period 2005-2016 with adjusted linear trend.

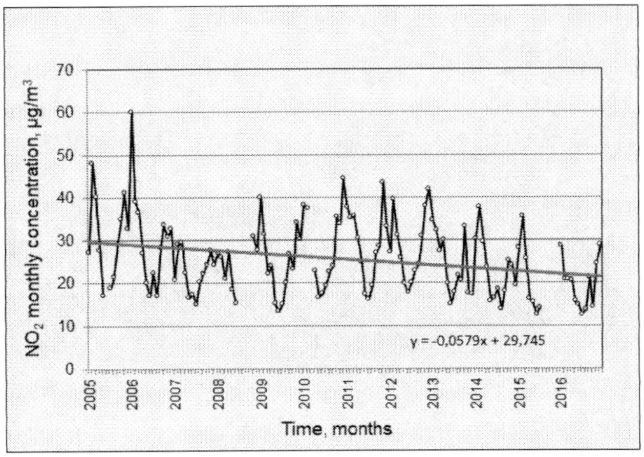

Fig. 7. The course of monthly average concentrations of NO_2 at the Zabrze station in the period 2005-2016 with adjusted linear trend.

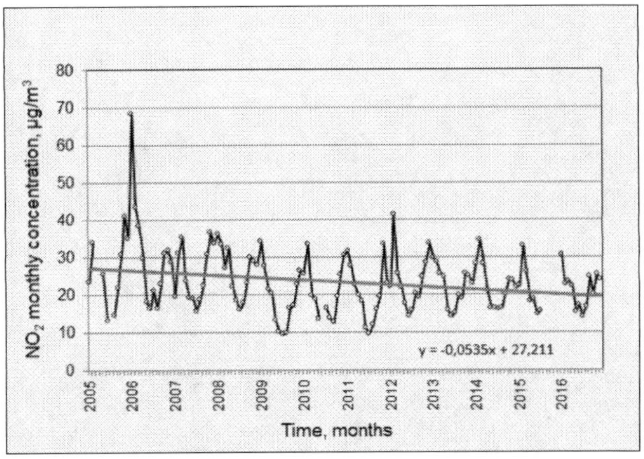

Fig. 8. The course of monthly average concentrations of NO_2 at the Rybnik station in the period 2005-2016 with adjusted linear trend.

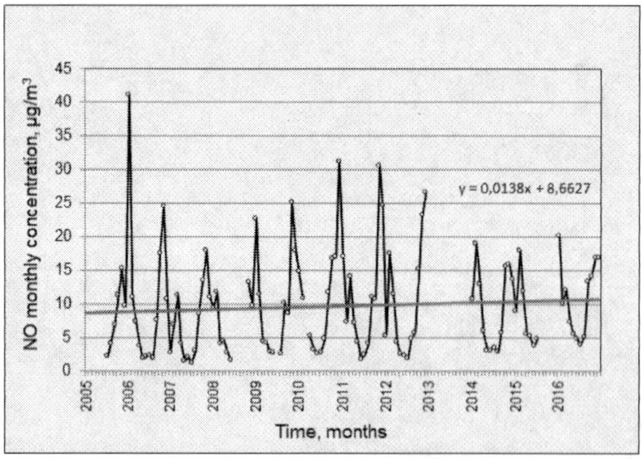

Fig. 9. The course of monthly average concentrations of NO at the Zabrze station in the period 2005-2016 with adjusted linear trend.

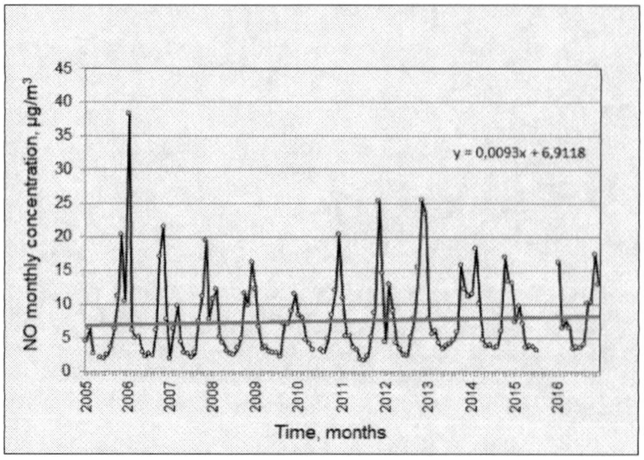

Fig. 10. The course of monthly average concentrations of NO at the Rybnik station in the period 2005-2016 with adjusted linear trend.

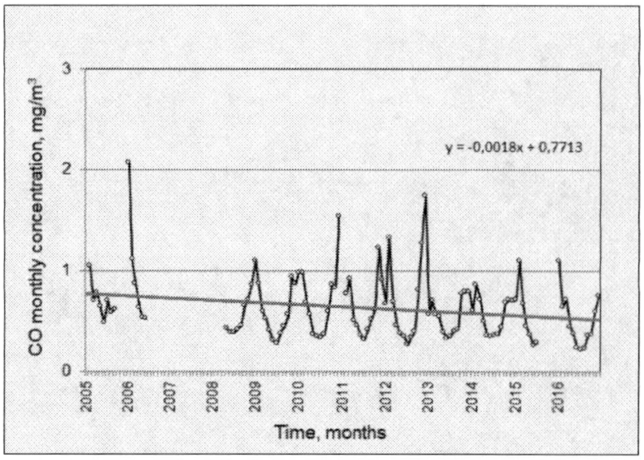

Fig. 11. The course of monthly average concentrations of CO at the Zabrze station in the period 2005-2016 with adjusted linear trend.

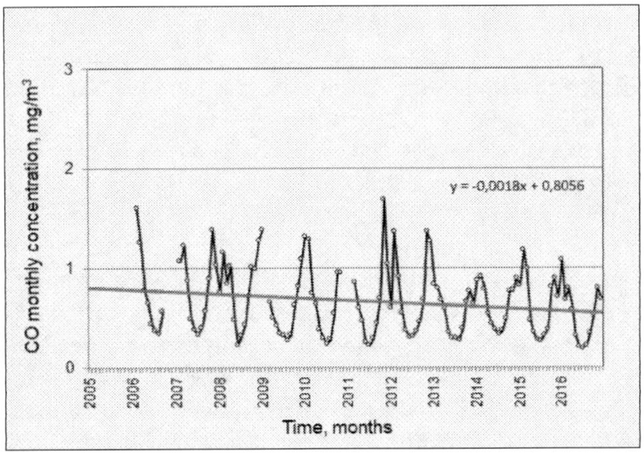

Fig. 12. The course of monthly average concentrations of CO at the Rybnik station in the period 2005-2016 with adjusted linear trend.

4 Summary and discussion

Summary of the results was shown in tab. 1. The direction coefficient values of the obtained trend equations were analysed to estimate the direction and strength of the concentrations changes. The strongest changes were observed in the case of SO_2, NO_2 and CO concentrations. In the both towns those concentrations have been decreasing. Concentrations of PM10 have been decreasing slower. O_3 concentrations have been decreasing in Zabrze but in Rybnik they were more stable. Only in the case of NO concentrations, the increasing was observed, however, at low concentration level.

Table 1. Long-term concentration changes – summary

Pollutant	Monitoring site	Trend equation (x – number of months)	Estimation of trend
O_3	Zabrze	y = -0.0185x + 43.2	Slight decrease
	Rybnik	y = -0.0007x + 44.0	Stabilization
PM10	Zabrze	y = -0.0148x + 50.7	Slight decrease
	Rybnik	y = -0.0514x + 59.3	Decrease
SO_2	Zabrze	y = -0.0884x + 26.1	Strong decrease
	Rybnik	y = -0.1053x + 26.8	Strong decrease
NO_2	Zabrze	y = -0.0579x + 29.7	Strong decrease
	Rybnik	y = -0.0535x + 27.2	Strong decrease
NO	Zabrze	y = 0.0138x + 8.66	Increase
	Rybnik	y = 0.0093x + 6.91	Increase
CO	Zabrze	y = -0.0018x + 0.771	Strong decrease
	Rybnik	y = -0.0018x + 0.806	Strong decrease

According to the European Environment Agency report, in the years 2000-2015 total emissions of air pollutants decreased in EU [4]. Similar trends have been observed in Poland. Since 2000 the following drops of emission has been estimated: 51% for SO_2, 25% for CO, 22% for particulates, 14% for NO_x [5].

Generally, the changes in pollution levels at stations Zabrze and Rybnik should be considered advantageous for the environment. The rapid fall in SO_2, CO and NO_2 levels indicates an effective policy of reducing air pollution. Several years ago Poland was classified as a country with exceptionally high SO_2 emission. An evident drop of SO_2 concentrations indicates that long-term programs for limitation emission of this gas are effective. Moreover, the favourable trends are noticeable in Silesia, the region where coal is mined and burned in large quantities. The increase of NO concentrations with the simultaneous decline of NO_2 concentration is not something dangerous. The biggest problem are PM10 concentrations. Although the gradual decrease in concentration can be noted, but still the levels of concentration during heating periods are too high. The reason is low emission from very common domestic installations burning coal waste. It can be assumed that smog episodes resulting from high concentrations of PM10 will continue to occur in many sites in Poland, because government policy is very hesitant in this regard.

Trends observed in Silesian cities coincide with European trends. Europe's air quality is slowly improving, but particulate matter and ground-level ozone continue to cause serious impacts on health [6]. Despite numerous actions to improve the air quality, the following issues still constitute an serious problem in Poland: exceedances of target value for tropospheric ozone during the summer season and exceedances of limit values for particulate matter PM10 and benzo[a]pyrene during the winter season [7].

5 Conclusions

The 12-year observations of the concentration levels of the basic pollutants registered at two air monitoring stations allow the following conclusions:

1. Concentrations of the most of main air pollutants tend to decrease. The strongest drops are observed in the case of SO_2, NO_2 and CO concentrations.
2. Concentrations of PM-10 decrease slower.
3. Exceptionally, NO concentrations increase.

This study was carried out within a research project of Czestochowa University of Technology BS-PB-402-301/15.

References

1. Inspekcja Ochrony Środowiska, Wojewódzki Inspektorat Ochrony Środowiska w Katowicach, *Stan środowiska w województwie śląskim w 2016 roku* (Biblioteka Monitoringu Środowiska, Katowice 2017)
2. *Ustawa z dnia 27 lipca 2001 r. Prawo ochrony środowiska* (Dz.U. 2001 nr 62, poz. 627).
3. Główny Inspektor Ochrony Środowiska, *Program państwowego monitoringu środowiska na lata 2003-2005* (Warszawa 2003)
4. European Environment Agency, *Air quality in Europe - 2017 report* (No 13/2017, Publications Office of the European Union, Luxembourg 2017)
5. Central Statistical Office GUS, *Environment 2017* (Warsaw 2017), available on website www.stat.gov.pl
6. European Environment Agency, *The European environment - state and outlook 2015*: synthesis report (Copenhagen 2015)
7. Główny Inspektorat Ochrony Środowiska, *Stan środowiska w Polsce - Raport 2014* (Biblioteka Monitoringu Środowiska, Warszawa 2014)

Analysis of the impact of the eastern ring road of Lodz on selected components of the environment

Monika Janas[*], *Alicja* Zawadzka

Lodz University of Technology, Faculty of Process and Environmental Engineering, 90-924 Lodz, Wolczanska 213, Poland

Abstract. The geographical location of Poland and the pursuit of economic growth make it necessary to build a sufficiently equipped and compact network of expressways and motorways. Such investments do not, however, leave the environment unaffected. The work analyzes changes that have occurred in the natural environment under the influence of the construction and operation of Lodz eastern ring road being a section of the A1 motorway commissioned in August 2016. Changes in the pollution of air, soil and acoustic climate are discussed in detail.

1 Road transport in Poland

In Poland, numerous actions have been taken over the last few years to develop the road network and to improve its technical condition. Traffic flow and road capacity are crucial for the development of modern civilization. At the end of 2015, the network of public roads in Poland covered 419.6 thousand km, of which roads with hard surface had a length of 290.9 thousand km that accounted for about 70% of the length of public roads. Despite the systematic increase in the length of motorways, Poland is still among the countries with the least developed network of expressways and motorways in Europe. By the end of 2015, the total length of motorways was 1559.2 km and expressways 1292.2 km [1].

Road investments are seen as a key instrument of socio-economic development. Construction of new roads, especially motorways and expressways, plays a very important role in stimulating local business [2-4]. Today, motorways serve not only transport functions but they become a driving force for boosting business growth, revitalizing selected economic areas and reducing unemployment, and at the same time they contribute to raising the level of road safety. Such investments, however, affect the natural environment and change space organization of terrain. The range of direct impact is relatively small (several hundred meters) and due to the linear nature and investment processes affects the environment on a regional and national scale [5-8].

The negative impact of motorways on the environment is very burdensome to people. Cars driven on these roads emit to the atmosphere harmful substances which directly affect human health. As a result of fuel combustion in vehicle engines, the following pollutants

[*] Corresponding author: monika.janas@dokt.p.Łódź.pl

are released into the air: carbon monoxide, nitric oxide, hydrocarbons including PAHs and solid particles which may contain heavy metals [6-10]. Transportation routes are also contaminated with lubricants, oils and petrol, and the poor technical condition of cars in Poland aggravates this phenomenon. Rain rinses these contaminants out of the roads and thus pollutes soil as well as surface and underground waters in the vicinity of roads. Heavy use of vehicles results in noise emissions and vibrations, worsening the acoustic environment in the vicinity of motorways. The noise level depends on traffic intensity, the weight of vehicles, technical solutions and topographical conditions. The location of communication routes affects changes in the location of fauna and flora habitats, modifies microclimate of the surrounding areas and causes environmental pollution [11-16].

The construction of new transport routes will always involve interfering with the natural environment. The purpose of environmental reports prepared by specialist teams is to answer the question: what road impacts are most likely to be significant and how they can be minimized. A well-designed and well-used road enables better city functioning by improving safety conditions, reducing road traffic and, above all, contributing to the development of the region [17-19].

2 The purpose and area of research

The aim of the study was to analyze the impact of the eastern ring road of Lodz on selected components of the environment. Its influence on atmospheric air, soil and acoustic climate was determined. The analyzed section of the motorway is a new Polish road investment functioning in the geographical space for a relatively short time (commissioned on 1 July 2016) and thus not all effects connected with its exploitation are already possible to assess thoroughly. Nevertheless, it was considered appropriate to undertake research to determine short-term results of the investment, which will make it possible to carry out comparative analyzes in the future.

The A1 motorway, also called the Amber Motorway, is the only motorway in Poland in meridian direction. Finally, it will be almost 570 kilometers long and will connect Trojmiasto, Torun, Lodz, Czestochowa and Upper Silesia with the state border with the Czech Republic. The A1 motorway is undoubtedly one of the pillars of the communication system in the country, because it is or will soon be connected to many other important fast traffic routes. In the north, in Rusocin, it connects to the S6 expressway, giving it a connection with the cities of the Trojmiasto agglomeration and finally with Szczecin, through Slupsk and Koszalin. In Strykow, it connects to the A2 motorway, which leads from the border with Germany to Warsaw. In the south of Lodz, the A1 motorway connects to the S8 expressway leading to Wroclaw. In Gliwice it crosses the A4 motorway, connecting the Polish border with Germany with the border with Ukraine in the east. On the southern border of the country, in Bohumín, the A1 connects to the Czech highway D1 [1]. The existing infrastructure of the motorway is shown below (Figure 1).

The analyzed section of the A1 motorway in the eastern ring road of Lodz is 30 km long. This part is located entirely in the Lodz Province, in the Lodz-East and Zgierz Districts and in the city of Lodz. The analyzed section of the motorway is of strategic importance for the Lodz Province, especially for the city. The A1 motorway along with the A2 motorway has taken over most of the transit traffic from the national roads in the Lodz area, which has a positive effect on their capacity and limits the transit traffic in Lodz itself. Less crowding in cities located on the outskirts of Lodz improves the safety of pedestrians, cyclists and motorists, and improves the environment.

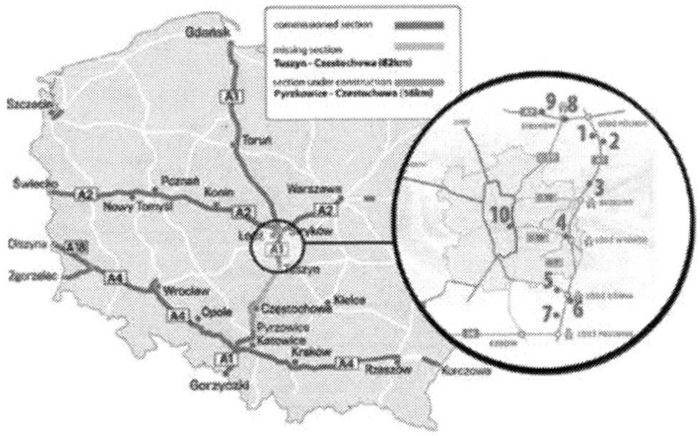

Fig. 1. A1 motorway and location of measurement points

Seven measurement points located along the eastern ring road of Lodz in the section from Skoszewy in the north (Lodz-East District) to Giemzowek in the south (Lodz-East District) were selected to analyze the environmental impact of the motorway. In two locations (Skoszewy-West and Wisniowa Gora-East) measurements were made only behind energy absorbing barriers. In addition, two measurement points were selected for comparison – one in Strykow, connecting the A2 and A1 motorways and the other one in the center of Lodz.

3 Research methodology

Analysis of the environmental impact of the eastern ring road of Lodz was carried out in accordance with the guidelines contained in the Ordinance of the Ministry of Environment on the requirements for measurements of substances or energy levels in the environment to be carried out by the manager of road, tram line, airport or port [20]. Periodic measurements were carried out on one of April Monday mornings with heaviest traffic under suitable meteorological conditions (wind speed below 5 m/s, air temperature above -10°C, no precipitation, no or low ground temperature inversion).

Air samples for analysis were taken along lines running perpendicular to the road axis. The samples were taken at points located 10 to 50 meters from the road edge. The concentrations of sulfur dioxide (SO_2) and nitrogen dioxide (NO_2) were determined by gas chromatography.

Soil samples in the immediate vicinity of the motorway were collected for analysis along lines running perpendicular to the road axis. The soil was collected from the surface level (humus level A, depth 5 cm) at points located 10 to 50 meters from the road edge (depending on the road infrastructure). The samples were dried at 105°C and then sieved through a sieve with 2 mm mesh size. The analyzes were performed in fraction <2 mm. Analysis of the content of nickel (Ni), lead (Pb), copper (Cu) and zinc (Zn) was carried out in solutions prepared on the basis of aqua regia (HCl: HNO_3 = 3:1 vol/vol). Copper (Cu) and zinc (Zn) content were determined by the AAS method on a Varian Spectr AA 220 spectrometer. The total content of cobalt (Co) and chromium (Cr) was determined by argon emission spectrometry [20].

Noise measurements on the analyzed section of eastern Łódź ring road were carried out with the use of an integrating SON-50 sound level meter. The meter was placed on a tripod, about 1.7 m above the ground. The adopted measurement level was in line with the recommendations of the International Society of Biometeorology for studies on the impact of atmospheric environment on humans. The impact of noise on the environment has been described by means of an average equivalent sound level, expressed in decibels. Measurement points were distributed along the profiles perpendicular to the main road. The distance of the measurement points from the edge of the main road was determined taking into account topography, plant cover and buildings. Measurement profiles were determined in places which enabled direct noise measurement from the main road burdened the least with local noise sources. The measurement point closest to the road was at a distance of 100 m, and the furthest one was at 300 m. The selection of measurement points was guided by the principle that they would best describe the acoustic climate created by traffic on a selected road section [20].

4 Results

Contamination of air emitted in the road transport sector depends on many factors including, among other things, fuel composition, engine characteristics and maintenance, vehicle type, vehicle speed or traffic jams. Figure 2 shows concentrations of sulfur and nitrogen dioxides in atmospheric air at specified measurement points.

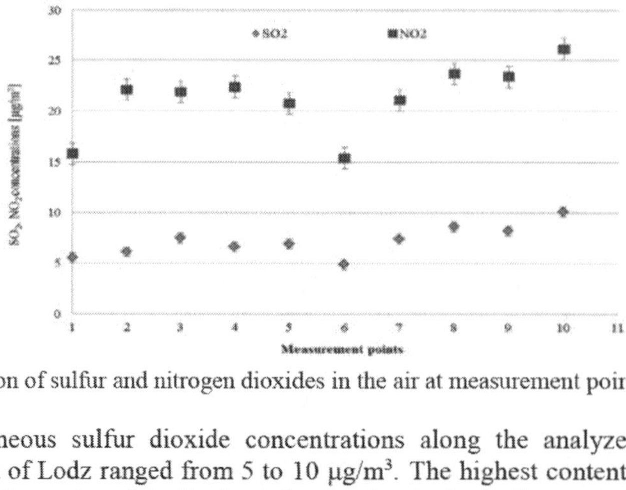

Fig. 2 Concentration of sulfur and nitrogen dioxides in the air at measurement points [µg/m³]

The instantaneous sulfur dioxide concentrations along the analyzed section of the eastern ring road of Lodz ranged from 5 to 10 µg/m³. The highest content of SO_2 in the air was recorded in the center of the Lodz agglomeration in the built-up area and in the region where the A1 motorway connects to the A2 motorway, where we observe increased traffic in comparison with the analyzed part of the eastern ring road of Lodz. The instantaneous concentrations of nitrogen dioxide outside the built-up area ranged from 15 to 24 µg/m³, while in the city center they were slightly higher, reaching 26 µg/m³. The lowest concentrations of SO_2 and NO_2 were recorded at measurement points located behind energy absorbing barriers at point 1 in Skoszewy West and at point 6 in Wisniowa Gora East.

The impact of road transport on soil contamination has been a subject of numerous studies conducted in different regions of the world. Heavy metal compounds released from vehicle operating processes and contained in particulate matter react with soil components or are adsorbed on humus colloids. Accumulation of heavy metals is observed mainly on the soil surface. Heavy metals can remain in the soil for a long time and affect the content of these compounds in plants. Figure 3 shows the content of selected heavy metals in soil samples collected in the vicinity of the analyzed Lodz ring road.

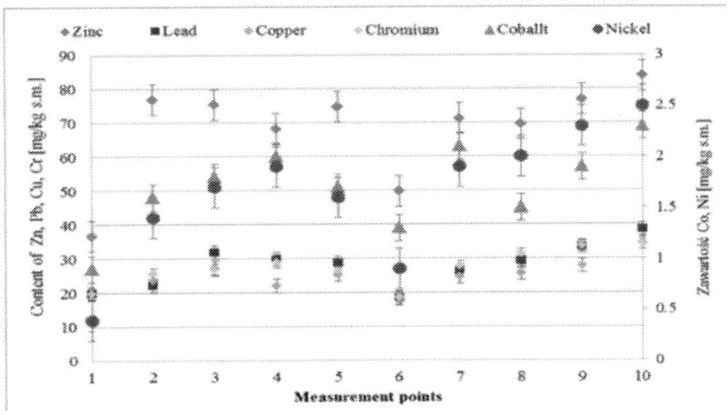

Fig. 3 Content of zinc, lead, copper, chromium, cobalt and nickel in the soil [mg/kg d.m.]

Soil samples taken along the eastern ring road of Lodz are characterized by different contents of heavy metals along the studied road section. Analysis of the samples shows that soil collected at the measuring points located in the immediate vicinity of the road is characterized by the highest zinc content equal to 35-85 mg/kg d.m. Zinc is commonly used in brake and clutch discs as well as in lubricating oils and tires. As a result, in places where cars need to brake more often and so wear off their tires and elements of the braking system, we can observe higher concentrations of this metal. Lead contaminants in the soil amounted to 20-40 mg/kg d.m. The concentration of cobalt and nickel is low and results from the natural content of these metals in the environment.

The content of heavy metals in the soil collected along the eastern ring road of Lodz is lower than in the area of the intersection of the A1 and A2 motorways and in the city center. The studied road section is the shortest one, and it can be concluded that for a short period of time the process of soil contamination is very slow.

Following changes in the noise level along the analyzed section of the eastern ring road of Lodz different noise intensities were observed. The analyzed section lies in the areas with different land cover (forests, fields, meadows, scattered farm buildings) and diverse housing density as well as the density of local roads. As can be seen from the studies conducted, in all noise profiles the permissible noise levels were exceeded (Figures 4-6). In assessing the level of road noise, the results obtained were compared with the permissible noise level specified in the Ordinance of the Ministry of the Environment on permissible noise levels in the environment [21]. These standards are dependent on the type of emission source, land use and time of day. The highest noise levels were recorded in the immediate vicinity of the road edge.

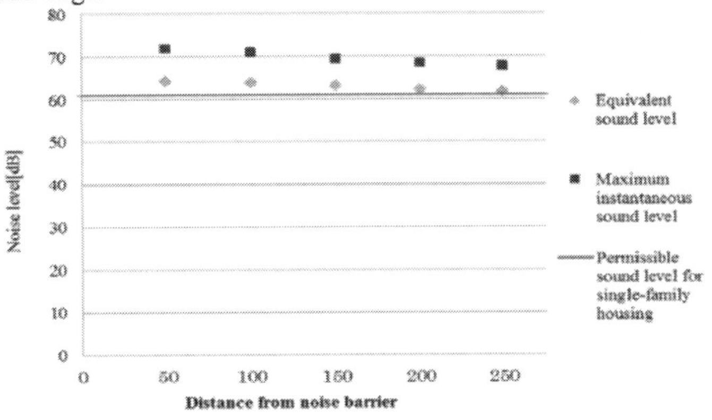

Fig. 4 Noise level at Rezerwat Wiączyń Street

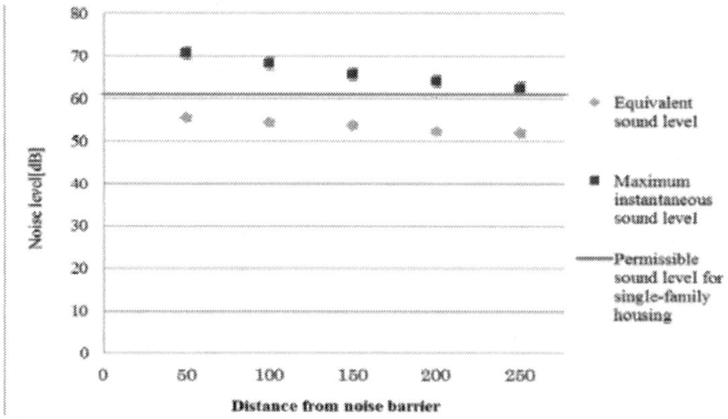

Fig. 5 Noise level at Nery Street

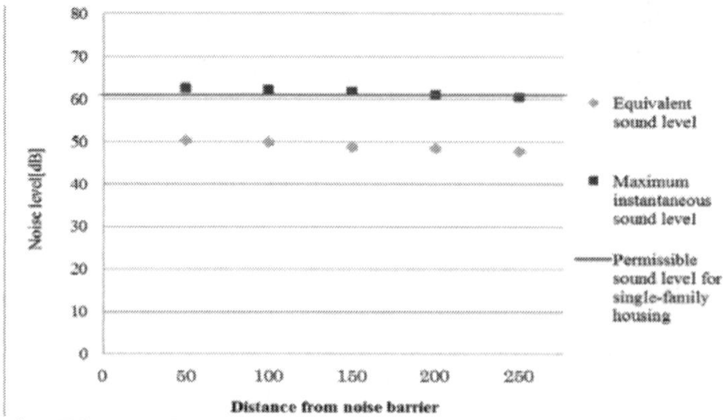

Fig. 6 Noise level at Taborowa Street

In the case of Wiaczyn Reserve Str. it was 71 dB (A) and a similar value was recorded on Nery road – 70 dB (A). The smallest exceeding of the allowed standards was noted at Taborowa Street. With the smooth movement of vehicles on motorways, the level of noise was lesser than the noise generated by frequent braking and acceleration (especially of heavy vehicles).

Studies show that the noise emitted by cars moving on the motorway decreases evenly as they move away from its edge. It has been found that a significant reduction in noise occurs at a distance of 250 m from the road. Moving away from the motorway, local factors gain importance in shaping the acoustic climate of the place, while traffic noise from the road is heard as steady humming noise.

5 Conclusions

The development of road transport is an important factor in the economic progress of the country and at the same time a significant source of environmental pollution and nuisance for man. The construction of expressways is a human intervention in ecological systems of the natural environment. It causes permanent occupation of the area under the planned route, temporary occupation of access roads to construction sites, warehouses for materials and machinery, aggregate extraction sites and others. From these areas the vegetation and soil layers are removed, the terrain is transformed and water relations changed. The development of expressway network has a significant impact on changes in the distribution of animal species, their habitats and causes environmental pollution.

Massive use of vehicles results in multilateral, negative environmental impacts associated with loading the atmosphere and land-water environment with contaminants, noise increase and barriers on animal migration routes [22- 26].

The expansion of the high-speed road network in Poland caused that road transport, as the most commonly used form of transport, has contributed significant to the environmental pollution. Car exhausts containing sulfur and nitrogen dioxides are much more harmful to people than industrial pollutants. Vehicle pollution spreads at low altitudes in the immediate vicinity of human habitats. These pollutants also cause damage to buildings, agricultural crops and forests.

On the analyzed part of the eastern ring road of Lodz, the concentrations of sulfur and nitrogen dioxides are significantly lower than the permissible levels of the analyzed substances in the air in accordance with the Ordinance of the Ministry of Environment on levels of certain substances in the air [27]. The value of a one-hour permissible level of nitrogen dioxide is 200 $\mu g/m^3$, and sulfur dioxide 350 $\mu g/m^3$. The emission of pollutants into the air coming from the car traffic is small and does not exceed the permissible values and at the same time does not significantly affect the environment.

Based on the analysis of soil samples, it can be concluded that not only traffic density but also duration of its impact on the soil is a very important element of the current sanitary condition. The analyzed part of the eastern ring road of Lodz is a relatively new investment, so the resulting heavy metal concentrations in the soil samples collected along this section are significantly lower than the admissible values described in the Ordinance of the Ministry of the Environment on the way of assessing the pollution of the earth's surface [28]. For example, the permissible content of zinc in the soil taken at a depth of 0 to 0.25 m is 300 mg/kg s.m., and it is almost three times higher than the determined concentration in the soil collected along the Lodz eastern ring road. Slightly higher concentrations, but also below the norm values were observed in soil samples taken in the city center. It should be noted, however, that such a trend may also be a consequence of the reduction of emissions due to technological changes (newer engine generations). At present, it cannot be confirmed that the state of soil pollution has decreased due to putting this motorway section into operation.

The conducted noise analyzes allowed us to determine the noise level along the eastern ring road of Lodz Its intensity depends on many factors such as topography, land use and weather conditions, local noise sources. In the immediate vicinity of the road, despite the use of energy absorbing barriers, there is no possibility of eliminating noise. In order to provide people with adequate living conditions, roads should be more than 200 to 300 meters away from residential buildings. Protection against roadway noise requires a wide range of activities from the planning stage to the proper traffic organization. It should be borne in mind that noise barriers are not the only way to reduce noise levels in road surroundings. It is necessary to use alternative methods of limiting undesirable noise emissions, including natural barriers in the form of broad green strips along the roads.

The conducted research allowed us to determine preliminarily the impact of the eastern ring road of Lodz on individual components of the environment. The results obtained constitute the first stage of the initiated monitoring of the environment on the analyzed road section and a full diagnosis of its environmental impact will be possible over the next few years.

5 Conclusions

1. D. Turek, *Road Transport in Poland in the years 2014, 2015* (Central Statistical Office, Warsaw, 2017)

2. M. van Poppel, L.I. Panis, E. Govarts, J. van Houtte, W. Maenhaut, Atmos. Environ., **60** (2012)
3. J.L. Durant, C.A. Ash, E.C. Wood, S.C. Herndon, J.T. Jayne, W.B. Knighton, M.R. Canagaratna, J.B. Trull, D. Brugge, W. Zamore, C.E. Kolb, Atmos. Chem. Phys., **10** (2010)
4. N.L. Gilbert, S. Woodhouse, D.M. Stieb, J.R. Brook, Sci. Tot. Environ., **312** (2003)
5. S. Hu, S. Friun, K. Kozawa, S. Mara, S.E. Paulson, A.M. Winer, Atmos. Environ., **43** (2009)
6. C. Hueglin, B. Nuchmann, R.O. Weber, Atmos. Environ., **40** (2006)
7. N.A.H. Janssen, P.H.N. van Vliet, F. Aarts, H. Harssema, B. Brunekreef, Atmos. Environ., **35** (2001)
8. M.P. Keuken, S. Jonkers, I.R. Wilmink, J. Wesseling, Sci. Tot. Environ., **408** (2010)
9. M. Kohler, U. Corsmeier, U. Vogt, B. Vogel, Atmos. Environ., **39** (2005)
10. M.C. Roorda-Knape, N.A.H. Janssen, J. de Hartog, P.H.N. van Vliet, H. Harssema, B. Brunkekreef, Sci. Tot. Environ, **235** (1999)
11. Y. Zhu, W.C. Hinds, S. Kim, S. Shen, C. Sioutas, Atmos. Environ., **36** (2002)
12. M.B.A. Dijkema, S.C. van der Zee, B. Brunekreef, R.T. van Strien, Atmos. Environ., **42** (2008)
13. N. Barros, T. Fontes, M.P. Silva, M.C. Manso, Transport. Res. A-Pol., **50** (2013)
14. P. Leisen, Sci. Tot. Environ., **134** (1993)
15. H. Pleijel, A. Ahlfors, L. Skärby, G, Pifl, G. Selldén, Å. Sjödin, Sci. Tot. Environ., **146-147** (1994)
16. L. Jiang, J. Kang, Sci. Tot. Environ., **543** (2016)
17. A. Pashkevich, M. Beliakova, A. Ivanov, A. Purju, Procedia Engineering, **178** (2017)
18. E.J. Judge, K. Werpachowski, M. Wishardt, J. Transp. Geogr., **12** (2004)
19. M. Karlson, U. Mörtberg, B. Balfors, Environ. Impact. Asses., **48** (2014)
20. Ordinance of the Ministry of Environment of 16 June 2011 on the requirements for measurements of substances or energy levels in the environment to be carried out by the manager of road, tram line, airport or port (Journal of Laws 2011 No. 140, item 824)
21. Ordinance of the Ministry of the Environment of 1 October 2012 on permissible noise levels in the environment (Journal of Laws 2012, item 1109)
22. A.H. Versluis, Sci. Tot. Environ., **146-147** (1994)
23. C. Beckx, W. Lefebvre, B. Degraeuwe, M. Vabhulsel, B. Kochan, T. Bellemans, S. Dhondt, L.I. Panis, Transport. Res. D-Tr E., **18** (2013)
24. T. Aljazzar, B. Kocher, Transport. Res. P., **14** (2016)
25. F. van der Hoeven, Tunn. Undergr. Sp. Tech., **26** (2011)
26. L.W. Gill, P. Ring, B. Casey, N.M.P. Higgins, P.M. Johnston, Sci. Tot. Environ., **601-602** (2017)
27. Ordinance of the Ministry of the Environment of 24 August 2012 on the levels of certain substances in the air (Journal of Laws 2012, item 1031)
28. Ordinance of the Ministry of the Environment of 1 September 2016 on the way of assessing the pollution of the earth's surface (Journal of Laws 2016, item 1395)

Determination of biomass burning tracers in air samples by GC/MS

Katarzyna Janoszka[1,*]

[1]Central Laboratory, Institute of Environmental Engineering Polish Academy of Science, M. Skłodowskiej-Curie 34 street, 41-819 Zabrze, Poland

Abstract. Levoglucosan (LG) as a main cellulose burning product at 300°C is a biomass burning tracer. LG characterize by relatively high molar mass and it is sorbed by particulate matter. In the study of air pollution monitoring LG is mainly analyzed in particulate matter, PM_1 and $PM_{2,5}$. The tracer create relatively high O-H···O bond and weaker C-H···O bond. Due to the hydrogen bond, LG dissolves very well in water. Analytical procedure of LG determination include: extraction, derivatization and analysis by gas chromatography coupled with mass spectrometry detector. In water samples levoglucosan is determined by liquid chromatography. The paper presents a methodology for particulate matter samples determination their analysis by gas chromatography coupled with a mass spectrometry detector. Determination of LG content in particulate matter was performed according to an analytical method based on simultaneous pyridine extraction and derivatization using N,O-bis (trimethylsilyl) trifluoroacetamide and trimethylchlorosilane mixture (BSTFA: TMCS, 99: 1).

1 Introduction

In the twenty first century more effort is put on finding an alternative sources of energy such as biomass burning. Biomass is an organic matter mostly referred to plant-based materials. Biomass combustion in fireplaces, residential, campfires and wildfires emits wood smoke which is a main source of soot, particulate matter (PM), free radicals, irritants and other chemicals which can be toxic or mutagenic [1-5]. The atmospheric aerosol formed can scatter or absorb incoming solar radiation, change visibility impairments and modify cloud formation processes [6-9].

Wood biomass is mainly consist of cellulose (40-50% of dry weight of wood, d.w.w) as well as hemicellulose (20-30% of d.w.w.), lignins (20-30% of d.w.w.) and additional components. The exact composition can differ depending on environmental aspects, type of tree (softwood or hardwood) and its genus. At temperature below 300°C cellulose depolymerize, water is eliminated, fragmentation and oxidation lead to formation of char. At the temperature above 300°C cellulose undergo pyrolysis involving bond cleavage, fission and disproportionation leading meanly to 1,6-anhydride of glucose, polar

* Corresponding author: katarzyna.janoszka@ipis.zabrze.pl

dehydrated glucose with a ketal functional group – levoglucosan, (LG; 1,6-anhydro-β-D-glucopyranose) [3-5,10-12]

Levoglucosan is thermal decomposition product form during thermal catalytic processing from plant biomass (cellulosic material) [13]. LG in a solid state has a crystal structure, it contain relatively strong O-H···O hydrogen bonds together with weaker interaction in C-H···O and because of the strong H bonds its molecules is linked into finite chain [14].

During biomass combustion there are also find isomers of levoglucosan, other monosaccharide derivatives like galactosan (GA) and mannosan (MN) (Fig. 1) but in minority because they are products of hemicellulose burning [5,10,11,15]. The ratio of levoglucosan to mannosan and/or galactosan can lead to the type of biomass burned by "chemical fingerprint" [16-19]. The marker that allow us to define biomass burning source should be a product of its degradation, emitted in majority at high concentration. It must be stable in environment and long-range transportation as well as easily detectable [1-3,10,20-22].

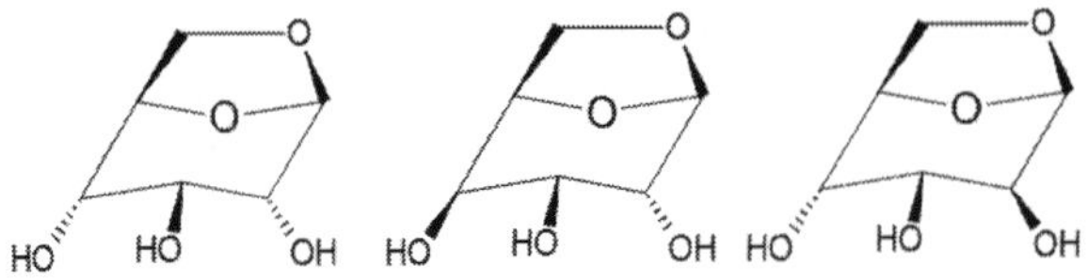

Fig. 1. Chemical structure of levoglucosan, galactosan and mannosan.

Levoglucosan is considered to be a biomass combustion marker because it meets the above requirements and as so it is determine. In atmospheric aerosol it shows resistance to photochemical oxidation, acid catalyzed hydrolysis and diagenesis in sediments. Because of its stability the same samples (e.g.: urban dust samples) can be determine in different laboratory conditions and use for comparative study [15,23]. LG is mainly entrapped in charcoal, ash and in urban dust and is associated with ultrafine particulate matter [10,15].

Gas chromatography coupled with mass spectrometry detector (GC/MS) is usually use to determine the LG in wood smoke, PM, ashes, soil and sediment samples [1-19,23] but other techniques as high performance liquid chromatography with aerosol charge detection (HPLC-ACD), high-performance anion-exchange chromatography with pulsed amperometric detection (HPAEC-PAD) or capillary electrophoresis with pulsed amperometric detection (CE-PAD) are also applied. The GC determination of levoglucosan consist of extraction, derivatization - mainly based on silylation giving trimethylsilyl ether derivative to reduce its polarity, same increase its volatility and analysis. The fragment ions used for identification was *m/z* 204, 217, 333 for electron impact ionization (EI). When levoglucosan was directly injected to GC/MS the results are unsatisfactory, chromatographic behavior was poor and response was low [1,10].

Extraction could be conducted in Soxhlet apparatus [6,24] or by pressurized fluid extraction with an accelerated solvent extraction (PFE ASE) [6]. However the most popular extraction technique is extraction in ultrasonic bath in different solvents or its mixture [1,2,5,11,18,25-28].

As for derivatizating agent different silylating reagents are used, like N,O-bis(trimethylsilyl)trifluoroacetamide (BSTFA), trimethylchlorosilane (TMCS), trimethylsilyl iodide (TMSI), trimethylsilylimidazole (TSIM) as well as N-methyl-N-(trimethylsilyl) trifluoroacetamide (MSTFA) in the presence or absence of solvent, like this use for extraction or e.g.: pyridine or DTE (1,4-dithioerythritol). When silylating agent or its mixture is added to prepared sample extract the mixture is allowed to react for specific time at adequate temperature.

2 Experimental

2.1 Chemicals

Levoglucosan (1,6-anhydro-β-D-glucose)was obtained from Sigma Aldrich Fine Chemicals with a stated purity of 99%. N,O-bis(trimethylsilyl)trifluoroacetamide (BSTFA) with trimethylchlorosilane (TMCS) (BSTFA:TMCS, 99:1) were obtained from Supelco. Pyridine, pure for analysis, were obtained from Avantor Performance Materials Poland S.A. (formerly POCH S.A.).

2.2 Sample collection

Samples of PM_{10} were collected on 47 mm Whatman QMA quartz filters at a 24-h cycle with the use of a low flow sampler with separating head PM_{10} Autoservice PNS3D15/LVS3d with a stabilized flow of 2.3 m^3/h. Aerosol samples were collected from March 2016 to April 2017 in Krynica Zdroj in the health resort - climatological station IMGW-PIB department in Cracow (Φ = 49°24'28", λ = 20°57'39", h = 582 m above sea level), near the border with Slovakia. The samples were sealed and transported to the laboratory at low temperature.

2.3 Extraction and Analysis

The quartz filters were cut into two parts and one half was placed in a reaction vial. Two milliliters of pyridine and 50 µl derivatizating mixture of N,O-bis(trimethylsilyl)trifluoroacetamide (BSTFA) and trimethylchlorosilane (TMCS) (BSTFA:TMCS, 99:1) were added in order to form silyl derivatives. The vial was sealed, vigorously shaken for 1 min. and put into an oven for 30 min. at 40°C in order to react. The 1 ml of extract were filtered through 0.22 µm syringe filter and analyzed by a Shimadzu GC-2010 gas chromatograph coupled with mass spectrometry detector (GC/MS) equipped with a 30 m; 0.25 mm i.d.; 0.25 µm film thickness HP-5MS capillary column.

3 Results and Discussion

During method development several reaction temperatures and reaction length were tested. Temperature at which reaction efficiency were tested was in a range from 40°C to 70°C (Fig.2.). Temperature increase caused the signal from levoglucosan derivative to decrease. At higher temperatures levoglucosan probably isomerizes and the reaction efficiency diminishes. Therefore reaction temperature was set for 40°C.

https://doi.org/10.1051/e3sconf/20182801015

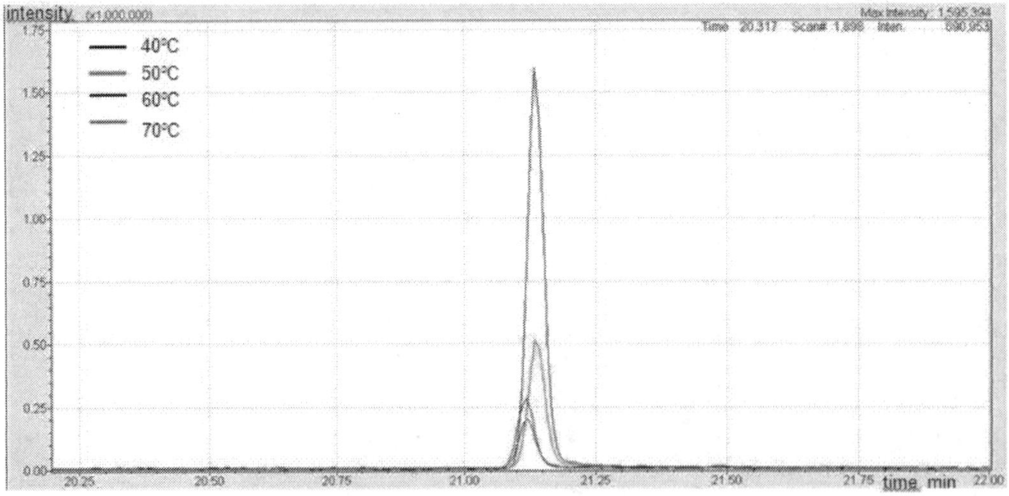

Fig. 2. Comparison of reaction temperatures.

The reaction length was also tested and there is no relevant differences between 30 minutes, 1 hour and 3 hours (Fig.3.). Because of that fact the reaction time was set for 30 min.

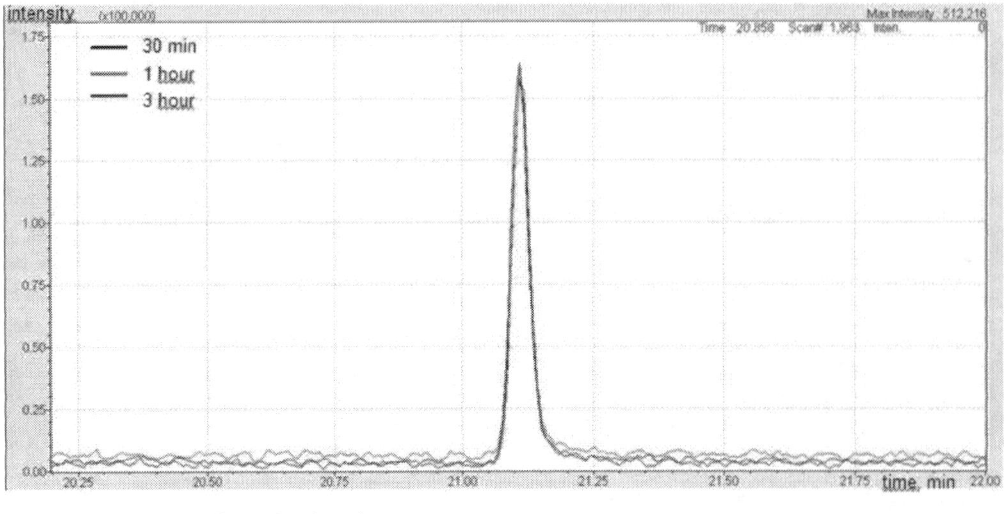

Fig. 3. Comparison of reaction length.

The set of 10 blank samples was analyzed. Blank filters were spiked with 1.5 μg of levoglucosan and the recovery was 95%. Precision - expressed as a relative standard deviation (RSD) - was equal 19%. Due to non-linear apparatus response the calibration range was divided into four calibration ranges. Calibration was performed in four concentration ranges: 0.02-0.2 μg/mL; 0.2-2.0 μg/mL; 2.0-6.0 μg/mL; and 5.0-35.0 μg/mL. The detection limit (LOD) was estimated as the concentration visible in the chromatogram with a peak height at least three times the signal to noise ratio, which was found to be 0.008 μg/mL and LOQ was expressed as 3 LOD amounts of 0.023 μg/mL. This corresponds to 0.001 $\mu g/m^3$ and 0.002 $\mu g/m^3$, respectively.

The method was applied to real samples of PM_{10} from Krynica Zdroj. The results of the levoglucosan concentrations are plotted at Fig. 3. Mean LG concentration during measurement campaign was 0.51 $\mu g/m^3$. The lowest LG concentration 0.01 $\mu g/m^3$ were

 https://doi.org/10.1051/e3sconf/20182801015

found in May (2016-05-23) and the highest concentration 3.15 μg/m³ in November (2016-11-24).

4 Conclusions

The method based on simultaneous extraction and derivatization of levoglucosan provide easy sample preparation and give low detection limit - 0.008 μg/mL and limit of quantification - 0.023 μg/mL. Analysis of PM_{10} samples shows the presence of levoglucosan. LG concentrations ranged from 0.01 up to 3.15 μg/m³.

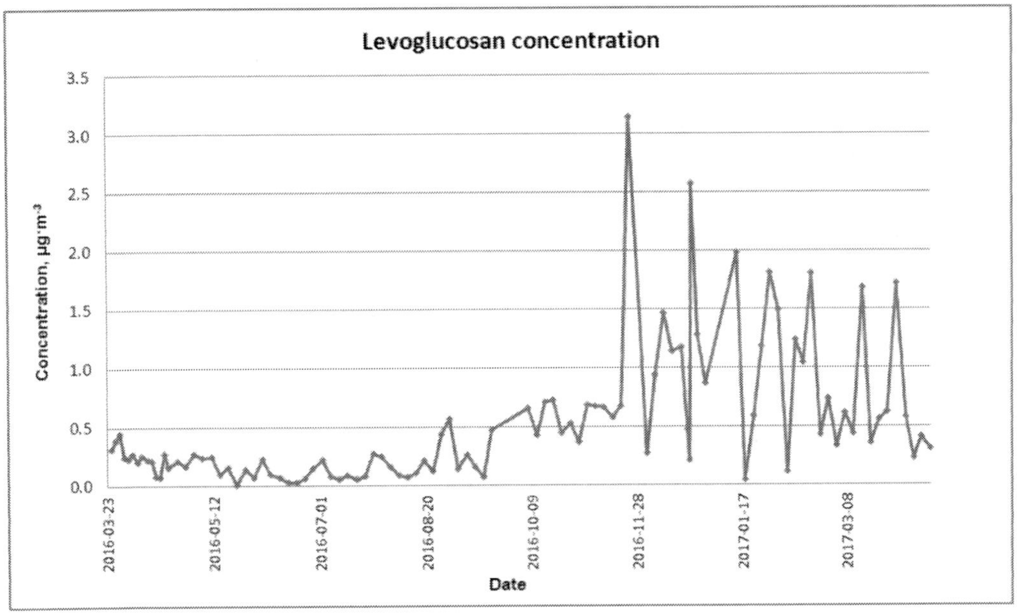

Fig. 4. Levoglucosan concentrations.

References

1. C.D. Simpson, R.L. Dills, B.S. Katz, D. A. Kalman, J. Air & Waste Manage. Assoc., **54**, 689-694 (2004)
2. O. Adetona, C.D. Simpson, G. Onstad, L.P. Naeher, Occup. Hyg.,**57**(8), 979-991 (2013)
3. B.R.T. Simoneit, J.J. Schauer, C.G. Nolte, D.R. Oros, V.O. Elias, M.P. Fraser, W.F. Rogge, G.R. Cass, Atmospheric Environment, **33**, 173-182 (1999)
4. M.-S. Bae, J.Y. Lee, Y.-P. Kim, M.-H. Oak, J.-S. Shin,K.-Y. Lee, H. Lee, S.Y. Lee, Y-J. Kim, Asian Journal of Atmospheric Environment, **6-1**, 53-66 (2012)
5. D.R. Oros, B.R.T. Simoneit, Applied Geochemistry, **16**, 1513-1544 (2001)
6. R.K. Larsen, M.M. Schantz, S.A. Wise, Aerosol Science and Technology, **40**, 781-787 (2006)
7. G. Schkolnik, Y. Rudich, Anal Bioanal Chem, **385**, 26-33 (2006)
8. S. Gao, D.D. Hegg, P.V. Hobbs, T.W. Kirchstetter, B.I. Magi, M. Sadilek, J. Geophys. Res., **108**(D13), SAF 27-1–SAF 27-16 (2003)
9. B. Graham, O.L. Mayol-Bracero, P. Guyon, G.C. Roberts, S. Decesari, C.M. Facchini, P. Artaxo, W. Maenhaut, P. Koll, M.O. Andreae, J. Geophys. Res., **107**(D20), LBA 14-1–LBA 14-16 (2002)

10. V.O. Elias, B.R.T. Simoneit, R.C. Cordeiro, B. Turcq, Geochimica et Cosmochimica Acta, **65**(2), 267-272 (2001)
11. M.R. Abas bin, D.R. Oros, B.R.T. Simoneit, Chemosphere, **55**, 1089-1095 (2004)
12. L.-J. Kuo, B.E. Herbert, P. Louchouarn, Organic Geochemistry, **39**, 1466-1478 (2008)
13. G.J. Kabo, Y.U. Paulechka, O.V. Voitkevich, A.V. Blokhin, E.N. Stepurko, S.V. Kohut, Y.V. Voznyi, J. Chem. Themodynamics, *85*, 101-110 (2015)
14. Ĺ. Smrčok, M. Sládkovičová, V. Langer, C.C. Wilson, M. Koóš, ActaCrys. *B62*, 912-918 (2006)
15. P. Louchouarn, L.-J. Kuo, T.L. Wade, M. Schantz, Atmospheric Environment, **xxx**, 1-7 (2009)
16. A. Otto, R. Gondokusumo, M.J. Simpson, Applied Geochemistry, **21**, 166-183 (2006)
17. R.L. Cordell, M. Mazet, C. Dechoux, S.M.L. Hama, J. Staelens, J. Hofman, C. Stroobants, E. Roekens, G.P.A. Kos, E.P. Weijers, K.F.A. Frumau, P.T. Panteliadis Delaunay, K.P. Wyche, P.S.Monks, Atmospheric Environment, **141**, 276-286 (2016)
18. R.L. Cordell, I.R. White, P.S. Monks, Anal Bioanal Chem, **406**, 5283-5292 (2014)
19. D. Fabbi, C. Torri, B.R.T. Simoneit, L. Marynowski, A.I. Rushdi, M.J. Fabiańska, Atmospheric Environment, **43**, 2286-2295 (2009)
20. M. Giannoni, T. Martellini, M. Del Bubba, A. Gambaro, R. Zangrando, M. Chiari, L. Lepri, A. Cincinelli, Environmental Pollution, **167**, 7-15 (2012)
21. C.D. Garcia, G. Engling, P. Herckes, J.L. Jr. Collett, C.S. Henry, Environ. Sci. Technol., **39,** 618-623 (2005)
22. B.Graham, O.L.Mayol-Bracero, P. Guyon, G.C. Roberts, S. Decesari, C.M. Facchini, P. Artaxo, W. Maenhaut, P.Koll, M.O. Andreae, J. Geophys. Res., **107**(D20), LBA 14-1–LBA 14-16 (2002)
23. K.E. Yttri, J. Schnelle-Kreis, W. Maenhaut, G. Abbaszade, M. Claeys, Atmos. Meas. Tech. **8**, 125-147 (2015)
24. E. Hedberg, C. Johansson, L. Johansson, E. Swietlicki, E. Brorstrom-Lunden, J. Air & Waste Manage. Assoc., **56**, 1669-1678 (2006)
25. C.-L. Hsu, C.-Y. Cheng, C.-T. Lee, W.-H. Ding, Talanta, **72**, 199-205 (2007)
26. Z. Zdráhal, J. Oliveira, D. Vermaylen, M. Claeys, W. Maenhaut, Environ. Sci. Technol, **36**, 747-753 (2002)
27. M. Mochida, K. Kawamura, P.Fu, T. Takemura, Atmospheric Environment, **44**, 3511-3518 (2010)
28. T. Zhang, M. Claeys, H. Cachier, D. Shuping, W. Wang, W. Maenhaut, X. Liu, Atmospheric Environment, **42**, 7013-7021 (2008)

Utilization of heat from High Temperature Reactors (HTR) for dry reforming of methane

Krzysztof Jastrząb*

Institute for Chemical Processing of Coal, Zamkowa 1 Str., 41-803 Zabrze, Poland

Abstract. One of the methods for utilization of waste carbon dioxide consists in reaction of methane with carbon dioxide, referred to as dry reforming of methane. It is an intensely endothermic catalytic process that takes place at the temperature above 700°C. Reaction of methane with carbon dioxide leads to formation of synthesis gas (syngas) that is a valuable chemical raw material. The energy that is necessary for the process to take place can be sourced from High Temperature Nuclear Reactors (HTR). The completed studies comprises a series of thermodynamic calculations and made it possible to establish optimum conditions for the process and demand for energy from HTR units. The dry reforming of methane needs also a catalytic agent with appropriate activity, therefore the hydrotalcite catalyser with admixture of cerium and nickel, developed at AGH University of Technology seems to be a promising solution. Thus, the researchers from the Institute for Chemical Processing of Coal (IChPW) in Zabrze have developed a methodology for production of the powdery hydrotalcite catalyser and investigated catalytic properties of the granulate obtained. The completed experiments confirmed that the new catalyser demonstrated high activity and is suitable for the process of methane dry reforming. In addition, optimum parameters of the were process (800°C, CO_2:CH_4 = 3:1) were established as well. Implementation of the technology in question into industrial practice, combined with utilization of HTR heat can be a promising method for management of waste carbon dioxide and may eventually lead to mitigation of the greenhouse effect.

Increasing demand for energy entails growth of interest in nuclear energy. In case of conventional electric power plants energy is produced by nuclear fission and then used for evaporation of superheated steam. Next, the utility is used to generate electricity in steam turbines, similarly to conventional thermal power plants. Nowadays the new generation of nuclear reactors is developed called High Temperature Reactors (HTR) and using neutrons decelerated in a graphic moderator [1]. Such generators produce the intermediate carrier of energy (helium) at the temperature of 600-800°C and are generally engineered for operation in the co-generation arrangement with simultaneous generation of high-temperature process heat and electricity. Thermal energy generated in HTR reactors can be used for a series of

* Corresponding author: kjastrzab@ichpw.pl

high-temperature technological processes, in particular the ones of chemical technologies. The most recent HTR projects are based on the following two key assumptions [2]:
- power is limited to several hundred of thermal megawatts (MWt);
- production of energy carrier with the temperature above 750°C.

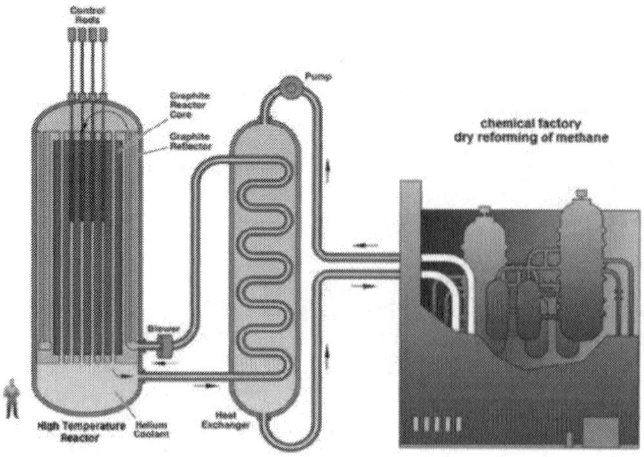

Fig.1. High Temperature Reactor (HTR), source: Wikipedia.

The first High Temperature Reactor (HTR) was operated from 1966 to 1988 at the scientific and research centre Forschungszentrum in Jülich, Germany [3]. Currently one HTR reactor is operated in Japan [4] and subsequent ones are being constructed in China [5]. Also in the USA and in Europe further research studies are in progress to launch HTR units (NGNP consortium and Nuclear Cogeneration Working Group). In Poland the HTRPL [6] project was being executed from 2012 to 2015 within the frameworks of a more extensive program sponsored by the National Centre of Research and Development. The key objectives of the project included examination of opportunities to implement HTR reactors in Poland. The group of industrial partners for the consortium comprises KGHM and TAURON, i.e. large Polish companies, where the first one consumes large amounts of energy whilst the second one is rated among the largest power generators in Poland. The project consortium was completed with such organizations as AGH University of Technology from Cracow, Institute for Chemical Processing of Coal (IChPW) from Zabrze, Central Mining Institute from Katowice, Silesian University of Technology from Gliwice and Institute of New Chemical Syntheses from Puławy.

The contribution of Institute for Chemical Processing of Coal (IChPW) consisted in tasks aimed at establishing of opportunities to utilize heat from HTR units to carry out strongly endothermic chemical syntheses that need application of carbon dioxide. These processes included:
- coal gasification with use of carbon dioxide [7];
- methane reforming with use of carbon dioxide [8];
- utilization of waste CO_2 for the Solvay technology [9].

The foregoing investigations referred to as a reasonable utilization of CO_2 may lead to substantial mitigation of greenhouse effect in future. Emission of CO_2 shall be reduced by both substitution of conventional coal-fired electric power plants with nuclear power plants as well as by conversion of already released carbon dioxide into new chemical products.

The class of strongly endothermal chemical technologies that are popular to enormous scale includes steam reforming of methane:

$$CH_4 + H_2O = 2\ CO + 3\ H_2 + 206\ kJ$$

https://doi.org/10.1051/e3sconf/20182801016

The synthesis gas originated from the foregoing process (mixture of CO and H_2) is then used for subsequent chemical reactions (production of ammonia, methanol, plastics, etc.).

The interesting modification of the foregoing process consists in reforming of methane by means of carbon dioxide [10] where steam is substituted with carbon dioxide:

$$CH_4 + CO_2 = 2\ CO + 2\ H_2 + 248\ kJ$$

It is also a catalytic process with high demand for energy. Similarly to the conventional steam reforming of methane the process balance is substantially shifted to the left side at the temperature below 600°C. It is also the case, i.e. at low temperatures, the hazard of thermal decomposition of methane exists with formation of carbon black, which leads to deactivation of the catalytic agent.

Although the foregoing process has been known for many years it is not implemented yet to the industry. The opportunity to develop the dry reforming technology lies in utilization of heat from HTR units. The additional benefit consists in the possibility to utilize streams of waste CO_2 that is released from conventional power engineering with coal-fired plants or from various chemical processes.

The studies completed within this project included a series of thermodynamic calculations intended to find out equilibrium composition for the post-reaction mixture with consideration of various process parameters. It was demonstrated that the optimum temperature for the dry reforming process of methane is 800°C since the reaction point is sufficiently shifted to the right-hand side. The risk associated with thermal decomposition of methane combined with deactivation of the catalytic agent is then minimized as well. Moreover, to achieve high degree of methane reaction the process should be carried out with the molar ratio between reagents, i.e. CO_2:CH_4 = 3:1.

Thermodynamic calculations made it possible to establish demand of the process in question for energy. The energy is consumed (adsorbed) during two phases (Fig. 2.):

- heating of substrates to the temperature of reaction,
- completion of the reaction at the assumed temperature.

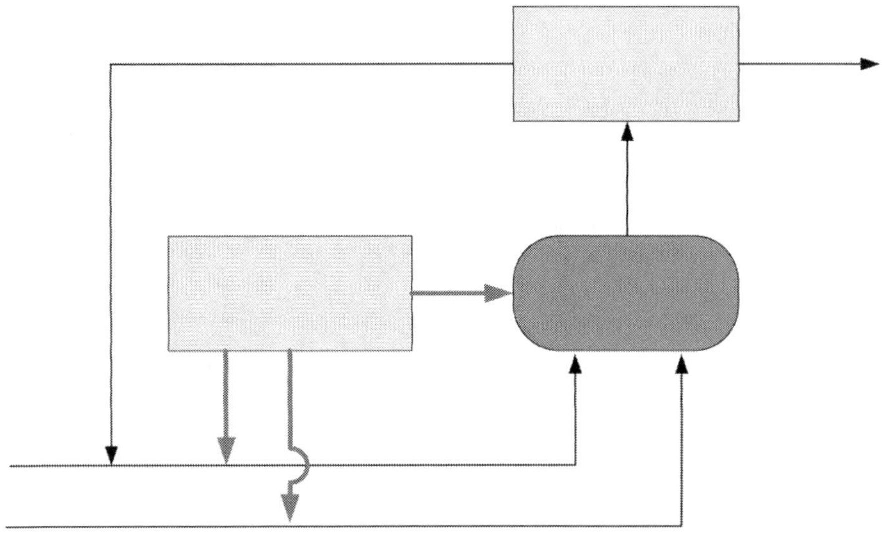

Fig.2. Synthesis gas production via dry reforming of methane using heat of HTR.

 https://doi.org/10.1051/e3sconf/20182801016

The completed calculations confirm that the demand for energy from the process in question is really high, which serve as the justification for use of heat released by HTR units. Results of calculation examples are disclosed in Fig. 3.

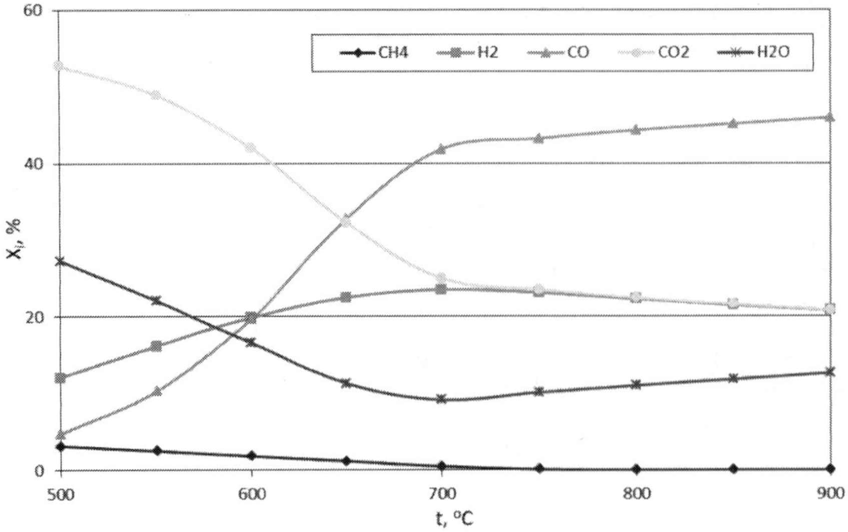

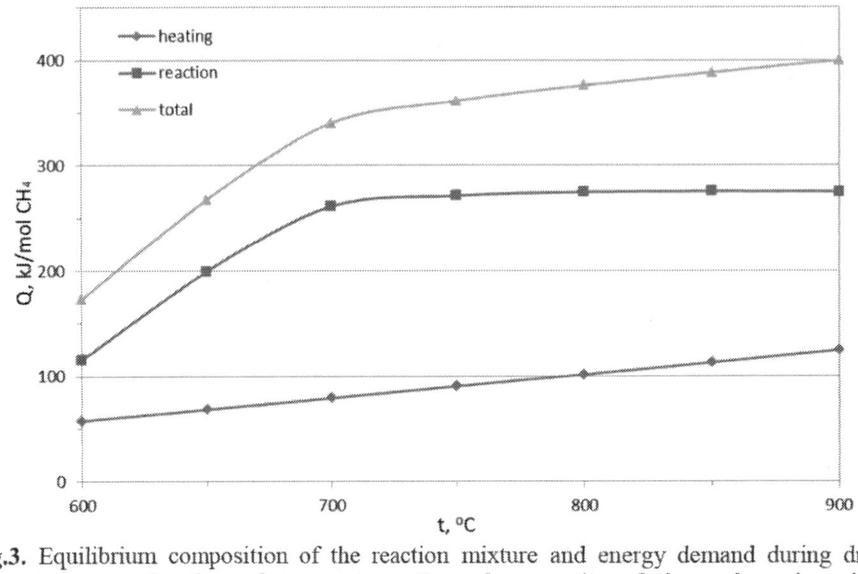

Fig.3. Equilibrium composition of the reaction mixture and energy demand during dry methane reforming as a function of temperature. Exemplary results of thermodynamic calculation at $CO_2:CH_4 = 3:1$.

Practical execution of the dry reforming of methane needs a catalytic agent with sufficient activeness. The appropriate catalytic agent enables high reaction rates, which significantly affects dimensions of possible industrial system for dry reforming and its performance. For instance, metallic nickel deposited on purposefully formed aluminium oxide can be used as conventional catalytic agent for the process of steam reforming. Unfortunately, in case of dry reforming such a catalytic agent fails to offer sufficient activeness. It is why seeking for an appropriate catalytic agent are still pending.

Promising properties were revealed for a hydrotalcite catalyser developed by scientists from the AGH University of Technology in Cracow [11]. Hydrotalcite is double-layered mineral with the structure of brucite where a part of magnesium cations is substituted with

https://doi.org/10.1051/e3sconf/20182801016

aluminium cations according to the isomorphic rule, which makes the general formula $Mg_6Al_2(OH)_{16}CO_3 \cdot 4H_2O$. To achieve the desired catalytic properties additional metallic promotors are introduced into the layered structure, for instance nickel or cerium.

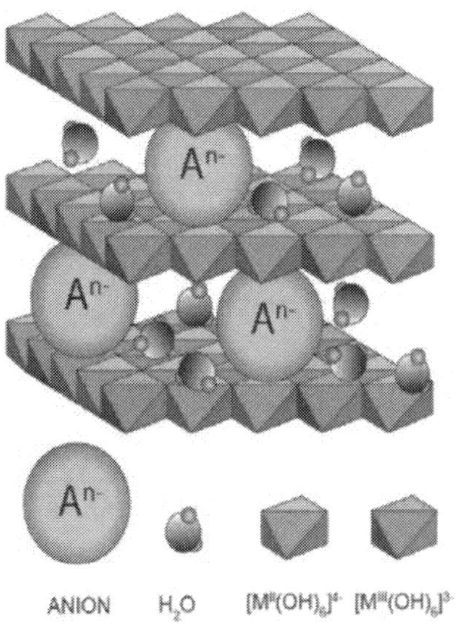

Fig.4. Structure of hydrotalcite.

Upon collaboration with the Faculty of Energy and Fuels at the AGH University of Technology (M. Motak, PhD.) the catalytic agent designated with the symbol of HT10 and with the content of 13% of nickel and 3% of cerium was selected since it proved to be the best catalytic agent for dry reforming of methane. To make its application easier, the technology for granulation of powdery was developed. The final product had five cylinders with the diameters of 5 mm and the length of 10 mm and with sufficient mechanical strength. Eventually, the granulate was used to determine catalytic properties of the catalytic material selected for tests.

The examinations were carried out under laboratory conditions inside a reactor with an immobile bed of catalytic material and a flowing gaseous phase. The reactor diameter was 35 mm whilst weigh of catalytic agent samples placed into the reactor was 50 g. The investigations were carried out for a mixture of methane and carbon dioxide with the total flow of 0.300 Nm^3/h.

The sample were investigated under the following range of parameters:
- temperatures: 600, 700 and 800°C;
- molar ratio between CH_4 and CO_2 in the inlet gas: - 1:3, 2:2 and 3:1.

The studies were focused on the analysis of the post-reaction mixture, i.e. the composition of gases at the outlet of the test setup. The particular attention was paid to key reagents: hydrogen and methane. Examples of the analysis results are shown in Fig. 5.

The achieved results served as a confirmation that efficiency of hydrotalcite catalyser and its suitability for the process of dry reforming of methane is really high. The analysis of these results was then used to define optimum conditions for the process execution. The overall objective of the studies was to achieve the maximum degree of methane reaction with high concentration of methane in the post-reaction gas mixture as well as the minimum possible temperature of the chemical process. It was demonstrated that the following parameters are the best suitable for dry reforming of methane:

Temperature of the dry reforming of methane: 800°C
Three times more of CO_2 then CH_4 in the reactive mixture

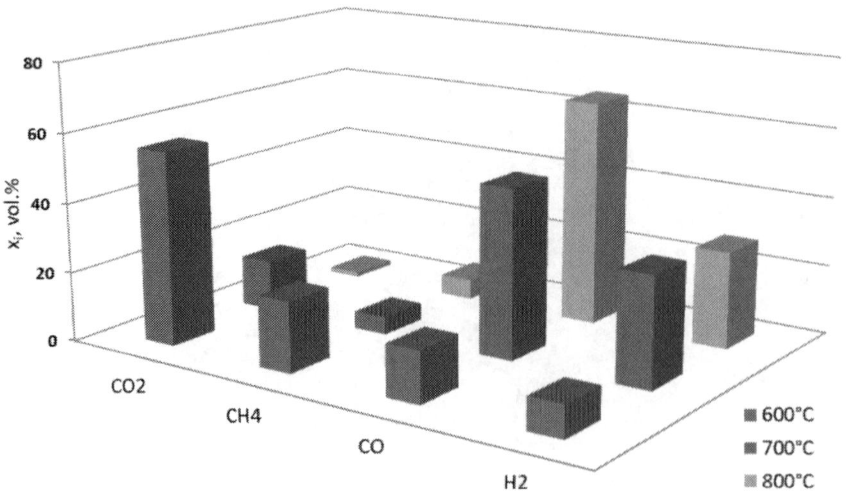

Fig.5. Equilibrium composition of the reaction mixture and energy demand during dry methane reforming as a function of temperature. CO_2:CH_4 = 3:1.

These conditions correspond to the optimum conditions determined on grounds of theoretical calculations. It should be emphasized that the established temperature of the chemical process is the boundary temperature for the circular /intermediate carrier of energy (i.e. helium), the maximum possible one that can be achieved in reactors of HTR type. The results achieved from examinations made it possible to formulate assumptions necessary to design a reference pilot plant for the process of methane reforming with use of carbon dioxide with the capacity of 1000 Nm^3 of gas per hour.

It was also demonstrated that the hydrotalcite catalyser activated with nickel and cerium is a material with high catalytic activeness and suitability for the dry process of methane reforming. The process running at the temperature of 800°C and with molar ratio CO_2:CH_4 = 3:1 makes it possible to achieve high degree of reaction with pretty high concentration of hydrogen in the post-reaction mixture (above 50% of H_2 V/V). The optimum temperature necessary for the chemical process and established from these studies enables utilization of heat sourced from nuclear reactors of HTR types is sufficient for both preheating of reagents and for execution of the dry process of methane reforming itself. Implementation of the technology in question into the industrial practice and combining the technology with utilization of heat from HTR units may prove an attractive approach to utilization of waste carbon dioxide and, in consequence, can enable mitigation of the greenhouse effect. Emission of CO_2 shall be reduced by both substitution of conventional coal-fired electric power plants with nuclear power plants as well as by conversion of already released carbon dioxide into new chemical products.

The work was performed under the strategic research project SP/J/1/166183/12 "*Development of high-temperature reactors for industrial applications*" acronym HTRPL, funded by The National Centre for Research and Development, Poland.

References

1. J. Vujića, R. Bergmanna, R. Škodab, M. Miletićc, Energy **45**, 288 (2002)
2. R. Kuhr, Nuclear Engineering and Design, **238**, 3013 (2008)
3. http://www.world-nuclear.org/reactor/default.aspx/AVR Juelich, recevied 2017-10-25
4. H. Hayakawa, Nuclear Engineering and Design **233**, 391 (2004)
5. Y. Sun, J. Xu, Z. Zhang, International Journal of Nuclear Hydrogen Production and Applications **1**, 104 (2006)
6. Strategic research project SP/J/1/166183/12 "*Development of high-temperature reactors for industrial applications*" acronym HTRPL, funded by The National Centre for Research and Development, Poland, 2012-2015
7. K. Jastrząb, O. Piotrowski, Energy Policy Journal, **16**, 309 (2013)
8. K. Jastrząb, S. Stelmach, *Cost analysis of using of HTR heat in the process of dry reforming of methane*, Proceedings of the 41th International Conference SSChE, Tatranske Matliare, Slovakia, 2014
9. K. Jastrząb, *Analysis of possibilities of using HTR heat in the Solvay process*, Proceedings of the 43th International Conference SSChE, Tatranske Matliare, Slovakia, 2016
10. T. Chmielniak, M. Ściążko, A. Sobolewski, G. Tomaszewicz, Energy Policy Journal **15**, 125 (2012)
11. R. Dębek, A. Gramatyka, M. Motak, P. Da Costa, Przemysł Chemiczny **93**, 2026 (2014)

Database for content of mercury in Polish brown coal

Krzysztof Jastrząb*

Institute for Chemical Processing of Coal, Zamkowa 1 Str., 41-803 Zabrze, Poland

Abstract. Poland is rated among the countries with largest level of mercury emission in Europe. According to information provided by the National Centre for Balancing and Management of Emissions (KOBiZE) more than 10.5 tons of mercury and its compounds were emitted into the atmosphere in 2015 from the area of Poland. Within the scope of the BazaHg project lasting from 2014 to 2015 and co-financed from the National Centre of Research and Development (NCBiR) a database was set up with specification of mercury content in Polish hard steam coal, coking coal and brown coal (lignite) grades. With regard to domestic brown coal the database comprises information on coal grades from Brown Coal Mines of 'Bełchatów', 'Adamów', 'Turów' and 'Sieniawa'. Currently the database contains 130 records with parameters of brown coal, where each record stands for technical analysis (content of moisture, ash and volatile particles), elemental analysis (CHNS), content of chlorine and mercury as well as net calorific value and combustion heat. Content of mercury in samples of brown coal grades under test ranged from 44 to 985 μg of Hg/kg with the average level of 345 μg of Hg/kg. The established database makes up a reliable and trustworthy source of information about content of mercury in Polish fossils. The foregoing details completed with information about consumption of coal by individual electric power stations and multiplied by appropriate emission coefficients may serve as the background to establish loads of mercury emitted into atmosphere from individual stations and by the entire sector of power engineering in total. It will also enable Polish central organizations and individual business entities to implement reasonable policy with respect of mercury emission into atmosphere.

1 Introduction

In August 2017 the so called Mercury Convention, also referred to a Convention from Minamata came into effect. The Convention, signed by Poland in September 2014, is aimed at comprehensive protection of environment and human health against release of mercury into atmosphere, water and earth. The regulations of the agreement govern issues associated with production of that metal, trading with mercury-containing products and application of mercury to products and industrial processes. In addition, the convention establishes rules

* Corresponding author: kjastrzab@ichpw.pl

for handling and disposal of mercury-containing wastes in the manner that is safe for natural environment, implementation of appropriate methods for storage of mercury as well as deals with issues related to reclamation of areas polluted with that metal. The subsequent legislative document that Poland has undertaken to obey is the Directive 2010/75EU of the European Parliament and of the Council on industrial emissions (integrated pollution prevention and control), so called IED Directive that is effective from 2010. On 17th August the document called Best Available Techniques (BAT) Reference Document for Large Combustion Plants, referred commonly as BAT Conclusions for Large Combustion Plants (LCP) was published in the Official Journal of European Union. The both documents impose really tough requirements to the permissible concentration of released pollutants and specify Best Available Techniques (BAT) necessary to keep the emission levels below the permissible limits.

Introduction of new legislative documents is particularly important since Poland is rated among countries with levels of mercury emission in Europe. The most recent survey carried out by the National Centre for Balancing and Management of Emissions (KOBiZE) revealed that emission of mercury from industrial processes and combustion of fossils on the area of Poland amounted in 2015 to 10,576 kg. Shares of individual sectors of the national economy in emission of mercury into atmosphere are depicted in Fig. 1. The breakdown of mercury emission as shown in the diagram is caused by the fact that power generation by Polish plants is predominantly based on combustion of hard and brown coal with substantial contribution from extensive industry of building materials. citation.

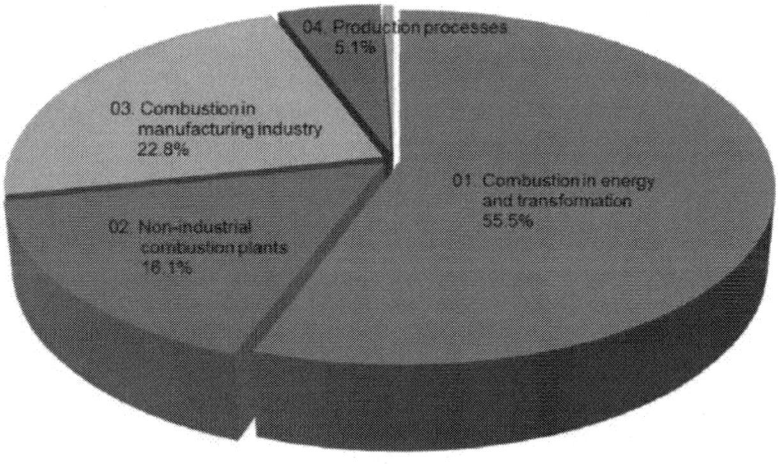

Fig. 1. Shares of sectors of the national economy in emission of mercury into atmosphere [1].

The balance presented herein indicates that the highest rates of mercury emission into atmosphere fall to the sector of power generation and conversion. Considerable portion of that emission is owed to power plants fuelled with brown coal (lignite) since these plants are rated among the largest ones in Europe with the total installed power of 9,220.5 MW at the end of 1st Q of 2015, which makes up 23.5% of total power available from all domestic electric power as well as heat and power plants. Only during the 1st half of 2015 the power plants fuelled with brown coal generated 13,518 GWh of energy. Unfortunately, such levels of power generation need huge consumption of brown coal. For long years production of these coal grades has been being maintained at the level of 60 million tons per year, which is currently only slightly less that than total production of hard coal. Only during first six months of 2015 as much as 31.6 tons of brown coal were mined with the largest share of the Brown Coal Mine (KWB) 'Bełchatów' amounting to more than 21 million tons, which makes 67.8% of total production of brown coal within the country. The shares of other open-pit mines in total production of brown coal were distributed between KWB Konin (14.2% /4.5 M tons), KWB Turów (11.1% /3.5 M ton) and KWB Adamów (6.3% /2.0 M ton).

Unfortunately, high consumption of brown coal always entails substantial emission of mercury and its compounds. First and foremost such emission levels result from high content of mercury in such a fossil like brown coal but lack of specialized systems or plants designed for mitigation of mercury emission into atmosphere is also quite important. Although some attempts to remove mercury from flue gas have already been undertaken at experimental plants operated at Electric Power Plant in Bełchatów (in collaboration between Polish Power Engineering Group (PGE) and Technical University of Łódź) and at 'Łaziska' Power Plant (jointly by Tauron, AGH University of Technology from Cracow and the Institute for Chemical Processing of Coal (IChPW) from Zabrze) but these studies are still on the level of tests.

The amount of mercury emitted into the atmosphere is mostly dependent on its content in fossils – hard and brown coal grades. It is estimated that the coefficient of mercury emission associated with combustion of brown coal is 0.15 g Hg/Mg, which makes 9.0 μg Hg/MJ [3]. Thus, the rough emission of mercury from a typical Polish power plant ranges from 52 to 57 kg of Hg per year [4]. According to the practice of recent years the content of mercury is determined in fossils and the results from obtained from investigation of mercury content in coal grades mined worldwide [4-9] indicated that content of that element ranges from 20 to1400 μg/kg. The lowest content of mercury was found in Japanese coal grades (30 to 100 μg/kg) and then Australian (30 to 400 μg/kg), New Zealand and Russian ones (20 to 600 μg/kg). The highest concentration of mercury was revealed in coal grades from Germany (700 to 1400 μg/kg) with direct Polish and South African grades (10 to 1000 μg/kg) right afterwards. Results of research studies carried out at the end of 20th century demonstrated that content of mercury in brown coal grades ranges from 81 to 1030 μg/kg with the average level of 323 μg/kg [5]. The least amount of mercury was found in coal from open pits of KWB Konin (200-250 μg/kg) whilst the most mercury-laden were coal grades from KWB Bełchatów (416 μg/kg).

The study [10] reports even higher content of mercury in brown coal grades from Polish mines, with the highest level of 2,600 μg/kg and the average of 1180 μg/kg. But on the other hand, analyses of 29 samples taken from 6 largest Polish deposits [11] revealed the range of mercury concentration from 60 to 665 μg/kg with the average of 197 μg/kg. At the same time content of mercury referred to the calorific value of fuel varied from 4.42 to 74.74 g Hg/TJ (21.38 g/TJ on average). Other studies completed at the Institute for Chemical Processing of Coal (IChPW) from 2009 to 2012 [12] made it possible to find out that content of mercury in brown coal grades ranged from 97 to 1300 μg/kg, whilst the average concentration of mercury amounted to 360 μg/kg and was nearly fivefold higher that the respective content of mercury in hard coal (73 μg/kg).

The foregoing review of literature references clearly shows that results disclosed by individual authors significantly differ from each other. The broad range of results from determination of mercury content in coal grades is associated with many of factors. First and foremost it is extremely difficult to determine such low content of mercury in coal samples (trace analysis). Secondly, different research centres apply different analytic methods (AAS, CVAAS, CVAFS, ICP and other). The next premise lies in various weighs of coal samples and the sampling methods. Variable content of mercury in a coal deposit within the same coal mine as well as sampling locations (samples taken from faces, from haulage conveyors or averaged samples at the power plant) also cannot be neglected. All these reasons contribute to variations in determination results.

2 Experimental

In 2013 a consortium was established by the Central Mining Institute (GIG), AGH University of Technology, Institute for Chemical Processing of Coal (IChPW) and Technical University of Częstochowa with the aim to launch the scientific and research project called BazaHg and financed by the National Centre for Research and Development (NCBiR). The project scope included comprehensive evaluation of mercury content in steam grades of hard coal (GIG in Katowice), coking grades of hard coal (AGH from Kraków) and brown coal (lignite) grades (IChPW from Zabrze). The achieved results made it possible to create a reliable and trustworthy database with specification of mercury contents in Polish fossils. The database shall enable reasonable and efficient management of mercury loads released into atmosphere from processes of fossil combustion and treatment.

To develop such a database with content of mercury in brown coal grades from Polish mines as many as 65 samples were taken from open pits on the area of Poland during the time period from 2014 to 2015 and parameters of these samples were determined. Each time samples with the weight of 50 kg were taken from open-pit faces. Then every single sample was disintegrated and averaged to obtain two representative analytic samples to determine physical and chemical properties of brown coal grades. The scope of investigations included:

- complete technical analysis (content of moisture, ash and volatile particles),
- complete elemental analysis (CHNS),
- determination of net calorific value and combustion heat,
- assay of chlorine and mercury content.

All analyses were carried out by the Accredited Laboratory of Fuels and Active Fuels operated by the Institute for Chemical Processing of Coal in Zabrze (accreditation from the Polish Centre of Accreditation, certificate number AB 081) pursuant to accredited standards and procedures summarized in Table 1.

Table 1. Applied accredited standards and procedures.

Analysis		Symbol	Standards/procedures	Error
Technical analysis	Total moisture, %	W_t^r	PN-80/G-04511	0.5
	Analytical moisture, %	W^a	PN-G-04560:1998	0.1
	Ash, %	A^a	PN-G-04560:1998	0.2
	Volatile particles, %	V^{daf}	PN-G-04516:1998	0.17
Elemental analysis	Content of carbon , %	C_t^a	PN-G-04571:1998	0.6
	Content of hydrogen, %	H_t^a	PN-G-04571:1998	0.27
	Content of nitrogen, %	N^a	PN-G-04571:1998	0.15
	Content of oxygen, %	O_d^a	z obliczeń	0.69
Heat of combustion, kJ/kg		Q_s^a	Q/LP/03/A:2011	83
Calorific value, kJ/kg		Q_i^a	Q/LP/03/A:2011	101
Chlorine content, %		Cl^a	PN-G-04534:1999	0.017
Total mercury content, µg/kg		Hg^d	Q/LP/32/A:2011	28

The content of Hg^d was determined according to the accredited procedure Q/LP/32/A:2011. The analyses were carried out with use of the automatic Japanese analyzer MA-2 from Nippon Instrument Corporation (Fig. 2.) designed for mercury assay. The determination process consists in thermal decomposition (850°C) of a sample under test and condensing of mercury on a gold trap with formation of amalgamate. Atomic mercury was then released from the collector by heating (600°C) and its amount was assayed in an absorption cell with use of the CVAAS method, at the wavelength of 253.7 nm. The error of the procedure is 28 µg Hg/kg.

Other determination procedures were carried out with use of the following instruments:

- thermogravimetric analyser TGA 701 from LECO with a full set of auxiliary equipment for determination of analytic moisture W^a and content of ash A^a
- CHNS 628 instrument from LECO to determine content of carbon, hydrogen and nitrogen C_t^a, H_t^a, N^a
- SC-632 instrument from LECO to determine total content S^a of sulphur S_t^a
- AC-350 or AC-500 calorimeters from LECO to determine calorific value and combustion heat Q_s^a and Q_i^a

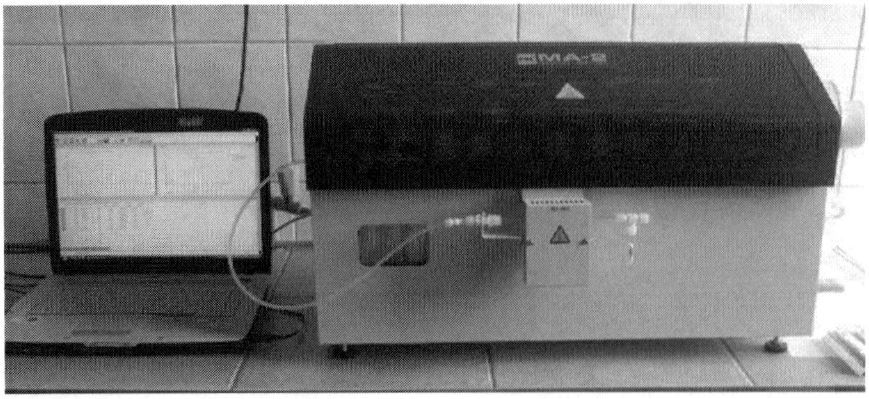

Fig.2. MA-2 mercury analyzer (NIC/Japan).

All completed measurements were subsequently used to set up a database with content of mercury in Polish mineable coal grades and products of processing thereof. Currently the database comprises 130 records related to brown coal, where each record stands for complete technical analysis (content of moisture, ash and volatile particles), complete elemental analysis (CHNS), content of chlorine and mercury as well as net calorific value and combustion heat. The database shall be consecutively supplemented with results of subsequent assays for brown coal samples.

3 Results and discussion

Table 2. Mercury content in Polish brown coal.

Brown Coal Mine (KWB)	Mercury content µg Hg/kg	Average mercury content µg Hg/kg
KWB Bełchatów, Bełchatów open pit	201 - 985	499
KWB Bełchatów, Szczerców open pit	175 - 922	439
KWB Adamów, Adamów open pit	251 - 619	442
KWB Adamów, Koźmin open pit	192 – 527	339
KWB Turów	144 – 601	287
KWB Sieniawa	44 - 310	125
Average	44 – 985	345

The analysis of mercury content in individual samples of brown coal (Table 2.) revealed that average content of mercury in Polish grades of brown coal makes 345 µg Hg/kg. The highest concentration of mercury was found for coal samples from Brown Coal Mine (KWB) in Bełchatów (the Bełchatów open pit) with the lowest content of 44 μg Hg/kg recorded for coal samples from KWB Sieniawa. The comparison made for average content of mercury in coal samples taken from individual open pits revealed that the highest content of mercury occurred in coal samples from the Bełchatów open pit, i.e. 499 μg Hg/kg. Slightly lower concentration of mercury was recorded for the Szczerców open pit also operated by KWB Bełchatów (439 μg Hg/kg) and from the Adamów open pit operated by KWB Adamów (442 μg Hg/kg). The lowest level of mercury was recorded for coal from KWB Turów (287 μg Hg/kg) and from KWB Sieniawa (125 μg Hg/kg). The foregoing details, completed with information on coal consumption by individual power plants and multiplied by emission coefficients may serve for determination of mercury load released into the atmosphere from individual power plants and from the entire sector of power engineering in total. In addition, after taking account of combustion parameters and other figures standing for characteristics of fossils, one can precisely evaluate how much mercury is released into the atmosphere in flue gas from power plants.

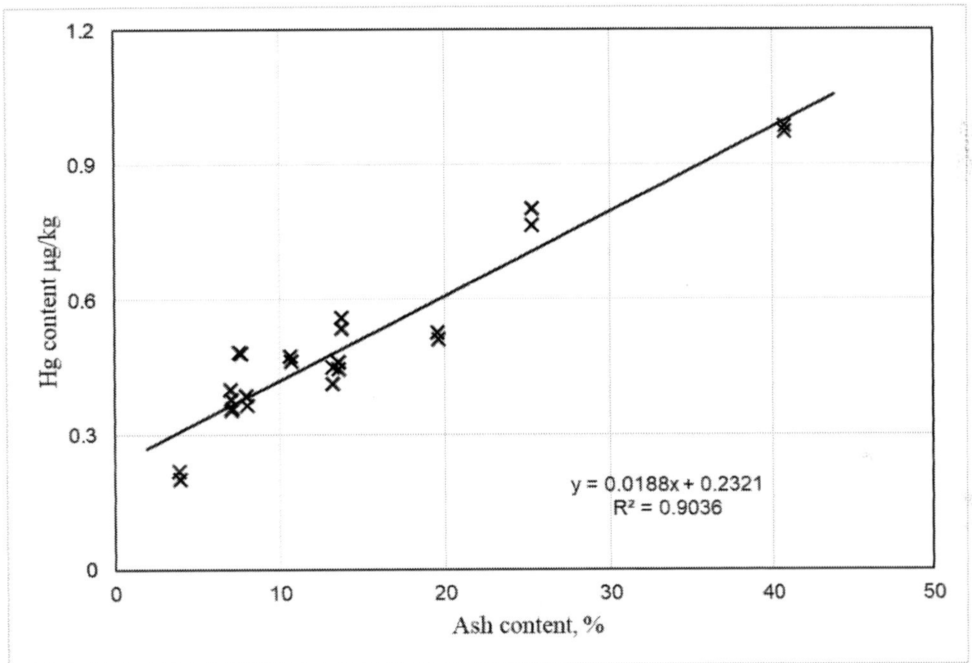

Fig.3. Exemplary dependence of mercury content on ash content. KWB Bełchatów.

Large amount of data enables also statistical analyses. The investigations were focused on correlations between individual physical and chemical parameters determined for fossils and including details on elementary composition or calorific value on one side and concentration of mercury on the other side. The calculations involved conventional analysis of regression (Statistica 6.0 from StatSoft). Sufficient amount of assay results enables also the statistical analysis with use of neural networks (Statistica Neutral Network from StatSoft). As a consequence, the following substantial relationships were established:

- content of mercury as a function of ash and volatile particles content,
- content of mercury as a function of sulphur content.

Examples of established correlations are shown in Fig. 3. and 4.

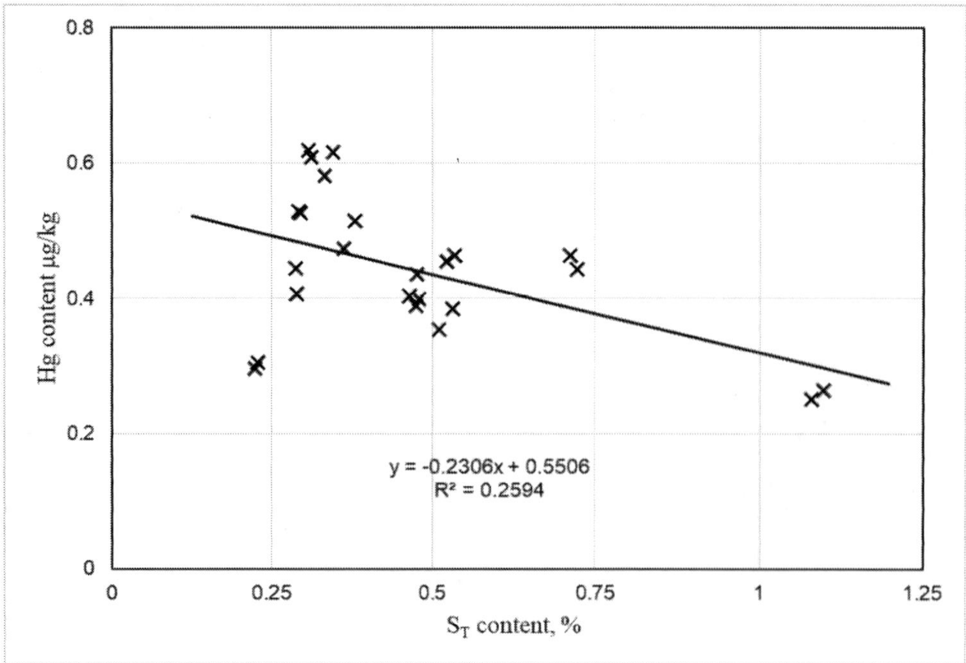

Fig.4. Exemplary dependence of mercury content on total sulphur content KWB Adamów.

4 Summary

Information disclosed in this study make up only a small portion of details comprised in the database described herein and set up for mercury content in domestic grades of brown coal. The developed database contains fully reliable and trustworthy knowledge about content of mercury in Polish fossils and enables reasonable and efficient management of combustion processes in the Polish sector of power engineering and associated release of mercury and its compounds into the atmosphere. The detailed analysis of coal grades shall also make it possible to predict emission of mercury from other processes associated with use of coal but different from simple combustion, for instance coal gasification, pyrolysis, coking technologies, production of activated charcoal and a long series of other products. It shall also enable Polish governmental organizations to cope with the exaggerated 'mercury problem' in Poland. The database with mercury contents shall be updated on current basis and shall be soon available on the project website.

The work was performed under the research project BazaHg no PBS2/A2/14/2013 founded in 2013-2016 by The National Centre for Research and Development, Poland.

References

1. *National emission balance of SO_2, NO_x, CO, NH_3, NMLZO, dust, heavy metals and TZO from 2014 - 2015 in the SNAP classification. Synthetic report.* KOBiZE Publishing House, Warszawa, 2017
2. A. Pietraszewski, Węgiel Brunatny **2**, 4 (2015)
3. L. Sloss, *Mercury emissions and effects-the role of coal.* IEAPER/19, London, England, August 1995

4. A. Głodek, J. Pacyna, Atmospheric Environment **43**, 5668 (2009)
5. I. Bojakowska, G. Sokołowska, Biuletyn PIG **394**, 5 (2001)
6. K. Wojnar, J. Wisz, Energetyka **4**, 280 (2006)
7. U. Lorenz, Z. Grudzinski, Gospodarka Surowcami Mineralnymi **24**, 1271 (2008)
8. S. Tewalt, L. Bragg, R. Finkelman, *Mercury in U.S. Coal. Abundance, distribution and modes of occurrence*, U.S.Geological Survey; Fact Sheet 095, 2001
9. E. Pacyna, J. Pacyna, J. Fudała, E. Strzelecka-Jastrząb, S. Hławiczka, D. Panasiuk, Sci. Tot. Environ. **370**, 147 (2006)
10. B. Bielowicz, Gospodarka Surowcami Mineralnymi **29**, 47 (2013)
11. J. Gołaś, A. Strugała, P. Burmistrz, T. Dziok, A. Rozwadowski, *The content of mercury in Polish coal and products of their enrichment. Mercury emission and possibilities of reducing it in the Polish energy sector*, AGH Publishing House, Kraków 2015
12. E. Misztal, T. Chmielniak, I. Mazurek, *The content of mercury in fuels used in the Polish energy sector and possibilities of its reduction by means of thermal treatment before combustion*, III International Scientific Conference, Mercury in Environment, Gdynia, May 8-10, 2013

E3S Web of Conferences **28**, 01018 (2018) https://doi.org/10.1051/e3sconf/20182801018
Air Protection in Theory and Practice

The identification of Volatile Organic Compound's emission sources in indoor air of living spaces, offices and laboratories

Beata Kultys[1,*]

[1] Jagiellonian University in Krakow, Faculty of Chemistry, Gronostajowa 2, 30-387 Krakow, Poland

Abstract. Indoor air quality is important because people spend most of their time in closed rooms. If volatile organic compounds (VOCs) are present at elevated concentrations, they may cause a deterioration in human well-being or health. The identification of indoor emission sources is carried out by comparison indoor and outdoor air composition. The aim of the study was to determinate the concentration of VOCs in indoor air, where there was a risk of elevated levels due to the kind of work type carried out or the users complained about the symptoms of a sick building followed by an appropriate interpretation of the results to determine whether the source of the emission in the tested room occurs. The air from residential, office and laboratory was tested in this study. The identification of emission sources was based on comparison of indoor and outdoor VOCs concentration and their correlation coefficients. The concentration of VOCs in all the rooms were higher or at a similar level to that of the air sampled at the same time outside the building. Human activity, in particular repair works and experiments with organic solvents, has the greatest impact on deterioration of air quality.

1 Introduction

Indoor air quality is an important issue for this reason that people spend most of the time in all kinds of rooms [1]. If the air in the room is contaminated, it can adversely affect the well-being and human health [2, 3]. The compounds that most often pose such a threat are volatile organic compounds (VOCs), which are defined as organic compounds at boiling point below 250 °C ant at the pressure 101.3 kPa [4].

The sources of indoor air pollution can be different, they may be both materials for furnishing the room and human activity or VOCs can get from the outside through ventilation or windows if the building is located near emitters such as roads or industrial plants [5-7]. The effects on well-being and human health depend on VOCs concentrations and its nature. Poor air quality due to the presence of VOCs, as well as biological or physical factors, may cause symptoms known as "sick building syndrome", that is the state of ill feeling with symptoms such as headache, nausea, eye or respiratory irritation, that

* Corresponding author: beata.kultys@uj.edu.pl

only appear and worsen while staying in this building and disappear almost immediately after leaving the building [8].

The emission sources may occur both in the room itself and outside. Considering that, in case of high concentration of VOCs, it is essential to first identify the location of emission sources, and then to take action to improve the air quality. When investigating indoor air, outdoor air tests should be carried out simultaneously. Then the first stage of identification the presence of emission sources may be the appropriate interpretation of the results, for example comparative analysis with the concentration of compounds outside or analysis of correlation coefficients of compounds indoor and outdoor.

2 Research methodology

2.1 Scope of research

In the years 2009-2016 in Krakow concentration of VOCs was investigated in several rooms, such as student houses during renovation, private apartment, laboratories and office spaces. These rooms have been selected for testing because of the expected elevated level of VOCs or have been indicated by users because of fragrancy nuisances or syndromes of the "sick building syndrome". The characteristics of individual rooms in which the indoor air quality was tested are given in Table 1. Part of the name marked in bold in Table 1 will be used in the further part of the work. Air samples were taken out of the building in parallel for each places.

Table 1. Place and date of research, kind of rooms and numbering the places.

The number of place	Name and location of the building	Type of rooms	Date	Number of samples
1	**Student house "Żaczek"**, the 3th May Avenue, Krakow	student rooms, kitchen, corridor	April 2009	9
2	**Nowohucka apartment**, Nowohucka 3, Krakow	private apartment	March 2014	2
3	**Faculty of Chemistry** of the Jagiellonian University, Ingardena 3, Krakow	laboratories, main hall, preparation room, doctoral students room,	March 2014	7
4a	**The III Campus UJ**: Faculty of Physics, Astronomy and Applied Computer Science of the Jagiellonian University, Łojasiewicza 11, Krakow	workshop, humidifier room	April 2015	3
4b	**The III Campus UJ**: Department of Environmental Chemistry, Gronostajowa 3, Krakow	office room, chromatographic lab	April 2016	3

Locations numbered 1 and 3 are situated in the centre of the city with compact building, 200 m from high traffic road - A. Mickiewicz Avenue. Private apartment (place No. 2) is located in Nowa Huta, 50 m from the crossing of Peace Avenue and Nowohucka Street. Location No. 4 are at the III Camus of the Jagiellonian University, at this place the air was investigated in two buildings (4a and 4b) located next to each other and that buildings will be described together. That buildings are located 100 m from Grota-Roweckiego Street, the road with much less traffic than in other places. This place has much better aerodynamic conditions, is more favourable for ventilation and distribution of pollutants in the atmosphere.

2.2 The method of determination

Air samples were collected into stainless steel canisters, reusable containers with a capacity of 6 dm^3. Canisters were evacuated to pressure 0.001 mBar before air sampling, when a cellulose filter was placed on the canister's inlet and air got into canister due to pressure difference after opening the valve. Canisters with air samples were transported to laboratory and chromatographic analysis were performed using gas chromatograph Varian Star 3600 CX equipped with flame ionization detector and sample preconcentration trap. The method of determination is based on the USA EPA method T0-14 [9]. The method detection limit was 0.2 μg/m^3. Precision was determined based on five real samples and ranged between 5% and 10% relative standard deviation (RSD) for peaks that are well separated, while partially overlapping peaks have RSD of about 20%.

The analysis consisted of several steps: rinsing the supply lines, cryogenic preconcentration of a sample of 160 cm^3 volume (preconcentration at -175°C for 4 minutes), thermal desorption of analytes at 120°C, separation of analytes on a capillary column DB-1. The column parameters were as follows: 60 m length, 0.32 mm diameter, stationary phase of polydimethylsiloxane 3 μm thick. The temperature column program was used: initial column temperature -60°C, when the temperature increased to 20°C at a rate of 8°C/min and 5°C/min to final temperature 250°C kept for 10 minutes. Helium was used as the carrier gas at a flow rate of 2.5 cm^3/min. The temperature of detector and valves were 300 and 160°C respectively. The gas flows in the detector were as follows: hydrogen 30 cm^3/min, air 300 cm^3/min, make-up 27.5 cm^3/min. Up to 60 compounds were identified belonging to the so-called ozone precursors, the number was the result of the standard mixture composition. Two standard mixtures have been used: the 27-component mixture manufactured by National Physical Laboratory and 25-component mixture of the Resteck catalogue. Compounds were identified by retention times and by Kovatč retention indexes and the propriety of identification was verified against the application note [10].

3 Results

The paper presents a discussion of the results of the analysis of 24 air samples collected in five different buildings. The number of identified compounds is not the same in subsequent samples, which is reflected in the charts for individual rooms. Table 2 gives the range of summary concentration of VOCs in indoor air and the summary concentrations of VOCs outdoors. In both cases total VOCs includes all concentrations of identified volatile organic compounds. A detailed discussion on identifying the presence of an emission source in a room is presented in the next subsections.

3.1 Student house "Żaczek"

The research was conducted to determine the magnitude of the impact of renovation work on air quality in the renovated rooms and in the whole building. Premises were tested before renovation, one year and two days after repairs. Rooms were closed before sampling. Four samples were taken in the new part of the student house (student rooms, common places - corridor and kitchen), three samples in the old part of the student house and two outside at the main entrance of the dorm and in the courtyard.

Concentrations in the renewed part are the highest, the mean aggregate concentration of identified compounds in the refurbished, old and outside are 2549, 763 and 224 μg/m^3 respectively. The ratio of average concentration in refurbished and old rooms to concentrations in outdoor air are 11.4 and 3.4 respectively.

Table 2. Characteristics of the air condition in the each building: indoor and outdoor concentration range and compounds with the highest concentrations in relation to outdoor air.

Building	The range of concentration of total VOCs in premises [µg/m³]	Total outdoor concentration of VOCs [µg/m³]	Compounds in elevated concentrations with respect to outdoor air
Student house „Żaczek"	285 - 4641	162	Compounds with a short carbon chain (C2- C5): 1-butene, 1-pentene, n-butane, acetylene, ethane, propane,
Nowohucka apartment	392	351	Saturated aliphatic compounds and compounds considered to be derived from biogenic sources: n-butane, 2-methylbutane, n-hexane, isoprene, 3-methylheptane, α-pinene 3-methylheptane, trimethylbenzene
Faculty of Chemistry	318 - 11401	133	cis-pentene, 2-methylbutane, n-hexane, cyclohexane, toluene, n-pentane, 2-methylpentane, 3-methylpentane
The III Campus UJ (4a and 4b)	58 - 120	61	cyclohexane, propylbenzene, toluene, benzene, 1-pentene

Figure 1 shows average concentrations of the individual compounds outside (blue), in the refurbished part (orange) and in the old part (grey). The compounds in the graph are shown in the order as they appear on the chromatogram, compounds at the lowest boiling point are on the left. The high concentrations are mainly attributed to the fraction of the most volatile hydrocarbons, with a carbon chain of 2 to 5 carbon atoms in the molecule. The compounds that were at the highest concentrations in the restoration are: 1-butene (943 µg/m³), n-butane (326 µg/m³), 2-methylpropane (197 µg/m³) and propane (171 µg/m³). The only compounds whose outside concentrations were higher were benzene, 2,3-dimethylpentane and o-xylene. Surprisingly, there was a relatively high concentration in parts of the building where renovation works were not carried out. The same compounds occur in elevated concentrations in both the old and the renewed part of the building. This proves that the air from the refurbished part penetrates into the not renewed part [11].

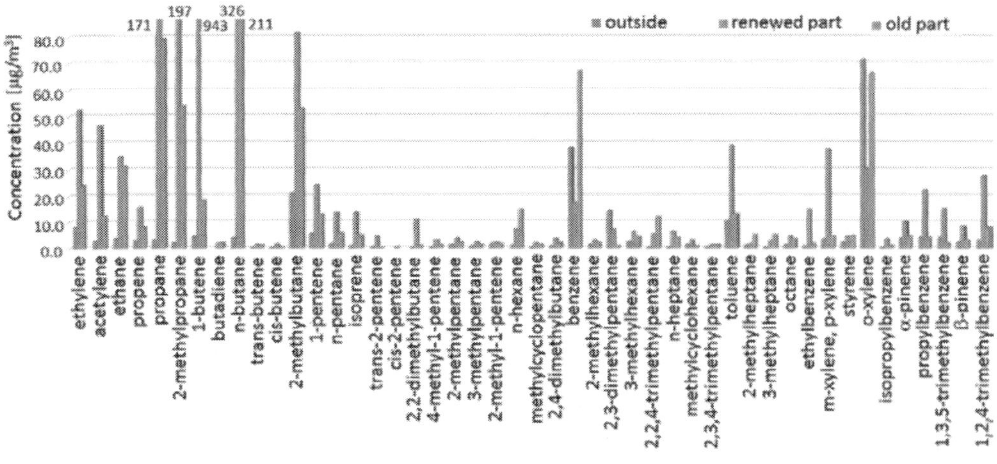

Fig. 1. Mean concentration of identified compounds in different parts of student house "Żaczek" and outside the building. Bars corresponding to the concentrations of the five compounds are shortened and the concentration values as written nearby.

3.2 Faculty of Chemistry

In 2014, the air was examined in several rooms of the Faculty of Chemistry of the Jagiellonian University, especially in the Department of Organic Chemistry: doctoral students' room, preparatory room and students' workshop during courses. Reference samples were taken at the main hall in the central part of the Faculty of Chemistry and before the main entrance to the building. All samples were taken in the morning within one hour and additionally in one of the rooms (lab 216) also in the afternoon after ventilation.

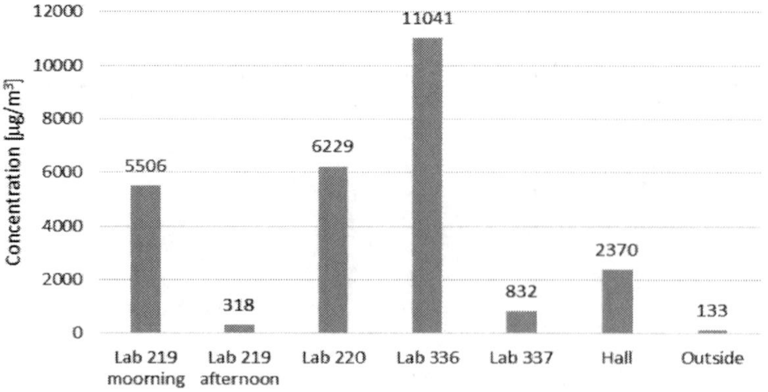

Fig. 2. Total concentration of identified compounds in different places of the Faculty of Chemistry.

The concentrations of the identified compounds in individual rooms and outside are presented in Figure 2. The concentration is higher in each room than in outdoor air (last bar). Several specific compounds are responsible for high concentration: 2-metylbutane, cis-2-pentene,, n-hexane, toluene, as illustrated in Figure 3, showing concentrations in the hall and outside the building together. The highest concentration is from the compounds identified as 2-methylbutane. But including the fact that the peak is unsymmetrical and acetone have the same retention time, it can be concluded that acetone, a compound commonly used in laboratories, is involved in high concentrations.

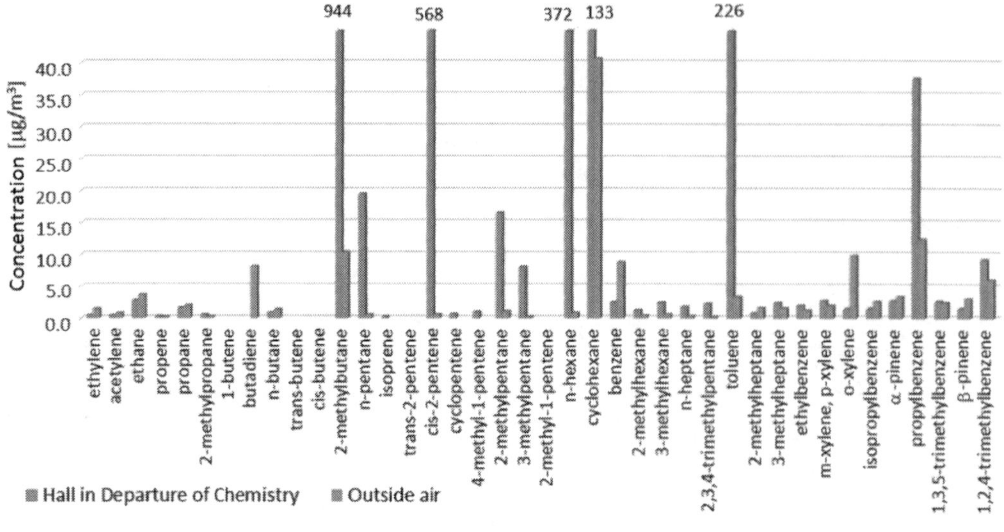

Fig. 3. Concentration of compounds in the main hall in the Faculty of Chemistry (▮) and outside the building (▮). Bars corresponding to the concentrations of the five compounds are shortened and the concentration values as written above.

It was noted that there is a correlation between concentrations in laboratories (Figure 4a) and also between concentrations in laboratories on the third floor and the hall on the second floor (Figure 4b). This demonstrates the penetration of pollutants from laboratories into other parts of the building. At the same time there is no correlation between indoor and outdoor air (Figure 4c). Only a similar profile of concentrations with outdoor air occurred in room 219 in the afternoon, where a few hours of airing of the room occurred (Figure 4d). Room ventilation caused a 17-fold decrease in concentration.

The results obtained in this part of the study show a strong influence of the way the room is used on the condition of air, the possibility of contaminating the rooms due to the movement of pollutants within the building [11].

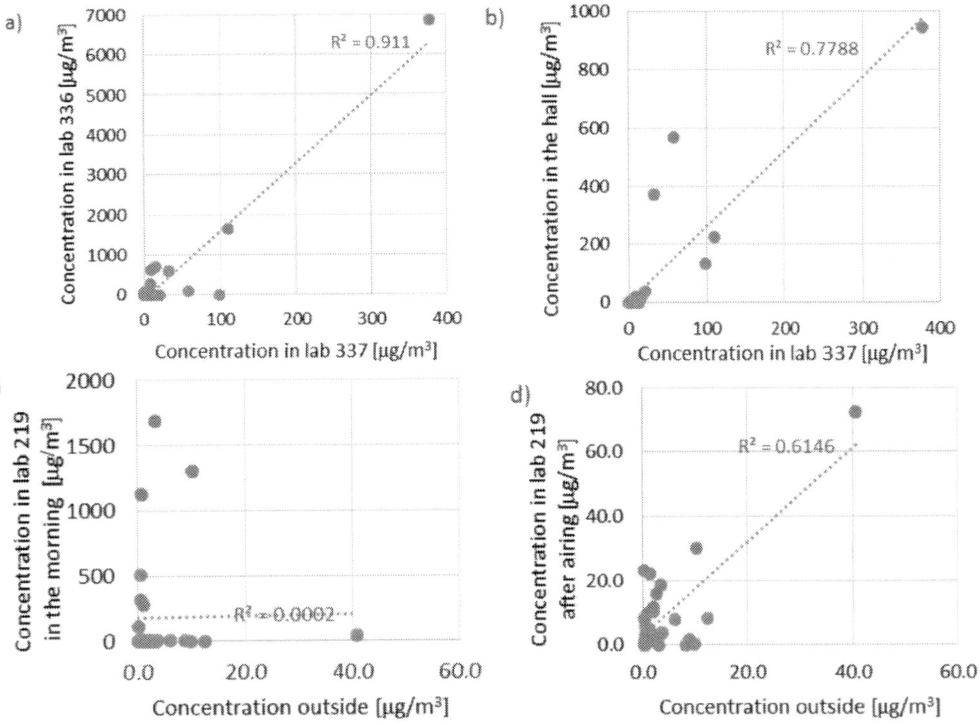

Fig. 4. Correlation between concentrations in: a) lab 336 and 337, b) hall and lab 337, c) lab 219 in the morning and outside, d) lab 219 in the afternoon and outside.

3.3 Nowohucka apartment

Research was conducted in a private apartment in a block at Nowohucka Street, a road with high traffic. Cleaners and air fresheners are used in the apartment; one of the rooms was repaired 3 years before the test. In this case, the total concentration of the identified compounds in the apartment is slightly higher than the concentration in the outdoor air, these values are 392 and 351 µg/m³ respectively. Compounds that have a higher concentration in the room are most of all iso-pentane, n-butane and some aromatic hydrocarbons. On the other hand, compounds that have higher concentrations outside are cyclohexane, 1-butene, o-xylene, ethylene and propane.

According to literature data, average concentration in homes is in the order of 200-500 µg/m³ and is usually two or several times greater than in outdoor air [5, 12, 13]. The value 392 µg/m³ is within the given range, however, after adding the concentration from unidentified compounds, the number 500 is exceeded. There is a fairly high correlation between indoor and outdoor concentrations (Figure 5), so it can be concluded that in this case air quality is more dominated by outside air than by endogenous emission [14].

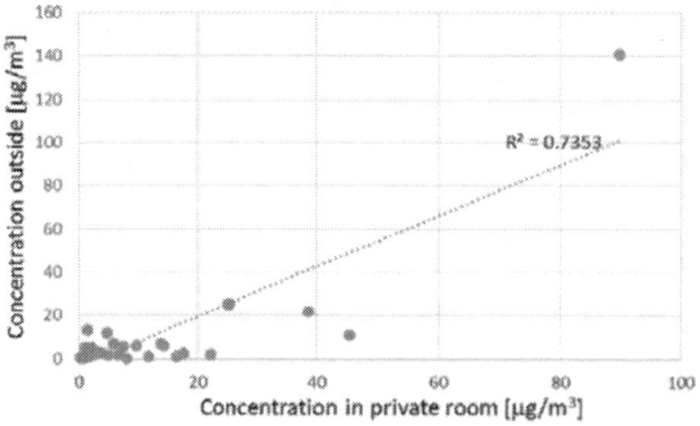

Fig. 5. Correlation between concentrations of identified compounds in private apartments and outdoor.

3.4 The III Campus UJ

The air in two buildings at the III Jagiellonian University Campus was examined in years 2015-2016: Faculty of Physics, Astronomy and Applied Computer Science (4a) and Department of Environmental Chemistry (4b). Tests in the first of the buildings was run about 1 year after completion of the finishing work, in two rooms: workshop and a room with air humidification equipment. The building of Department of Environmental Chemistry was commissioned in 1999.

The total concentration of the identified compounds in place "4a" was 98-120 µg/m³. Also it was stated the presence of many unidentified compounds giving a total concentration of more than 500 µg/m³. In both rooms, the concentrations of compounds are lower than in the previously tested rooms and lower than the average concentration of hydrocarbons in Krakow, but they are almost twice as high as in atmospheric air in that part of Krakow (60 µg/m³). An important factor affecting the air condition in this part of the city is the fact that the III Campus UJ is in a well-ventilated area, there are no traffic jams and in this situation the pollution from car communication is much less affected the air than in the city centre. There are only several compounds responsible for higher concentrations: mainly cyclohexane, toluene, propylbenzene.

A clear answer to the question whether emission sources are present in the examined rooms, give correlation graphs of concentrations in rooms with outdoor air. The correlation coefficient of concentration in both rooms, located in the separate wings of the building and at a different level, ie in the basement and on the second floor, and have different destinies, is about 0.77 while the correlation coefficient of indoor and outdoor concentrations is only 0.10. At the same time, there is another concentration profile for compounds above 5 carbon atoms in the molecule for the air in the rooms compared with outside. Hydrocarbons having a carbon chain of 2 to 4 carbon atoms in the molecule are at a very similar concentration. The final conclusion is that the emissions from finishing materials occur but are rather low and do not cause significant deterioration to indoor air quality.

The air from chromatography laboratory and office room in Department of Environmental Chemistry (place "4b") was also examined. The average concentration of the total identified compounds is 58 µg/m³. This value does not differ significantly from the concentration in atmospheric air in this part of the city and at the same time, the concentration profile is similar and the correlation coefficient is 0.81. In this case, there are no sources of emission in the room.

4 Conclusion

It is essential to conduct indoor and outdoor air research simultaneously. However, to find out if the sources of emissions are inside the building, it is necessary to compare the concentrations of individual compounds, not just the total concentration of VOCs. Very helpful in identification the presence of endogenous emission sources is the analysis of the correlation coefficient between indoor and outdoor concentrations of compounds as well as between rooms in one building. Appropriate interpretation of the results allows us to say whether there are or are no sources of emission in the examined room, but they do not indicate what the source is.

The VOCs concentration in the rooms were higher or at a similar level to that of the air sampled at the same time outside in all tested rooms. Most of the study areas found endogenous sources of emissions, but their impact on air quality was very differential. Human activity, in particular repair works (the use of paints and varnishes) and experiments with organic solvents have the greatest impact on the deterioration of the air quality in the examined premises. Significant improvement in air condition was achieved after room ventilation. The only room where the endogenous source did not exist was the chromatographic laboratory on the III Jagiellonian University Campus.

References

1. P.L. Jenkins, T.J. Phillips, J.M. Mulberg, S.P.Hui, Atmos. Environ. **26A**, 2141 (1992)
2. M. Kampa, E. Castanas, Environ. Pollut. **151**, 362 (2008)
3. M. Lippmann, *Environmental toxicants: human exposure and their health effects* (John Wiley & Sons Inc., 2008)
4. *Directive 2004/42/CE of the European Parliament and of the Council of 21 April 2004 on the limitation of emissions of volatile organic compounds due to the use of organic solvents in certain paints and varnishes and vehicle refinishing products and amending Directive 1999/13/EC*
5. A.P. Jones, Atmos. Environ. **33**, 4535 (1999)
6. P.K. Srivastava, G.G. Pandit, S. Sharma, A.M. Mohan Rao, Sci. Total Environ. **255**, 161 (2000)
7. A. Boudehane, A. Lounas, Y. Moussaoui, C. Balducci, A. Cecinato, Atmos. Environ. **144**, 266 (2016)
8. W.E. Goldstein, *Sick Building Syndrome and Related Illness* (Taylor & Francis Group, 2010)
9. *Determination of Volatile Organic Compounds in Ambient Air Using SUMMA® Passivated Canister Sampling and Gas Chromatographic Analysis* (US EPA,1999)
10. *Continuous Monitoring of C2 to C10 Ozone Precursors in Air Using Cryogenic Concentration and Single Capillary Column.* (Hewlett Packard Application Note 228 – 218. Gas Chromatography, 1993)
11. S. Batterman, C. Jia, G. Hatzioasilis, Environ. Res. **104**, 224 (2007)
12. B. Zabiegała, Pol. J. Environ. Stud. **15**, 383 (2006)
13. H. Guo, S.C. Lee, W.M. Li, J.J. Cao. Atmos. Environ. **37**, 73 (2003)
14. E. Ilgen, N. Karfish, J. Angerer, P. Schneider, J. Heinrich, H.E. Wichmann, L. Dunemann, J. Begerow, Atmos. Environ. **35**, 1235 (2001)

E3S Web of Conferences **28**, 01019 (2018) https://doi.org/10.1051/e3sconf/20182801019
Air Protection in Theory and Practice

Application of headspace for research volatile organic compounds emitted from building materials

Beata Kultys[1,*], *Karolina* Wałąg[1]

[1] Jagiellonian University in Krakow, Faculty of Chemistry, Gronostajowa 2, 30-387 Krakow, Poland

Abstract. Headspace technique and gas chromatography method with mas detector has been used for the determination of volatile organic compounds (VOC) emitted from various building and finishing materials, such as sealing foams, mounting strips, paints, varnishes, floor coverings. The tests were carried out for different temperatures (in the temperature range of 60 to 180 °C) and the time of heated vials with tested materials inside. These tests were conducted to verify the possibility of use this method of determination the VOC emission. Interpretation of chromatograms and mass spectra allowed to identify the type of compounds emitted from the tested materials and the optimum time and temperature for each type of material was determined. The increase in heating temperature of the samples resulted in increase the type and number of identified compounds: for four materials the increase was in the whole temperature range, for others it was from 90 °C. On the other hand, emission from mineral wool was low in whole temperature range. 30-minutes heating of the samples was sufficient to identify emitted compounds for most of tested materials. Applying a longer time, i.e. 24 hours, significantly increased the sensitivity of the method.

1 Introduction

Volatile organic compounds (VOC) may be one of the factors that may adversely affect indoor air quality and in the extreme case they can lead to symptoms of the so-called "Sick building syndrome". Sources of emissions of these compounds can be both indoor and outdoor [1-4]. Parallel testing of indoor and outdoor air only allows you to statement if the source of the emission is in the test room. But if the research is conducted to improve the quality of indoor air it is necessary to identify the source of the emission. VOC emissions may result from human activities in the room (for example the use of solvents, doing experiments) or may come from the elements of the room, finishing and construction materials. Emissions test from materials are typically carried out in emission chambers at a steady gas flow and humidity [5-9]. They allow to determine the magnitude of emission of test material in order to verify compliance with emission standards, e.g. before

* Corresponding author: beata.kultys@uj.edu.pl

introduction into usage. VOC emission tests also use thermal desorption, ie the release of volatile compounds from the solid under the influence of heat [10, 11].

In this study, the headspace technique was used for examination VOC emissions from construction and finishing materials. Headspace is based on the analysis of the gas phase in equilibrium with the condensed phase [12-15]. That method has many uses, in particular for complex matrix samples, such as environmental and biological samples [14-16]. The gas phase composition was determined by means of a gas chromatograph with a mass detector, this made it possible to identify the compounds emitted from the materials. The research was conducted for building and finishing materials used in the construction of the Faculty of Chemistry of the Jagiellonian University, such as sealing foams, mounting strips, paints, varnishes and floor coverings. The appropriate time and temperature of the material being heated in the vial is essential for the headspace technique to be effective. The paper presents the results of studies on the influence of temperature and time on the type and amount of emitted compounds from tested materials.

2 Materials and research methods

In order to determine the type and amount of compounds emitted from materials, a static headspace method was used, in which the gaseous phase in equilibrium with the condensed phase (liquid or solid) is analyzed. In this method, a sample with an analyte content of C_S^0 is placed in a vial (stage I), where the analyte passes to the gas phase according to the partition coefficient $K = C_S/C_G$ at a set temperature (stage II). The gas phase is taken for chromatographic analysis using a syringe (stage III) or a dosing loop [12, 13]. The principle of the headspace method is illustrated in Figure 1.

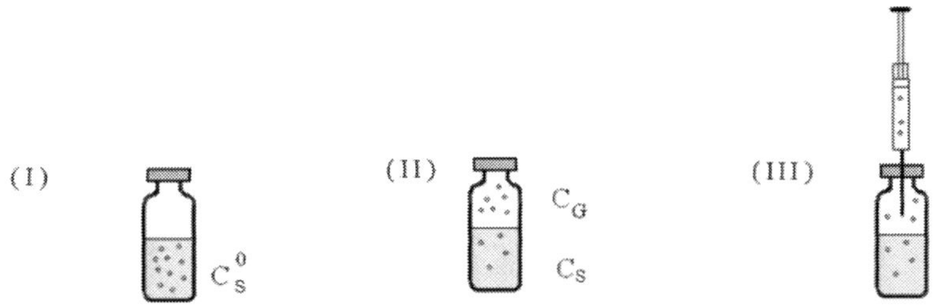

Fig. 1. Stages of headspace technique: (I) – placement of the test material in the vial, (II) – establishing the equilibrium, (III) – taking the gas phase for chromatographic analysis.

The basis for calculating the amount of analyte in the headspace method is dependence the concentration of the analytes in the gas and condensed phase from the initial concentration in condensed phase. Mass balance of the system is expressed by the equation:

$$C_S^0 \cdot V_S = C_G \cdot V_G + C_S \cdot V_S \qquad (1)$$

where: C_S^0 – is concentration of the analyte in the sample; V_S – volume of condensed phase; C_G – concentration of the analyte in the gas phase at steady state; V_G – volume of the gas phase above the condensed phase; C_S – concentration of the analyte in condensed phase at steady state. Calibration is necessary for quantitative determination but in this case, such calibration was not performed because the purpose of the study was merely to identify the type of compounds emitted from the tested materials.

Thirteen types of building and finishing materials were tested. This materials were used during building a new building of the Faculty of Chemistry of the Jagiellonian University at Gronostajowa 2 street. The characteristics of the tested materials, i.e. the commercial name

Table 1. Characteristics of the tested materials.

	Trade name	Chemical compozition	Average weight [g]	Use of building material
A	PVC cable trays (perforated)	polyvinyl chloride	6.82	Designed for carrying cables, for example in control cabinets
B	corrugated PVC pipe	polyvinyl chloride	1.63	Shielded plastic tube for electrical cables
C	styrofoam EPS	expanded polystyrene	0.36	Designed for thermal insulation of buildings
D	mineral wool, on the outside are covered with aluminum foil (CONLIT PLUS system, Rockwool producer)	silica 34 - 45%, alumina 15 - 24%, iron(III) oxide 2 - 1%, calcium oxide + magnesium oxide 20 - 35%, sodium oxide + potassium oxide 1 - 7% (aluminum foil not included in the samples tested) [17]	1.77	Designed for fire protection insulation, for example isolation of ventilation ducts, air conditioning ducts and smoke extraction ducts
E	fireproof polyurethane foam	polyurethane	0.75	Designed to fill, for example, the ceiling contact and the wall in walls of the fire zone
F	blue styrodur STYROFOAM XPS 300	extruded polystyrene	0.62	Designed for example for warming up the floor on the ground, on the ceiling or as a foundation warming
G	PVC Tarkett iQ Granit	polyvinyl chloride	11.99	Used in the corridors
H	polyurethane foam	polyurethane	0.85	Used for insulation
I	dispersion latex white paint for interior (Acryl 3000 PS Dekoral professional)	polymer emulsions, it also contains: 5-chloro-2-methyl-2H-isothiazol-3-one (2-methyl-2H) isothiazol-3-one (CMIT/MIT) 3: 1 and 1,2-benzisothiazol-3(2H)-one	5.45	Designed for decorative painting of walls and ceilings indoors
J	Ecophon Focus E - suspended ceiling (filled with glass wool, painted outside)	glass wool: silicon(IV) oxide 45 - 70%, aluminum(III) oxide 0 - 7%, iron(III) oxide 0 - 5%, calcium oxide 5 - 13%, magnesium oxide 0 - 5% [18], Akutex ™ FT coating	1.54	Used for example in the office rooms. It has an aesthetic function and above all an acoustic function
K	polyurethane semi-matt parquet lacquer (Supramal 303)	polyurethane, it also contains N, N-dimethylethanolamine	2.53	Designed for parquet flooring
L	perforated profile	lack of information	6.39	Many uses, e.g. reinforcement for structural elements
M	gray styrodur XPS	extruded polystyrene	0.38	e.g. for warming up the floor on the ground, on the ceiling or as a foundation warming

of the material, its main chemical composition and application, and the average mass of the samples, are presented in Table 1. Photographs of the tested materials are shown in Figure 2.

Fig. 2. Photos of the tested materials. Indication of materials according to Table 1.

The preparation of samples for analysis consisted of: excision the right amount of material, placement in the vial, tight vial closure and finally heating the vials with the tested material for a fixed time and at a certain temperature. Samples of material were placed in vials of 20 cm^3. In each case, the volume of the condensed phase was half the volume of the vial. Materials were heated at 60, 90, 120 and 150 °C. Heating time was 30 minutes, 2 and 24 hours. After heating of the sample, the gas phase was introduce to the gas chromatograph using a loop. Chromatographic analyses were performed by gas chromatograph (HP 6890) coupled to a mass detector (HP 5973). A capillary column with low polarity phase was used (methyl-5%-phenylsiloxane, film thicknes 0,25 mm) with a length of 30 m and a diameter of 0.25 mm, purchased from J&W Scientific.

The temperatures of the Headspace-GC transfer line and the injector were 160 °C and 200 °C, respectively. Helium was used as the carrier gas at a flow rate of 1 cm^3/min Chromatographic analysis conditions were as follows: the column was initially maintained at 50 °C for 3 min, then the temperature increased to 200 °C at a rate of 15 °C/min, which then was held for 10 min.

The analytes after separation in the chromatographic column were directed into mass detectors where the compounds were ionized with an electron beam of energy 70 eV and ions were separated in quadrupole. The MS was operated in the SCAN mode with the m/z range of 30 - 250 amu. The temperature of ion source and quadrupole was 230 °C and 150 °C respectively.

3 Results and discussion

The studies presented in this paper have been performed to verify the applicability of the Headspace method in combination with GC-MS for identification VOC emitted from interior materials. Such tests could supplement indoor air studies. Examining the indoor air it can be stated only that there are emission sources in the room. But if the purpose of research is to improve air quality, it is necessary to identify the source of emission, i.e. to identify the VOC emitting material, to be able to remove the source of emission from the room. Testing of the method was made by determining the gas phase composition under various operating conditions of the headspace. The type and amount of emitted compounds was investigated according to temperature and the time of heating the samples. 30 minute sample healing was applied during the temperature studies, while the effects of time were investigated at 60 and 90 °C.

3.1 Emission dependence from time and temperature

For four of the tested materials (G, I, K and L), the area of chromatographic peaks increases with increasing temperature, i.e. the quantity and number of emitted compounds were increased over the whole temperature range. This is illustrated in Figure 3 and 4a. For the next four materials (A, B, C and J), the growth of the peak area is shown with increasing temperature, but only at a temperature of 90 °C (Fig. 4b). On the other hand, for samples of materials H, F and M, the emission levels in the range of 60 - 120 °C show slight fluctuations and the emission increase is observed only at 150 °C (Fig. 4c). While for the sample of mineral wool no significant increase in emissions was observed in the temperature range of 60 - 150 °C (Fig. 4d).

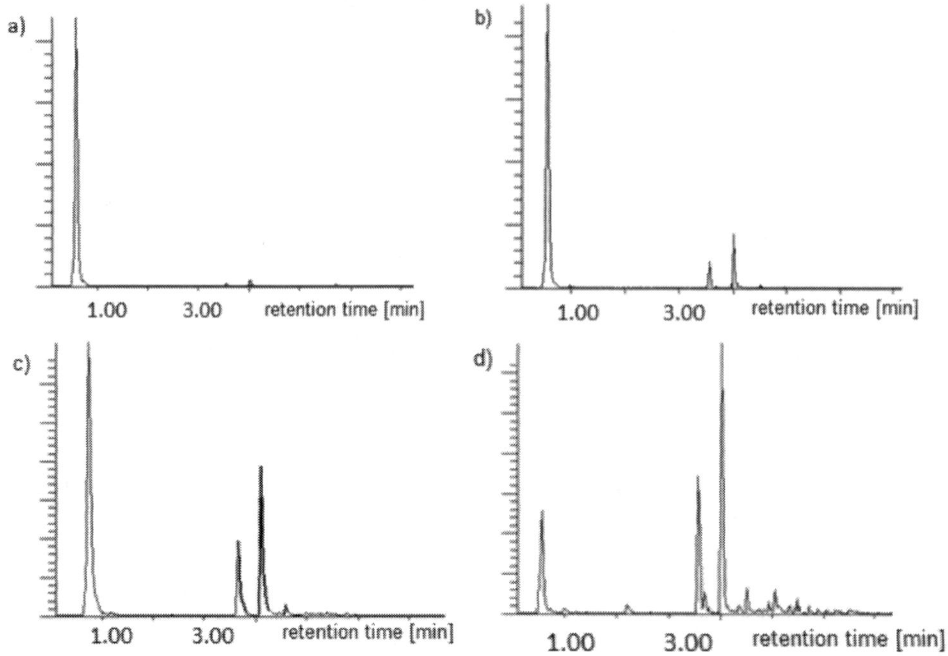

Fig. 3. Chromatograms for blue styrodur STYROFOAM XPS 300 heated at: a) 60 °C, b) 90 °C, c) 120 °C, d) 150 °C.

In conclusion, it can be stated that it is difficult to indicate the preheating temperature of the sample which will be most optimal for all building materials. On the one hand, to simulate more real VOC emission conditions, low temperatures that are closest to room temperature should be used. The lowest temperature at which the samples were analysed was 60 °C because it was the temperature that allowed stable operation of the headspace. Raising the temperature increased the emission of organic compounds, for all materials except mineral wool. However, these are not the conditions that occur in buildings every day.

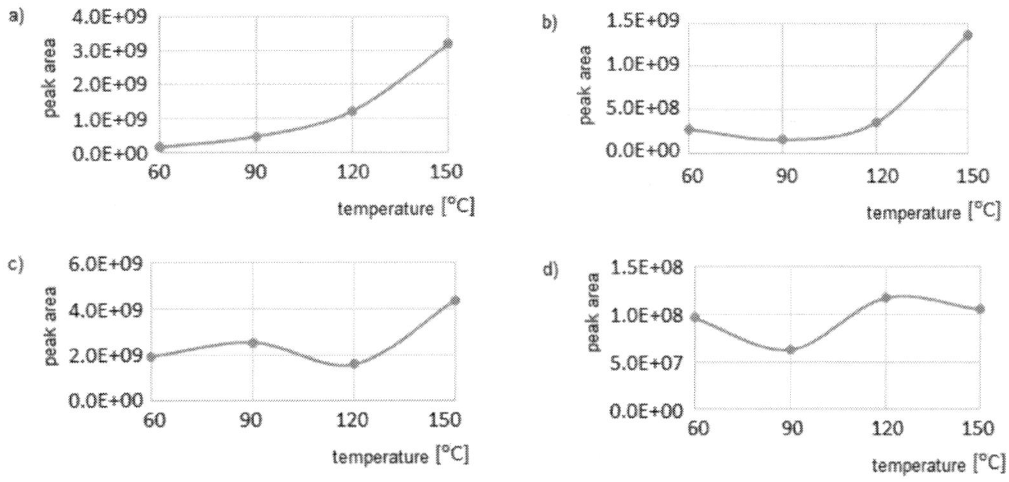

Fig. 4. Dependence of the total surface area of the peaks on the temperature for selected materials: a) perforated profile, b) corrugated PVC pipe, c) gray styrodur XPS, d) mineral wool.

Prolongation of the heating time from 30 min to 120 min does not cause significant changes in type and amount of emitted compounds. While 24-hour sample heating resulted in almost fourfold increase the peak area (Fig.5). The increase in the total surface area of the peaks is mainly due to the increase in the surface area of the peaks. Changes in the profile of released compounds are small. Probably during long-term heating VOCs are released to the gas phase from both the surface layer and also from inside of the material.

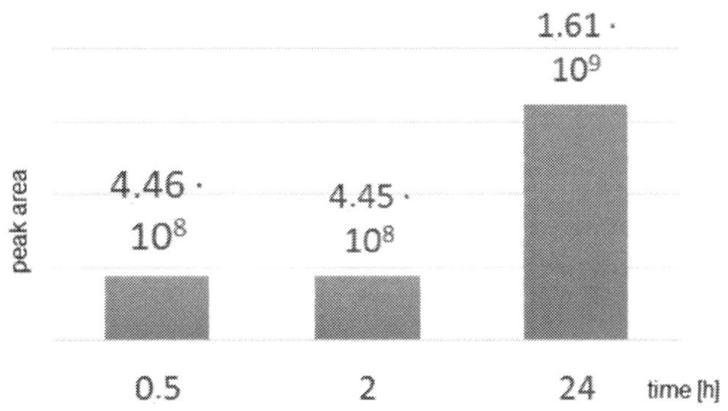

Fig. 5. Dependence the total peak area on the heating time for styrofoam.

For most tested materials, 30-minute sample heating was sufficient to determine which compounds emitted the material. The longer time, i.e. 24 hours, is recommended only for low emission materials.

3.2 Type of compounds emitted from tested materials

The number of emitted compounds was very different for each material, from several for mineral wool or polyurethane foam to several dozen for perforated profile. Table 2 lists the groups of emitted compounds in the most abundant and in parentheses specific examples.

The most secure material among the selected for analysis was mineral wool due to the amount and type of emitted compounds. No emissions of aromatic hydrocarbons and formaldehyde from this material were recorded. The greatest risk to health due to the large number of emitted organic compounds potentially creates a corrugated tube. On the other hand, such materials as blue styrofoam (F) and fire-retardant polyurethane foam (E) may pose a risk due to the high participation of aromatic hydrocarbons.

Comparing the obtained results with literature reports [2, 10], the compatibility of the group of emitted compounds can be found, while there are differences in the emission of specific compounds. This partial accordance may result from testing one specific material and the fact that other substances may now be used in the production than it was a dozen or so years ago.

Table 2. List of the most common VOCs emitted from the tested materials. Indication of materials according to Table 1

	The most common VOCs emitted from the tested building and finishing materials
A	alcohols (especially: 2-(2-ethoxyethoxy) ethanol), saturated aliphatic hydrocarbons, carboxylic acid esters, ketones
B	aliphatic saturated hydrocarbons, alcohols (e.g. 2-(2-ethoxy-ethoxy)ethanol), carboxylic acid esters (e.g. 1,2-benzene-dicarboxylic acid diethyl ester), amines, ketones
C	aromatic hydrocarbons (especially: ethylbenzene, styrene, propenylbenzene, propylbenzene, xylene), alcohols (among others: 2-(2-ethoxyethoxy)ethanol), aldehydes (among others: benzaldehyde), saturated aliphatic hydrocarbons (especially: 2-methylbutane)
D	alcohols (especially: 2-(2-ethoxyethoxy)ethanol, 2-[2-(2-ethoxyethoxy)ethoxy]ethanol)
E	aromatic hydrocarbons, carboxylic acid esters, alcohol: 2-(2-butoxyethoxy)ethanol, saturated aliphatic hydrocarbons
F	aromatic hydrocarbons (especially: xylene, ethylbenzene, styrene, saturated aliphatic hydrocarbons (especially: 2-methyl-propane), alcohols (e.g. 2-(2-ethoxyethoxy)ethanol), aldehydes (e.g. benzaldehyde)
G	alcohols (especially: 2-(2-ethoxyethoxy)ethanol), amines, ketones, aliphatic hydrocarbons (especially: 2,2,4,6,6-penta-methylheptane), aldehydes
H	alcohols, saturated aliphatic hydrocarbons, ketones, aldehydes, amines
I	alcohols (especially: 2-(2-ethoxyethoxy)ethanol, 2-(2-butoxyethoxy)ethanol), carboxylic acid esters (especially: butanoic acid butyl ester), amines, aldehydes (especially: benzaldehyde), acids (e.g. acetic acid)
J	alcohols (especially: 2-(2-butoxyethoxy)ethanol, 2-(2-ethoxyethoxy)ethanol), carboxylic acid esters (especially: butanoic acid butyl ester), ketones, aldehydes
K	alcohols (especially: 2-(2-ethoxyethoxy)ethanol), carboxylic acid esters (especially: butanoic acid butyl ester), amines, ketones, saturated aliphatic hydrocarbons
L	aliphatic saturated hydrocarbons (e.g. 2,2,4,6,6-pentamethyl-heptane), carboxylic acid esters, alcohols (especially: 2-(2-ethoxyethoxy)ethanol), aromatic hydrocarbons (especially: xylene, ethylbenzene), ketones, amines, amides
M	aromatic hydrocarbons (especially: ethylbenzene, styrene, propylbenzene), saturated aliphatic hydrocarbons, aldehydes (especially: benzaldehyde)

4 Conclusion

Headspace method, in combination with gas chromatography and mass spectrometry enables the identification of organic compounds emitted from finishing and building materials. Such tests may be complementary to indoor air quality research that identify VOC emission sources.

It is important to determine the optimum working conditions for the tested materials. By heating the sample at a lower temperature, you can simulate measurement conditions that are close to the real conditions of use the tested materials. The increase in sample temperature results in an increase in the type and amount of identified compounds for most tested materials. But the sensitivity of the method can also be increased by prolongation the time of heating the sample.

References

1. A.P. Jones, Atmos. Environ. **33**, 4535 (1999)
2. B. Zabiegała, Pol. J. Environ. Stud. **15**, 383 (2006)
3. *Organic Indoor Air Pollutants: Occurrence, Measurement, Evaluation* (Wiley Online Library, 2009)
4. *Indoor air quality & its impact on man, EC Raport No.19* (Luxembourg, 1997)
5. A. Colombo, M. De Bortoli, E. Pecchio, H. Schauenburg, H. Schlitt, H. Vissers, Sci. Total Environ. **91**, 237 (1990)
6. A. Niesłochowski, Izolacje **1/2011**, 40 (2011)
7. I. Pecka, R. Wilglusz, E. Sitko, G. Nikiel, Roczn. PZH **55**, 181 (2004)
8. J.M. Hugo, M.W. Spence, T.D. Lickly. Appl. Occup. Environ. Hyg. **15**, 512 (2000)
9. T.J. Kelly, J.D. Myers, M.W. Holdren, Indoor Air **9**, 117 (1999)
10. A. Niesłochowski, Prace ITB **165**, 9 (2013)
11. C. Rodríguez-Navas, R. Forteza, V. Cerdà, Chemosphere **89**, 1426 (2012)
12. B. Kolb, L.S. Ettre, *Static Headspace-Gas Chromatography: Theory and Practice* (Wiley, 2006)
13. N.H. Snow, G.C. Slack, Trends Analyt. Chem. **21**, 608 (2002)
14. N.H. Snow, G.P. Bullock, J. Chromatogr. A **1217**, 2726 (2010)
15. J.C. Flórez Menéndez, M.L. Fernández Sánchez, J.E. Sánchez Uría, E. Fernández Martínez, A. Sanz-Medel, Anal. Chim. Acta **415** 9 (2000)
16. C. Ubeda, R.M. Callejón, C. Hidalgo, M.J. Torija, A. Mas, A.M. Troncoso, M.L. Morales, Food Res. Int. **44**, 259 (2011)
17. M. Półka, P. Sulik, Zeszyty Naukowe SGSP **40**, 99 (2010)
18. W. Śmiszek-Lindert, A. Bajorek, J. Kubacki, Chemik **69**, 411 (2015)

Particulate matter and polycyclic aromatic hydrocarbons in a selected athletic hall: ambient concentrations, origin and effects on human health

Karolina Kuskowska[1], *Wioletta* Rogula-Kozłowska[2,3], *Patrycja* Rogula-Kopiec[3]

[1]The Main School of Fire Service, Faculty of Civil Safety Engineering, Slowackiego 52/54, 01-629 Warsaw, Poland
[2]The Main School of Fire Service, Faculty of Fire Safety Engineering, Slowackiego 52/54, 01-629 Warsaw, Poland
[3]Institute Of Environmental Engineering, Polish Academy of Sciences, M. Sklodowiskiej-Curie 34, 41-819 Zabrze, Poland

Abstract. The paper presents the results of research on the concentration of 16 polycyclic aromatic hydrocarbons (PAHs) associated with total suspended particles (TSP) and their respirable fraction (PM_4) in one of the Warsaw sports halls. Samples of dust were collected for 15 days simultaneously inside and outside (atmospheric air) of the sports hall. The obtained data allowed calculating diagnostic ratios indicating the origin of the PAHs and selected indicators of cumulative exposure to the PAH mixture. Both PM and PAH in the surveyed area were derived from the combustion of solid fuels; this involved pollution both in the atmospheric air and in the air inside the sports hall. It has been shown that the exposure of sports hall users to PAHs is significantly higher than the exposure resulting from concentrations recorded in the atmospheric air.

1. Introduction

In recent years society's awareness of the necessity of maintaining one's physical condition as the most important parameter regulating the proper functioning of the body has increased considerably. As a result, the use of various sports facilities, both open (stadiums, playgrounds, outdoor tennis courts) and closed (gyms, gymnasiums, sports halls) has also increased. The quality of air in sports facilities is particularly important for regular trainers, who spend several hours per day in these buildings, several times a week.

Meteorological parameters and the air quality in open facilities are almost the same as in atmospheric conditions, so the air quality in open facilities can be assessed by comparing the concentrations of the different pollutants to the relevant standards (reference values, limit values, etc.).

1 Corresponding author: kkuskowska@sgsp.edu.pl

But in closed sports facilities there are no standards for air quality. Here, the sanitary condition of the air is regulated mainly by the efficiency of the air-supply system, which affects the values of many parameters, including the concentrations of volatile organic compounds (VOCs) in the air, airborne biological concentrations, and particulate air pollution [1, 2]. This last parameter seems to be the most important in the assessment of the health exposure because quantitative and qualitative analysis of dust provides additional information on other dust-bound substances (e.g. polycyclic aromatic hydrocarbons (PAHs), heavy metals) which may have adverse health effects.

The results of epidemiological studies show that both short-term and long-term exposure to high concentrations of particulate matter (PM) increase the risk of respiratory, blood and nervous system diseases [3]. PM particles settling on the walls of the pulmonary alveoli impede the gas exchange. For those who train in the room, this means that high PM concentrations during training will decrease the athletic performance [4]. In addition, oral breathing usually occurs during intense exercise [5–7]. Nasal respiration allows stopping most of the inhaled PM particles with diameters above 2 μm. Breathing through the mouth causes the omission of the natural respiratory filtration mechanisms in the upper part of the respiratory system. As a result, the amount of particles reaching the deeper parts of the respiratory system is increased [8, 9]. The high intensity of breathing while exercising causes not only fine (i.e. less than 2 μm in general) but also larger particles to be transferred into the air passages. The increased intensity of breathing occurs in people who regularly train not only during training. Oxygen consumption in people who regularly train is considerably higher than in the average person. Moreover the per minute breathing volume in physically active persons is greater than the per minute breathing volume in non-athletes [10]. This means that trainees will be exposed to more PM and PM-bound substances and pollutants in all conditions than those who do not regularly practice a sport.

A brief overview of enclosed sports facilities shows that most sports buildings in Poland are buildings equipped with gravitational ventilation and lacking internal facilities to improve indoor air quality. These are most frequently urban sports halls and halls of various sports clubs. New buildings equipped with modern installations to replace and clean the air are reserved mostly for professional athletes; their use by amateur athletes is associated with high (too high) costs. Due to this, people prefer sports facilities free of charge but of poorer quality. Regular school workouts and a variety of sports activities for children and young people are offered in public sports halls. These are usually rented by the instructors conducting the classes and are heavily used from early morning to late evening.

The first research in Poland into PM concentrations and 16 PM-bound PAHs within a Warsaw sports hall is that presented here. PAHs are classified as priority substances in the air due to their proven carcinogenic and mutagenic effects [11–14]. The main sources of PAHs in the air in urbanized areas are the following: processes related to energy consumption (burning solid and liquid fuels), incomplete combustion of organic compounds, chemical processes, waste incineration, forest fires, tobacco smoke, and chemical reactions in the atmosphere [15]. The most common and the most often examined PAHs in the environment are: acenaphtene (Acn), acenaphthylene (Acnl), anthracene (An), benzo[a]anthracene (BaA), benzo[a]pyrene (BaP), benzo[b]fluoranthene (BbF), benzo[ghi]perylene (BghiP), benzo[k]fluoranthene (BkF), chrysene (Ch), dibenzo[ah]anthracene (DahA), fluoranthene (Flt), fluorene (Flu), indeno[1,2,3-cd]pyrene (IP), naphthalene (Na), phenanthrene (Phen) and pyrene (Pyr) [16, 17].

 https://doi.org/10.1051/e3sconf/20182801020

2. Experimental

2.1. Description of the sampling sites

Samples of dust were collected inside and outside of the sports hall of the Sports and Recreation Center (OSiR) of the Targówek district located in Warsaw at 25 Ossowskiego Street (52◦16'19 N, 21◦02'42 E) (Fig. 1). Targówek district is split into two parts: industrial and residential. This hall is located in the residential part of the district. It adjoins on its south side a one-story residential building; on the north, the buildings of a school complex; on its west side is an Orlik-type playground; on its east side is Ossowski Street, with relatively low traffic due to the closure of part of the street in May 2016 to build a subway. Approximately 100 meters in a straight line to the west there passes Saint Vincent Street with the highest traffic in the area (according to the Association for the integration of capital communications (SISKOM) data, about 1900 cars/hour). There is a lot of open space around the hall. In the surroundings of the hall, especially on the northeast side, low-rise buildings are dominant.

OSiR is a multifunctional facility that allows training in many sports, such as handball, basketball, football, volleyball, table tennis, etc. It has two stories. On the ground floor is a sports hall with a parquet floor (measuring 44 m x 24 m) and washrooms. On the upper floor is an auditorium for 200 persons and offices. Only gravity ventilation is used in the building. During working days, the hall is used by students of the neighboring school and in the afternoons and on weekends the hall is used for training and inter-school competitions.

Fig. 1. Location of sampling place.

2.2. PM sampling

Measurements were carried out from December 2016 to January 2017. Fifteen parallel measurements of the concentrations of PM_4 and TSP were made simultaneously inside and outside the sports hall. The average time of the PM sampling was 436 minutes. The PM concentration was examined by the gravimetric method (PN-91/Z-04030/06, PN-91/Z-0430/05). For this purpose, two GilAir 3 aspirators and glass filters (Cat. no. GF1-025, 25 mm in a diameter) were used. The PM sampling was performed at a height of 2 m from the floor/ground level. The procedures for conditioning and weighing of pure filters and PM sample filters were the same as in [18,19].

2.3. PAH analysis

The samples prepared for the PM_4 and TSP analysis were extracted using dichloromethane in an ultrasonic bath. The resulting extract was filtered, washed and concentrated in a helium atmosphere. The crude extract was diluted with propanol-2 and redistilled water until a propanol-2:water equivalent ratio of 15:85 (V/V) was reached. Next, the samples were subjected to an extraction to solid phase (SPE) using the columns c-18 with stationary phase. A Perkin Elmer Clarus 500 gas chromatograph with flame-ionisation detector (FID) was used to analyse the extracts. An RTE-5 capillary column (30 m x 0.32 mm x 0.25 μm) from the Restek company was used for the separation of the components. Helium was used as the carrier gas with a flow rate of 1.5 cm^3/ min. The quantitative analyses were based on the calibration curves for 16 standard PAHs. The linear correlation of dependence between the peak area and the PAH concentrations was tested for a range of concentrations from 1 to 20 ng/μl. The correlation coefficients for this concentration range were between 0.98 and 0.99. The samples were fed into the dispenser. The evaporation temperature was 240°C and the detector temperature was 280°C. The analysis was carried out using a programmable temperature increase in the operating column. The initial temperature of the analysis for up to 4 minutes was 60°C, then increased by 10°C/min to reach 280°C. At this temperature the column was soaked for 14 minutes. The total analysis time was 40 minutes. The FID was powered by hydrogen (flow rate 45 cm^3/min) and air (450 cm^3/min) [14].

3 Results and discussion

The average concentrations of PM_4 and TSP in the indoor and outdoor air of the sports hall during the whole of the sampling period are presented in Table 1.

Table 1. The average concentrations of PM_4 and TSP indoor and outdoor the hall.

Concentration	Indoor [μg/m³]	Outdoor [μg/m³]	I/O
PM_4	74.2 ± 21.5	22.3±8.2	3.3
TSP	115.1 ± 9.6	65.2±34.7	2.1

The concentrations of PM_4 and TSP were significantly higher (three times for the PM_4 and two times for TSP) inside the hall than in its surroundings. In outdoor air only about 33% of the TSP mass was PM_4 particles whereas in the indoor air of the sports hall the analogue PM_4 proportion exceeded 60%. It seems that the reason for maintaining high concentrations of PM inside the hall is the accumulation of PM particles in the indoor air. Probably PM particles reach the indoor air of hall with atmospheric air and due to the limited ventilation they are not removed to an adequate extent to balance the PM migration.

A similar phenomenon was observed in air quality studies in university teaching rooms in Gliwice and Warsaw [20, 21]. Over time, the particles inside the hall create larger agglomerates and they are added to the mass of the coarse dust (in this case, the aggregate mass of the particles in the 4100 μm range). Then the particles settle on the ground and undergo permanent resuspension during the movement of the air in the hall until they are mechanically removed (cleaning, vacuuming). For these reasons the difference in PM_4 concentration indoor and outdoor of the hall were clearly higher than the difference in TSP concentrations.

Fig. 2 shows that the concentrations of 16 PM_4- and TSP-bound PAH both in the indoor and outdoor air of the sports hall recorded during the study period were very high. In atmospheric air they exceeded values previously quoted in Warsaw several times [22], and were similar to those that were recorded only in heavily polluted areas of Poland in the heating period (increased emissions from burning coal and biomass in household furnaces) [15].

The main reason for the high concentrations of PAHs was doubtless the season in which the measurements took place: the heating season. The research was conducted in the Targówek district, where over 1000 buildings (according to the statistical data of city office of Warsaw are not connected to the municipal central heating network. These buildings are individually heated, usually with solid fuels (coal, wood). The quality of the air in the district is also affected by air migration from neighboring districts (Białołęka, Marki, Ząbki), which are characterized by low buildings with individual fireplaces also.

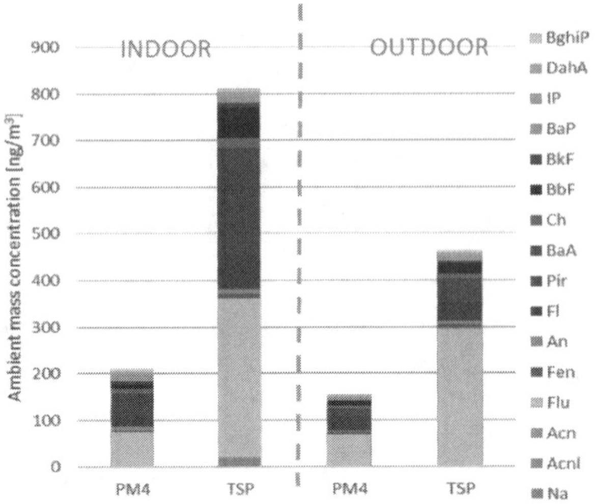

Fig. 2. Average ambient concentration of 16 PAH related to PM_4 and TSP in indoor (athletic hall) and outdoor (atmospheric) air.

To confirm the above theses and establish the origin of the PM_4- and TSP-bound PAHs, we used Diagnostic Ratios (DR). Diagnostic ratios are the mutual proportions of the ambient concentrations of a single PAH or groups of PAHs that have similar physicochemical properties and impact in the environment [14, 17–19]. From the many known diagnostic ratios [17], several were selected (Table 2). Their values indicate that the PAH mixture both in the indoor and outdoor air of the sports hall was mainly connected with the burning of solid fuels (coal, wood) and with engine exhaust gases (with the predominant ones being diesel engines).

Table 2. PAH diagnostic ratios for the indoor and outdoor sampling points.

Diagnostic ratio	Indoor		Outdoor		Source of PAHs [17]
	PM_4	TSP	PM_4	TSP	
CPAH/TPAH	0.820	0.845	0.748	0.866	~ 1 - combustion
Flu/(Flu+Pir)	0.620	0.659	0.948	0.905	> 0.5 - petrol emissions
An/(An+Fen)	0.692	0.523	0.322	0.298	> 0.1 - petrogenic; 0.24 - coal burning
Flt/(Flt+Pir)	0.323	0.312	0.886	0.571	< 0.4 - unburned petroleum; 0.57 - coal burning; > 0.5 coal, grass, wood
BaA/(BaA+Ch)	0.434	2.145	2.181	1.918	0.46 - coal burning
BaP/(BaP+Ch)	0.558	0.405	0.547	0.447	0.5 - diesel
IP/(IP+BghiP)	0.536	0.517	0.427	0.452	0.2–0.5 - petroleum combustion; > 0.5 - grass, wood, coal combustion
IP/BghiP	1.156	1.070	0.746	0.824	~1 - diesel
BaP/BghiP	3.057	3.342	3.125	1.832	> 1.25 - coal/coke

The measured concentrations of 16 PAHs clearly showed that the average concentration of the PAH mixture (ΣPAH) inside the sports hall is almost twice as high as the concentration of the PAH mixture in the surroundings of the hall. This applies both to PM_4- and TSP-bound PAHs. Particularly high contents of PAHs were found in the TSP fraction inside the sports hall (809.435 ng/m^3). The values of the concentrations of individual compounds from the group of TSP-bound PAHs in the indoor air of the hall were within the range of values from those values not exceeding the detection threshold (Na, Acnl, Acl) to the value 340 ng/m^3 (Flu). The concentrations of individual compounds from the group of TSP-bound PAHs in the outdoor air was as follows: from the detection limit (Na, Acnl, Acl) to 296 (Flu) ng/m^3. In the indoor air the concentration of individual PM_4-bound PAHs ranged from the detection limit (Na, Acnl, Acl) to 75 ng/m^3 (Flu) and outside from the detection limit (Na, Acnl, Acl) to 70 ng/m^3 (Flu).

Table 3 presents the individual compounds that had the largest contribution to the average ambient concentrations of PM4.0- and TSP-bound PAHs in the indoor and outdoor air. Similar PAH profiles in the indoor and outdoor air also indicate the infiltration of atmospheric air (by slots, open windows and doors) as the main source of PAHs inside the sports hall.

An important substance, due to its adverse health effects, is BaP, for which its PM_4-bound average concentration in the sampling period in the indoor air (12.668 ng/m^3) was two times higher than in the atmospheric air (6.929 ng/m^3). This result may indicate an internal source of this compound in the sport hall.

Table 3. The contribution of individual compounds to the average ambient concentrations of PM-bound PAH in indoor and outdoor air

Position	Indoor		Outdoor	
	PM_4	**TSP**	**PM_4**	**TSP**
1	Flu – 36%	Flu – 42%	Flu – 45%	Flu – 64%
2	Pir – 22%	Pir – 22%	Fl – 20%	Fl – 9%
3	Fl – 10%	Fl – 10%	BaA – 8%	Pir – 7%
4	BaP – 6%	Bbf – 8%	BBf – 6%	Bbf – 5%

The high concentration of PM-bound PAHs in the indoor air of the sports hall certainly causes high health risks especially for its users who train in this facility systematically. The level of exposure of persons actively using the analysed sports facility to hazards from the high PM and PAH concentrations can be expressed by indicators of the exposure such as CEQ, MEQ and TEQ [14]. Regardless of the PM fraction, these indicators were significantly higher for the mixture of PAHs in the indoor air (athletic hall) than in the atmospheric air (surroundings). Carcinogenic inhalation exposure to the PAH mixture

(CEQ) in the indoor air reached the values 40 ng/m^3 for the PM_4-bound PAH and 53 ng/m^3 for the TSP-bound PAH (Fig. 3.). These are values several or dozens of times higher than previously recorded in urbanized areas of Europe, including Polish cities [14]. Similar (high) values of the risk indicators were reached during studies in the indoor air of a sawmill in Biskupice (Zabrze; Upper Silesia Agglomeration) where the CEQ for the PM_4-bound PAHs ranged from 6.6 to 69 ng/m^3, MEQ: 6.5–61.39 ng/m^3, TEQ: 53.4–293 pg/m^3 and the CEQ for the TSP-bound PAH mixture, ranging from 41.5 ng/m^3 to 114.98 ng/m^3, MEQ: 32–61.4 ng/m^3, and TEQ: 138–293 pg/m^3 [16]. However, one must keep in mind that almost all activities performed in a sawmill are involved with dust emission and high concentrations of PM and PM-bound PAHs were to be expected; but sports facilities are designed for recreation and improving of one's physical condition, to which is associated the expectation of the appropriate quality of the air.

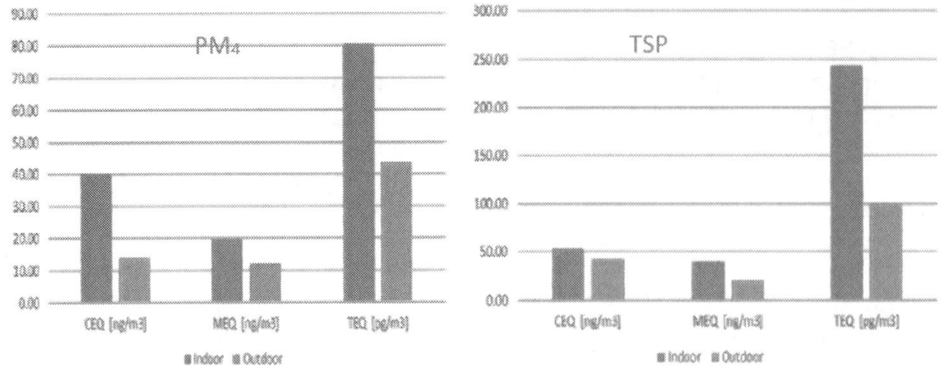

Fig. 3. Mean values of the risk indicators (CEQ, MEQ, TEQ) for PM_4-bound and TSP-bound PAH in indoor (athletic hall) and outdoor (atmospheric) air.

Other important parameters for the assessment of the exposure to PAH are the relations between the average ambient concentration of carcinogenic PAHs and the average mass concentration of the PAH mixture ($\sum PAH_{canc}/\sum PAH$). The average mass share of the sum of carcinogenic compounds in the average ambient concentration of PAH in the indoor air was 25% for PM_4-bound PAH and 20% for PM-bound PAH.

4 Conclusions

The main source of fine PM (PM_4) and PM-bound PAH in the sports hall was atmospheric air. However the poor air exchange in the hall and the related accumulation of PM_4 dust as well as the accumulation of PM-bound PAH resulted in the values of risk indicators for sports hall users as expressed by the selected indicators CEQ, MEQ, and TEQ, being almost twice as high as the values determined at the same time for atmospheric air. On the basis of the selected diagnostic ratios it can be assumed that the source of PM-bound PAHs both in the air of the sports hall and in its surroundings was the burning of fuels, viz. diesel and coal. It seems that the observed high concentrations of PM and PM-bound PAHs in the air of the sports hall are specific to the winter period, when in the vicinity of the hall many different sources of PM and PM-bound PAH emissions associated with the heating of homes and flats are active and in general the consumption of energy is intensified during this period. To draw more general conclusions about the exposure of users of the hall to PM-bound PAH, similar studies should be conducted during the non-heating season.

The article was realized within the statutory research no. S/E 422/8/16/17 co-financed by funds of the Ministry of Science and Higher Education provided to the Faculty of Civil Safety Engineering of The Main School of Fire Service in 2017.

References

1. J.S. Pastuszka, *Higiena pracy w pomieszczeniach nieprzemysłowych*, Higiena Pracy, Łódź, J.A. Indulski (ed.), 1-514 (1999)
2. T. Nathanson, *Indoor Air Quality in Office Buildings: A Technical Guide*, Report of the Federal-Provincial Advisory Committee on Environmental and Occupational Health, Minister of National Health and Welfare, Canada (1995)
3. B. Brunekreef, S.T. Holgate, Lancet, **360,** 9341, 1233-1242 (2002)
4. S. Menteşe, M. Arisoy, A.Y. Rad, G. Güllü, Clean, **37,** 6, 487-493 (2009)
5. V. Niinimaa, P. Cole, S. Mintz, Respir Physiol, **42,** 1, 61-71 (1980)
6. C.C. Daigle, D.C. Chalupa, F.R. Gibb, P.E. Morrow, G. Oberdorster, M.J. Utell, Inhal. Toxicol. **15,** 539–552 (2003)
7. California Environmental Protection Agency, Air Resources Board, *Short–Term Health Effects Associated with Exposure to Air Pollution During Exercise* (2008)
8. Abelsohn, D.M. Stieb, Can. Fam. Physician, **57,** 8, 881-887 (2011)
9. G. Lippi, G.C. Guidi, N. Maffulli, Int. J. Sports Med., **29,** 696-698 (2008)
10. P. Srikanth, S. Sudharsanam, R. Steinberg, Indian J. Med. Microbiol. **26,** 4, 302-312 (2008)
11. J.L Durant, W.F. Busby, Jr., A.L. Lafleur, B.W. Penman, C.L. Crespi, Mutat. Res. Gen. Toxicol., **371**, 123–157 (1996)
12. K.H. Kim, S.A. Jahan, E. Kabir, R.J.C. Brown, Environ. Int., **60**, 71–80 (2013)
13. D.S. Jyethi, P.S. Khillare, S. Sarkar, Environ. Sci. Pollut. Res., **21**, 366–378 (2014)
14. Kozielska, W. Rogula-Kozłowska, P. Rogula-Kopiec, I. Jureczko, Ecol. Eng. **49,** 25-32 (2016)
15. Kozłowska, N. Pawlas, M. Zaciera, L. Kapka-Skrzypczak, R. Jasiński, Environ. Med., **14,** 3, 28-38 (2011)
16. P. Rogula-Kopiec, W. Rogula-Kozłowska, B. Kozielska, I. Sówka, Pol. J. Environ. Stud. **24, 4,** 1867-1873 (2015)
17. W. Rogula-Kozłowska, *PAH and Heavy Metals in Ambient Particulate Matter: A Review of Up-to-Date Worldwide Data,* Synergic Influence of Gaseous, Particulate, and Biological Pollutants on Human Health, J.S. Pastuszka (ed.), CRC Press, 68-108 (2015)
18. Dvorská, G. Lammel, J. Klánová, Atmos. Environ., **45,** 2, 420-427 (2011)
19. F. Mirante, C. Alves, C. Pio, O. Pindado, R. Perez, M.A. Revuelta, B. Artiñano, Atmospheric Res., **132–133**, 345–361 (2013)
20. G. Majewski, K. Kociszewska, W. Rogula-Kozłowska, H. Pyta, P. Rogula-Kopiec, W. Mucha, J.S. Pastuszka, Atmosphere, **7,** 9, 1-12 (2016)
21. K. Kociszewska, P. Rogula-Kopiec, G. Majewski, W. Rogula-Kozłowska, W. Walter Mucha, B. Mathews, Sci. Rev. Eng. Env. Sci., **26,**1, 108-124 (2017)
22. W. Rogula-Kozłowska, B. Kozielska, G. Majewski, P. Rogula-Kopiec, W. Mucha, K. Kociszewska, J. Environ. Sci., DOI 01238 (2017)

E3S Web of Conferences **28**, 01021 (2018) https://doi.org/10.1051/e3sconf/20182801021
Air Protection in Theory and Practice

Environmental issues and process risks for operation of carbon capture plant

Radosław Lajnert[1,*], *Martyna* Nowak[1] and *Jolanta* Telenga-Kopyczyńska[1]

[1]Institute for Chemical Processing of Coal, Zamkowa 1 St., 41-803 Zabrze, Poland

Abstract. The scope of this publication is a presentation of environmental issues and process risks connected with operation an installation for carbon capture from waste gas. General technological assumptions, typical for demonstration plant for carbon capture from waste gas (DCCP) with application of two different solutions – 30% water solution of monoethanoloamine (MEA) and water solution with 30% AMP (2-amino-2-methyl-1-propanol) and 10% piperazine have been described. The concept of DCCP installation was made for Łaziska Power Plant in Łaziska Górne owned by TAURON Wytwarzanie S.A. Main hazardous substances, typical for such installation, which can be dangerous for human life and health or for the environment have been presented. Pollution emission to the air, noise emission, waste water and solid waste management have been described. The environmental impact of the released substances has been stated. Reference to emission standards specified in regulations for considered substances has been done. Principles of risk analysis have been presented and main hazards in carbon dioxide absorption node and regeneration node have been evaluated.

1 Introduction

Implementation of carbon dioxide (CO_2) capture technology into existing coal power plants is more and more real, because of European Comission decisions in the field of greenhouse gases emission reduction. The main part of such systems is a CCP (Carbon Capture Plant) installation, in which the CO_2 is captured. One of the technologies of carbon dioxide reduction commonly applied in the industry, is amine absorption process [1]. The technology of CO_2 capture and storage (CCS) is a key element, which helps to fulfill the requirements imposed by current legislation (EU ETS directive, CCS directive, RED II directve, non-ETS decision) in the field of CO_2 emission reduction into atmosphere. This solution fits in the CCS directive. Implementation of CCS technology allows for significant carbon dioxide emission reduction, and hence reduction of CO_2 emission costs within the European Union Emission Trading Scheme, established by the EU ETS directive [2]. This article presents the environmental conditions and process risks of such installation.

* Corresponding author: rlajnert@ichpw.pl

2 General technical and technological assumptions

The base used for analysis in the article is a demonstration installation of CO_2 reduction for exhaust gases (DCCP - Demo Carbon Capture Plant) [3]. The installation has to capture CO_2 from exhaust gases, which are generated in a reference 900 MW power plant. It is predicted, that the installation will remove about 90% of the generated CO_2 per year, which is about 1.5 million tons of CO_2/year. Thanks to the research conducted on pilot installations [4,5], carried out by research team from the Institute for Chemical Processing of Coal, a concept of DCCP installation was prepared, for Łaziska Power Plant in Łaziska Górne, belonging to the TAURON Wytwarzanie S.A. The basic technological element of the DCCP installation is a CO_2 removal module with amine solution (30% water solution of ethanolamine (MEA)) [6] or water solution with 30% AMP (2-amino-2-methyl-1-propanol) and 10% piperazine (PZ) [7]. The CO_2 removal module consists of a few process parts: amine solution storage, CO_2 absorption, heat exchange and solution regeneration. The most important part is absorpion system with absorber. The exhaust gas introduced into an absorption column is in contact with the amine water solution. Heat recovery system, located between the absorber and desorber (regenerator) is also an essential part. Part of the heat present in the hot solution, which leaves the desorber is transferred into the saturated solution, directed for regeneration. The main element of the absorption solution regeneration part is a desorption column (regenerator), in which the CO_2 capture from solution takes place, thanks to the delivered heat [8]. Absorption processes are the most often selected technologies for CO_2 removal from exhaust gases, generated in fossil fuels power plants and are currently found as best for this application [9].

3 Environmental protection and hazards

DCCP installation is a potential source of hazardous substances. Most of the substances are periodically managed as waste, part of them, which circulates in the closed system, can be dangerous only in the case of accident. However, some of the substances are transferred into the exhaust gases and are emitted into the atmosphere. Among the wastes are also hazardous wastes, which are dangerous to heatlth and life or to environment, because of their origin, chemical or biological composition and other properties. These are substances which have at least one the the hazardous properties [10]. In this installation following types of such wastes can be present: flammable, harmful, toxic for reproduction, sensitizing, ecotoxic and corrosive. Specification of main hazardous materials typical for amine-based carbon capture plant together with risk type is shown in Table 1.

Table 1. Specification of main hazardous materials typical for amine-based carbon capture plant [11].

Waste code	Material	Type of risk in the case of accident
16 10 01*, 16 10 02	Residues from reclaimer	Flammable, toxic for reproducion, sensitizing, ecotoxic, corrosive
	Amine solution	Toxic for reproduction, sensitizing, ecotoxic, corrosive
	Amine decomposition	
07 01 10*	Residues from filters	
06 13 02*	Used bed, active carbon	Harmful, toxic for reproduction, sensitizing, ecotoxic, corrosive
06 02 04*	Solution of NaOH	Corrosive, harmful
16 10 03*, 16 10 04	Condensate	

Emissions from the installation include solid, liquid and gaseous susbtances as well as noise emission. All solid products, if they are not recycled to the process, must be transported into proper storage and processing site. In the case of off-site processing, the substances must be properly stored in the DCCP installation site, until they are transported off-site. Spent active carbon can be used in the plant as a fuel, together with spent active carbon used in the power plant. Therefore, it can be qualified as a side product in the administrative procedure. In such situation, the DCCP installation will not cause a significant increase of waste amount from the whole plant. All wastes generated in the DCCP installation must be included in the waste production permission update. Such permission is required, if the plant generates more than 5000 Mg of wastes per year [12].

Gaseous emission includes: emissions level directly fom the installation stack and concecntration values off-site. From the absorpion column are released NO_X, SO_X and particulate matter [11]. These are the same substances, which are emitted into the atmosphere from power plant without CCS installation. The main difference in the released gas into atmosphere is lack of CO_2. Gaseous amines degradation products are the characteristic substances for CO_2 removal process. Therefore, small amounts of amine vapors and degradation products, are transferred to the cleaned exhaust gases. Togehter with amine absorption process development by selection of new sorbents, which increases the CO_2 removal efficiency and decreases the energy consumption, attention is paid on amines and their degradation products emissions (Table 2). During the installation exploitation losses of sorbent (amine) occur, which results from thermal and oxidative degradation, and also physical phenomena, like evaporation or entraintment by flowing gas stream. As a result of sorbent degradation, compounds, which lower its active concentration, are formed. Besides, a lot of the substances are highly volatile and are released with the cleaned gas into atmosphere and cause environmental pollution.

Table 2. Main compounds formed as a result of degradation of described amines [13-15].

MEA degradation	AMP degradation	Piperazine degradation
– ammonia – ethanolamine – 3-hydroxyethylamine – N-hydroxyethyl-propanamide – 4-hydroxyethyl-2-piperazine – 2-hydroxyethylamine – N-hydroxyethyl-acetamide – metal cations – salt anions	– 2-methyl-2-(methylamine)-1-propanol – 4,4-dimethyloxazolidine – 3,4,4-trimethyloxazolidine – 1-(2-hydroxy-2-methylpropyl)-4,4-dimethylimidazolidine	– ethylenediamine – formic acis – N-nitroso-piperazine – N,N-di nitroso-piperazine

During DCCP installation exploitation emission level of the amines into atmosphere will not have a significant impact on the environment. It is proved by amine installation of this type commonly applied in refinery and petrochemical industries. According to literature [13], the amine concentration in the emitted exhaust gases after CCP proces can be lower than 0.2 ppm.

The allowable emission levels is determined by emission standards. DCCP installation is integrated with power plant and that is why the emission standards [16] are refered to the whole plant, as industry object and its all emitors. The emission exhaust gases without CO_2 is not a stream which DCCP installation adds to the general balance, but emission from the power plant, and cannot exceed the values determined in regulations. According to the

regulation [16], emission standards for existing power plant, for which the building permit was established before 1st of July 1987, with nominal thermal power of Łaziska power plant – 196 MW are (for hard coal combustion, with 6% O_2):
- 250 mg/m_u^3 for SO_2,
- 25 mg/m_u^3 for PM10

and for NO_X the individual value for Łaziska power plant is 600 mg/m_u^3 (up to 31st of December 2017). The emission values, which are adequate for allowable emission standards for hard coal combustion for each emitors are presented in an integrated permit. For Łaziska power plant the permissible emission values of SO_2 and PM10 are less rigorous in the permit than in the emission standards [16]. The regulation does not define emission standards for MEA, AMP, PZ and ammonia, but their emission values can be assessed by actual emissions from insdustrial objects. The biggest installation of CO_2 removal by chemical absorption method in Europe - „CO_2 Technology Centre Mongstad" in Norway has a permit for following emissions [17]:15 and 23 mg/m_u^3 for MEA and NH_3 respectively. According to de Koeijer [18], for the abovementioned instalation following design emissions were defined respectively 2.5 and 3.6 mg/m_u^3. It is expected, that the actual emissions will be 1.3 and 1.8 mg/m_u^3. A simulation prepared for the abovementioned installation shows, that for 25% of AMP and 15% of piperazine solution, following emission values are obtained [13]:
- for NH_3 – 0,012 ÷ 1,3 g/Mg CO_2,
- for AMP – 16 ÷ 1340 g/Mg CO_2,
- for PZ – 0,6 ÷ 25 g/Mg CO_2,

depending on gas temperature and the CO_2 molar charge nad removal efficiency. Additionally, in Commission Recommendation [19] information about piperazine monitoring are presented: *„[...] The competent authorities in the Member States concerned lay down conditions, emission limit values or equivalent parameters or technical measures regarding piperazine in the permits issued under Council Directive 96/61/EC [...]"*. This term recommends inclusion of piperazine emission in the integrated permit, but the act includes only the concentration values in working environment. According to the directive [20] and regulation [21], the highest allowable levels in working environment of piperazine are (for 293.15K and 101.3 kPa):
- 0.1 mg/m_u^3 – the highest allowable concentration referred to an average working time,
- 0.3 mg/m_u^3 – the highest temporary allowable concentration.

Similarly, the values for ammonia are 14 and 28 mg/m^3 and for MEA 2.5 and 7.5 mg/m^3.

Although, the regulation [16] does not define emission standards for amines and ammonia. To determine an impact on the environment off-site, pollutants propagation calculation should be performed and should include the background of considered pollutants. The regulation [22] does not define allowable amines concentration, but in the regulation [23] there are presented subtances, characteristic for CO_2 removal by absorption method – ethanolamine (for MEA solution) and ammonia (for both solutions). According to the regulation [23], the reference levels of substances concentration of-site are:
- for ethanolamine – 30 µg/m_n^3 average for one hour and 1.6 µg/m_n^3 average for the year,
- for ammonia – 400 µg/m_n^3 average for one hour i 50 µg/m_n^3 average for the year.

Corresponding standards for the basic substances are:
- for NO_X – 30 µg/m_n^3 average for the year,
- for NO_2 – 200 and 40 µg/m_n^3 average for one hour and for the year, respectively,
- for SO_2 – 350, 125 and 20 µg/m_n^3, average for one hour, day and the year,
- for PM10 – 50 and 40 µg/m_n^3, average for the day and the year.

Because of the wastewater, generated in the installation, all process areas in which amines are applied, must be equipped in trays and each area should have its own drainage

 https://doi.org/10.1051/e3sconf/20182801021

with wastewater tank. The assessment procedure for the liquid utilization is based on a list from annex to the regulation [24], which contains substances, which are particularly harmful for water environment, which introduction into sewage system requires water permit or inclusion in the integrated permit.

During CCP installation exploitation there will be new noise sources, such as a compressor, a pump room with amine pumps and chemicals pumps (in this rooms, the noise should not exceed 85 dB (A)) and electrical appliances in a electric building (65 dB (A)). Predicted noise emission of the DCCP installation during exploitation will be negligible for the surrounding.

4 Hazard studies

Process hazards are an inherent part of technological processes. Chemical or physical phenomena present during the installations exploitations can cause damage to property and environment. Assessment of potencial hazards identification is targeted at limiting the possibility of occurence and propagation of hazards from the experimental installation, appliances, objects etc. The identificaion procedures are focused at [25]:
- identification of all process hazards, which have an impact on technical safety,
- analysis of all elements, which have an impact on technical safety level,
- finding solutions in order to eliminate the occurence of lapses and errors.

Hazard studies give the possibility of identification and assessment of hazards, which can occur during the installation designing, building, commissioning and exploitation. Information in process design are the basis for hazard study development. Depending on the investment hazard study, persons responsible for its realization undertake appropriate safety measures, in order to avoid, reduce, accept or transfer the hazard.In the following part, the hazard study with HAZID, HAZOP and SIL methods is shortly described.

HAZID (Hazard Identification) study is a method, which aims at full understanding of basic risk related to the process. In this method, potential events, frequency of their appearance and hierarchical control, which manages this risk are considered [26]. The aim of the HAZID analysis is identification of key hazards and thanks to it, there is a possibility of their elimination at the design step. It should be underlined, that the HAZID analysis is a node-by-node study, not line-by-line assessment. The nodes are selected in a way to include a big area of the installation, in which different chemical processes in similar conditions are conducted instead of analyzing similar areas.

HAZOP (Hazard and Operability) study is a method intended to provide a detailed assessment of the proposed design, such that suitable and sufficient prevention, control and mitigation measures can be included to reduce the risk to a tolerable level [26]. The benefits of this method are, increase of installation work reliability, lower exploitation costs, reduction of losses associated with downtime, full range of identified hazards, high effectiveness and early detection of hidden hazards.

SIL (Safety Integrity Level) assessment is a structured, logical, team-based approach to the determination of the requirement for safety instrumented functions necessary to ensure that risks from industrial activity are tolerable. Safety integrity is a probability, that the system related to safety will properly perform the required safety functions in particular conditions and time interval [26].

5 Hazards assessment

The following Table 3 presents scenarios of main and characteristic for DCCP installation hazards during accidents in key nodes, such as absorption and regeneration nodes.

Table 3. Assessment of the main risks in absorption and regeneration nodes [26].

Risk scenario	Causes	Potential impact on humans	Potential impact on environment
Absorption node			
CO_2 emission to the environment	Installation shut-downs, emergency venting	Potential suffocation	Minimal
Loss of tank/pipeline tightness – amines leakage	Corrosion, tank or pipeline failure, collision, tank overflow, pump seal failure	Vapors can irritate skin, eyes and the respiratory system. Contact can additionaly cause serious skin burns and eye damages	Can have a significant impact on the environment, if amines are released outside the tank/pipeline/ installation. Potential contamination of groundwater
Absorption column failure	Incorrect structural design, column overflow	Potential fatalities caused by installation failure and hazardous materials release	
Clogged gas inlet - implosion	Gas congestion during fast cooling and auxiliary fan operation		
Absorber leakage - hot exhaust gases and amine derivatives release	Corrosion caused by accumulated corosive substaces, expansion joint failure	Exposure to hot exhaust gases, potential suffucation or thermal shock. Injuries caused by falling parts	Limited impact on the environment, if the acids derivatives are released into drainage
Regeneration node			
Amines release into condensate during starts or shut-downs. Corrosion of turbine/boiler	Leakage in heat exchanger, malfunction of vapor valves system, which cause vacuum	Staff exposure to vapor leak – potentially fatal	Potential leak into sewage system. Possibility of amine contamination of the local environment
Vacuum on vapor, condensate and heat exchanger nodes	Extraction on low and medium pressure below 25% of efficiency, water vapor condensation in tanks	Vapors can cause skin and eyes irritation, respiratory sensitization. Contact with hot amine can cause serious skin burns and eye damage	Potential impact on local environment
Leakage of pipeline with cold or hot amine	Regeneration column or pipeline/heat exchanger damage, pump or reclaimer	Exposure to hot materials and vapors. Can cause skin and eyes irritation, respiratory	Potentially significant impact, if amines are released off-site. Potential

	leakage, corrosion, amine pipeline failure	sensitization	contamination of ground waters

6 Summary

Air emissions from described installation will not be harmful for the environment. Process waters and waste waters are also not problematic, because the process areas in which amines are applied are equipped with trays and each are will have their own drainage with waste water tank. In the worst, but less likely case of such leak, there can be soil contamination and hence amine penetration into ground water. Wastes generated in the installation will be segregated and safely stored on the DCCP site until they are managed in appropriate place. Predicted noise emission of DCCP installation during exploitation will not have a significant impact on the surrounding.
According to rules applied in HAZID, HAZOP and SIL studies, the most important technological nodes of the DCCP installation were analyzed and risks to humans occurs only during emergency situation. Most of the identified hazards are typical for industrial installations, such as fire, electrocution, fall from a height, unsealing because of appliances structural integity decrease, resulting in vapor or water leak and splinter falling. There are also hazards characteristic for the DCCP installation, related to amine solutions and their deivatives presence. Hazardous substances forming during the installation exploitation are: amine solution, amine degradation products, residues from the reclaimer, residues from filters, sodium hydroxide and condensate. The described hazards can cause ground water contamination, if the amines are released off-site. Environmental protection and hazards issues in case of AMP and PZ solutions application do not differ significantly in case of MEA solution application.

Presented research results were obtained during realization of research project no 11.16.008 „Monitoring of national and community regulations and actions supporting municipalities and industry within environmental protection", financed by the Ministry of Science and Higher Education within dotation for research potential maintenance.

References

1. L. Więcław-Solny, A. Tatarczuk, A. Krótki, A. Wilk, Karbo, **57**, 62 (2012)
2. L. Więcław-Solny, M. Ściążko, A. Tatarczuk, A. Krótki, A. Wilk, Polityka Energ., **14/2**, 441, (2011)
3. A. Tatarczuk, M. Stec, M. Ściążko, L. Więcław-Solny, A. Krótki, Przem. Chem., **95/11**, 2322, (2016)
4. M. Stec, A. Tatarczuk, L. Więcław-solny, A. Krótki, T. Spietz, A. Wilk, Clean Technol. Environ. Policy, **1**, (2015)
5. L. Więcław-Solny, A. Tatarczuk, M. Stec, A. Krótki, Energy Procedia, **63**, 6318, (2014)
6. A. Tatarczuk, M. Ściążko, M. Stec, S. Tokarski, Chemik, **67/5**, 407, (2013)
7. D. Śpiewak, A. Krótki, T. Spietz, M. Stec, L. Więcław-Solny, A. Tatarczuk, A. Wilk, Chem. Process Eng., **36/1**, 39, (2015)
8. T. Spietz, L. Więcław-Solny, A. Tatarczuk, A. Krótki, M. Stec, Chemik, **68/10**, 884, (2014)
9. Technology Roadmap: Carbon Capture and Storage (2013), International Energy Agency, https://www.iea.org (accessed 14.11.2017)

10. Act of 14th December 2012 on waste (Journal of Laws of 2013 item. 21, as amended)
11. UK Carbon Capture and Storage Demonstration Competition - Project HSE Report. ScottishPower CCS Consortium, (2011) https://pl.scribd.com/document/72781748 (accessed 14.11.2017)
12. Act of 27th April 2001 Environmental Protection Law (Journal of Laws of 2001 no 62 item. 627 as amended)
13. N. Dave, T. Do, P. Jackson, P. Feron, M. Azzi, M. Attalla, CSIRO, Australia, (2010)
14. A. B. Rao, E. S. Rubin, i M. B. Berkenpas, Department of Engineering and Public Policy, **1**, (2004).
15. T. Do, D. Narendra, P. Feron, M. Azzi, CSIRO, (2012)
16. Regulation of the Minister of Environment of 4t November 2014 on emission standards for installations, combustion sources and combustion appliances or waste incineretion (Journal of Laws of 2014 item. 1546)
17. Y. Maree, S. Nepstad, i G. de Koeijer, Energy Procedia, 37, 6265, (2013)
18. G. de Koeijer et al., Energy Procedia, **4**, 1207, (2011)
19. Commission communication on the results of the risk evaluation and the risk reduction strategies for the substances: Piperazine; Cyclohexane; Methylenediphenyl diisocyanate; But-2yne-1,4-diol; Methyloxirane; Aniline; 2-Ethylhexylacrylate; 1,4-Dichlorobenzene; 3,5-dinitro-2,6-dimethyl-4-tert-butylacetophenone; Di-(2-ethylhexyl)phthalate; Phenol; 5-tert-butyl-2,4,6-trinitro-m-xylene
20. Commission Directive 2000/39/EC of 8 June 2000 establishing a first list of indicative occupational exposure limit values in implementation of Council Directive 98/24/EC on the protection of the health and safety of workers from the risks related to chemical agents at work
21. Regulation of the Minister of Labour and Social Policy of 6th June 2014 on maximum admissible concentrations and intensities for agent harmful to health in the working environment (Journal of Laws of 2014 item. 817, as amended)
22. Regulation of the Minister of Environment of 24th August 2012 on admissible levels of some substances in the air (Journal of Laws of 2012 item. 1031)
23. Regulation of the Minister of Environment of 26th January 2010 on reference values of some substances in the air (Journal of Laws of 2010 no 16 item. 87)
24. Regulation of the Minister of Environment of 15th December 2008 amending the regulation on substances that are particularly harmful to the aquatic environment, which introdution into sewage system requires water permission (Journal of Laws o 2008 no 229 item. 1538)
25. T. Podczaszy, Opinia w zakresie konieczności wykonania analizy identyfikującej potencjalne zagrożenia dla bezpieczeństwa technicznego instalacji doświadczalnych i stanowisk testowych oraz środowiska naturalnego dla CCTW (2011) – unpublished
26. Kingsnorth CCS Hazard Identification (HAZID) Report (2011), http://webarchive.nationalarchives.gov.uk/20121217151014/https://www.decc.gov.uk (accessed 14.11.2017)

Investigation of air pollutants in rural nursery school – a case study

Anna Mainka[1,*], *Elwira* Zajusz-Zubek[1], *Barbara* Kozielska[1], and *Ewa* Brągoszewska[1]

[1]Department of Air Protection, Silesian University of Technology, 22B Konarskiego St., 44-100 Gliwice, Poland

Abstract. Children's exposure to air pollutants is an important public health challenge. Indoor air quality (IAQ) in nursery school is believed to be different from elementary school. Moreover, younger children are more vulnerable to air pollution than higher grade children because they spend more time indoors, and their immune systems and bodies are less mature. The purpose of this study was to evaluate the indoor air quality (IAQ) at naturally ventilated rural nursery schools located in Upper Silesia, Poland. We investigated the concentrations of volatile organic compounds (VOCs), particulate matter (PM), bacterial and fungal bioaerosols, as well as carbon dioxide (CO_2) concentrations in younger and older children's classrooms during the winter and spring seasons. The concentration of the investigated pollutants in indoor environments was higher than those in outdoor air. The results indicate the problem of elevated concentrations of $PM_{2.5}$ and PM_{10} inside the examined classrooms, as well as that of high levels of CO_2 exceeding 1,000 ppm in relation to outdoor air. The characteristics of PM and CO_2 levels were significantly different, both in terms of classroom occupation (younger or older children) and of season (winter or spring).

1 Introduction

One of the five EU benchmarks for Education and Training (ET 2020) is that by 2020, at least 95% of children between the ages of four and starting compulsory primary education should participate in early childhood education [1]. In the context of expanding early childhood education, the EU should be active in reviewing research pertaining to the quality of care afforded to the system's youngest participants, which is linked in particular with the microenvironment of nursery schools. There, they are exposed to unknown levels of indoor pollutants. Children are particularly vulnerable to air pollutants, because they spend more time indoors than active, healthy adults, moreover they breath higher volumes of air relative to their body weights and their tissue and organs are growing [2].

Many research that consider children underline the relation between IAQ (indoor air quality) and health effects [3].

Among indoor pollutants researchers point out volatile organic compounds (VOCs), particulate matter (PM), especially finer particles (particles with an aerodynamic diameter smaller than 2.5 μm, $PM_{2.5}$ and smaller than 1 μm, PM_1), bioaerosols as well as

* Corresponding author: anna.mainka@polsl.pl

temperature and relative humidity to provide thermal comfort [4]. Carbon dioxide (CO_2), although not defined as an air pollutant, is an indicator of low ventilation rates [5]. Increased levels of CO_2 led to a decrease in pupils' learning abilities of approximately 5% and communicate with respiratory illnesses [6].

In addition to air pollutants emitted indoors and penetrating from outdoors, ASHRAE Standard [7] points that some materials act as sinks for emissions and then become secondary sources of VOCs and PM as they reemit adsorbed pollutants. The sink materials include fabric partitions and other fleecy materials. The specific sorptive properties of soft materials are particularly relevant in the case of nursery schools, especially in younger children's classrooms, where except for carpets, there are many sorptive toys and additional materials such as bedcovers for the duration of an afternoon nap [8].

The aim of the present study is to characterize IAQ in rural nursery school located near Gliwice, in southern Poland. The study carried out simultaneous measurements of:

- VOCs (particularly benzene, toluene, ethylbenzene and xylenes - BTEX),
- PM (indoor: PM_1, $PM_{2.5}$, PM_{10} and total - TSP; outdoor: $PM_{2.5}$ and PM_{10}),
- bioaerosols (culturable bacterial and fungal aerosols),
- CO_2 concentration (outdoors and indoor in two classrooms) with physical parameters (temperature and relative humidity)

in naturally ventilated classrooms occupied by younger and older children classrooms during winter and spring seasons. The emphasis of this article will be on the difference between winter and spring seasons as well as younger and older children classrooms.

2 Materials and methods

The study was carried out rural nursery school situated about 10 km north of the city of Gliwice (Figure 1), which is located in the west district of the industrial region of Upper Silesia, Poland. The major activities influencing the ambient air quality of this site are agricultural activities (during non-heating periods) and burning of biomass and fossil fuels for domestic needs (during heating periods).

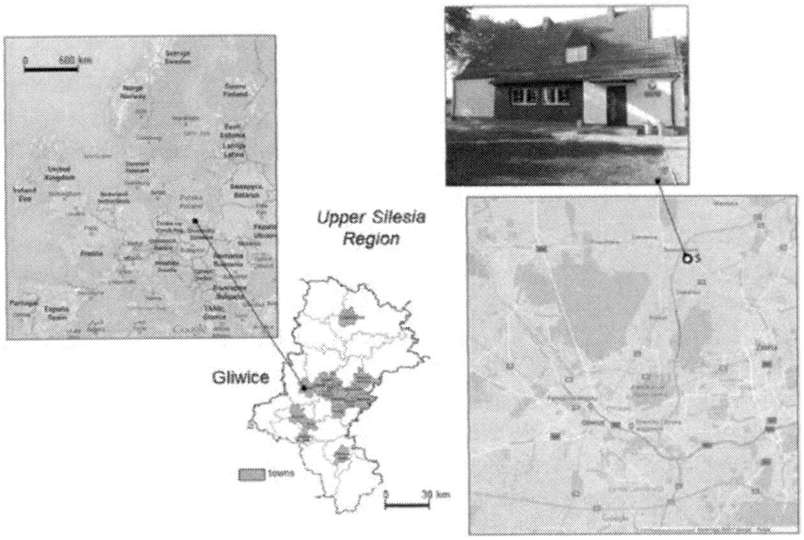

Fig. 1. Location of the investigated nursery school (Map data: 2017© Google, ORION-ME).

The building is located 50 m from the A1 highway. It is separated from the highway by highway screens. It is a detached building (Figure 1) that underwent thermal efficiency improvement processes, completed in the summer of 2013. During the thermal insulation process, natural ventilation using the buildings' air duct systems was left unchanged. Consequently, the IAQ is primarily ensured by means of stack ventilation and airing through open and unsealed windows. The measurements were conducted in both the classrooms of older (five- to six-year-old, I) and younger (three- to four-year-old, II) children. Table 1 summarizes the specification of the two sampling sites, including basic IAQ parameters and the occupancy of each classroom during the winter and spring seasons.

Table 1. Summary of the occupancy and IAQ parameter characteristics for each studied classroom, as well as for the outdoor air.

Parameters	Classroom of Children	
	Older (I)	**Younger (II)**
Children's Age, Years	5–6	3–4
Floor	Ground floor	Ground floor
Volume, m^3	169.7	118.6
Period of Occupation	6:30–16:00	8:00–12:30
Number of children in group	24	21
Median occupancy of children - winter	16	14
Median occupancy of children - spring	18	14
Indoor temperature, °C - winter	27.8±3.6	26.2±2.7
Indoor temperature, °C - spring	23.6±2.7	24.2±1.9
Indoor relative humidity (RH), % - winter	24.5±6.7	26.0±6.6
Indoor relative humidity (RH), % - spring	46.5±5.6	42.5±4.8
Indoor CO_2 concentration, ppm - winter	1122.7±571.0	1265.4±652.6
Indoor CO_2 concentration, ppm - spring	1374.5±825.5	1090.8±630.8
Outdoor parameters, mean ± SD		
Season	winter	spring
Temperature, °C	13.4±9.2	24.4±11.6
Relative humidity (RH), %	47.0±28.8	54.3±34.0
CO_2 concentration, ppm	374.0±17.1	353.5±16.2

The indoor and outdoor concentrations of selected VOCs, different fractions of PM (indoor: PM_1, $PM_{2.5}$, PM_{10} and TSP; outdoor: $PM_{2.5}$ and PM_{10}) and bioaerosols, as well as CO_2 concentrations, were measured in the classrooms of younger and older children in the selected building. The sampling position in classrooms was set at the height of an average child's head (i.e., about 0.8 to 1.0 m above the floor) and away from the door, thus avoiding disturbances resulting from air currents.

Sixteen VOCs (benzene, toluene, ethylbenzene, m-xylene, p-xylene, o-xylene, styrene, isopropylbenzene, n-propylbenzene, 1,3,5-trimethylbenzene, tert-butylbenzene, 1,2,4-trimethylbenzene, sec-butylbenzene, 4-isopropyltoluene, n-butylbenzene and naphthalene) were actively sampled using Perkin Elmer stainless steel tube samplers containing Tenax GR according to the US EPA TO-17 method [9]. Indoor and outdoor fractions of PM were actively sampled and determined following the reference procedure PN-EN 12341 standard [10]. Bioaerosols, defined as bacterial and fungal colony-forming units per cubic metre of air (CFU/m^3), were collected using a six-stage Andersen cascade impactor (with aerodynamic cut-size diameters of 7.0, 4.7, 3.3, 2.1, 1.1 and 0.65 μm). Continuous measurements of CO_2 concentrations both inside classrooms and outside the building were performed using automatic portable monitors (model 77535, Az Instruments). A detail procedures concerning determination of selected air pollutants has been presented in our previous articles [8,11–13].

Statistical analyses were performed using the statistical package Statistica 13 (StatSoft). The non-parametric Mann–Whitney U and Wilcoxon matched-pairs tests for VOCs, PM fractions, and bioaerosols were performed in order to test whether outdoor and indoor concentrations differed significantly, as well as whether the concentrations of compounds in older (I) and younger (II) children's classrooms differed significantly from one another. While for CO_2 levels the parametric t-test was used. A statistical significance level of $\alpha = 0.05$ was used throughout the study.

3 Results and discussion

3.1 Volatile organic compounds

The average concentrations of 16 determined VOCs outdoors 2.39 and 1.14 μg/m^3 during winter and spring, respectively, as well as average ΣVOC_{16} indoors was 10.6±4.9 μg/m^3 point to low concentrations of VOCs in the selected site. Analogically to outdoor concentration levels, indoors higher concentrations were observed in winter 16.1 μg/m^3 (I) and 14.3 μg/m^3 (II), with correspondence to 1.1 μg/m^3 (I) and 1.5 μg/m^3 (II) in spring. Figure 2 to presents the average indoor and outdoor concentrations of BTEX in older (I) and younger (II) children's classrooms as well as outside the building.

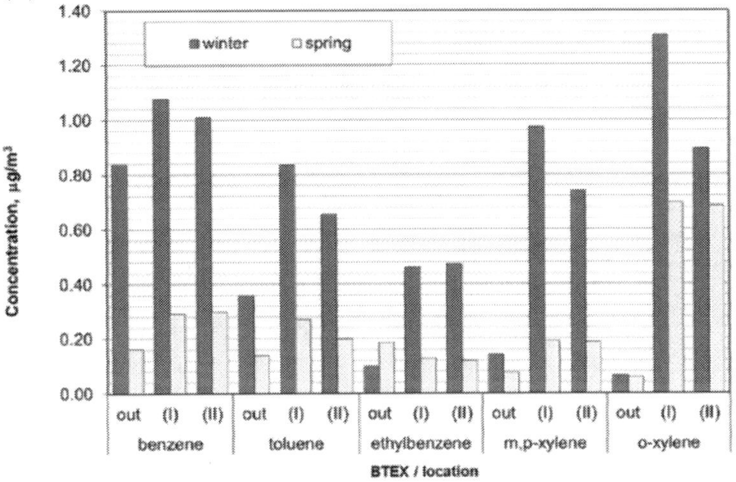

Fig. 2. The average concentrations of BTEX (μg/m^3) during winter and spring campaigns

The concentrations of BTEX found inside the classrooms were higher than outdoor concentrations. The levels of many VOCs are typically higher inside the residences compared to outdoors because indoor VOC source emissions are stronger than the infiltration of outdoor air [14]. As shown in Figure 2, during winter season the concentrations of xylenes were significantly different ($p > 0.05$) between older (I) and younger (II) children classrooms. During the spring season indoor levels of BTEX in both classrooms are on the similar levels except toluene ($p < 0.05$).

Additionally toluene and xylenes are well correlated R^2>0.7, which point to common emission sources. Toluene and xylenes are strong compounds that are used in many household and industrial products. In the nursery schools children may breathe air contaminated with toluene and xylenes by use of glues, paints, rubbers and plastic-modelling cements. The other source of toluene and xylenes might be cleaning solvents used by nursery staff. Along with other solvents, common sources of ethylbenzene and o-xylene might be paints, varnishes and to a lesser extent plastics, and synthetic fibre products e.g. in the coating of fabrics and papers [15].

3.2. Particulate matter

Polish legislation [16] specifies a 24-hour mean concentration of PM_{10} in ambient air, which is 50 μg/m³. For $PM_{2.5}$, there is no corresponding short term (24-hour) limit, but there is an annual level of 25 μg/m³. The World Health Organization (WHO) is more strict in this regard; for $PM_{2.5}$, it recommends a 24-hour average standard of 25 μg/m³ [17] and recommends applying to indoor spaces the same guidelines as for ambient air.
Simultaneously with indoor samples, outdoor samples of $PM_{2.5}$ and PM_{10} were collected on playground areas. The average concentrations of $PM_{2.5}$ and PM_{10} were 19.37 μg/m³ and 22.73 μg/m³ during spring, and did not exceeded WHO guidelines. While the concentrations during winter were 49.40 μg/m³ and 59.00 μg/m³, and exceeded WHO guidelines at 98% and 18% for $PM_{2.5}$ and PM_{10}, respectively.
The concentrations were significantly different between seasons as well as between older (I) and younger (II) children classrooms (Wilcoxon test, $p<0.05$). PM concentrations were higher in older children's classrooms (Table 2). The TSP, PM_{10} and $PM_{2.5}$ levels were significantly higher in older children classrooms ($p < 0.05$), while the PM_1 levels were higher but only in winter. Other research reported an analogous relationship between the classrooms of younger and older children [18], and indicated a potential reason for this being the cumulative effect of three major conditions: high occupancy, poor ventilation and the intensive activities of children resulting in the PM resuspension phenomenon [19].

Table 2. Average levels of PM fractions (μg/m³) measured during occupancy periods inside the classrooms of younger and older children, as well as indoor/outdoor ratios.

PM fraction	Winter				Spring			
	(I)	(II)	Indoor/Outdoor $(I/O)_I$	$(I/O)_{II}$	(I)	(II)	Indoor/Outdoor $(I/O)_I$	$(I/O)_{II}$
PM_1	102.11	49.04	—		45.62	49.04	—	
$PM_{2.5}$	125.69	67.65	3.30	1.11	69.31	67.65	4.78	3.85
PM_{10}	166.12	81.49	3.79	1.10	112.26	81.49	6.85	4.10
TSP	184.24	91.19	—		131.38	91.19	—	

The concentrations of $PM_{2.5}$ (67.65 to 125.69 μg/m³) and PM_{10} (81.49 to 166.12 μg/m³) in all classrooms exceeded WHO guidelines, indicating the low quality of air in all classrooms (Table 2).

The average outdoor concentrations of $PM_{2.5}$ and PM_{10} were typical for the Upper Silesia region [20–23]. Indoor mean concentrations of samples collected in Portuguese preschools were found to be at a similar level, e.g., PM_1, $PM_{2.5}$, PM_{10} and TSP were 33.08, 34.69, 50.94 and 85.81 μg/m³, respectively [4].

The I/O ratios were higher for larger fractions of PM_{10}. The larger the particles are in optical diameter, the heavier they are and the more easily they are able to be deposited on floors and furnishing. Consequently, the influence of re-suspension on indoor particle concentrations increases with particle size. To understand the effect of size distribution on the measured PM concentrations, three different PM size ratios were used to characterize indoor air: $PM_1/PM_{2.5}$, $PM_{2.5}/PM_{10}$ and PM_{10}/TSP. The ratios calculated for each classroom were 0.72, 0.75 and 0.88 for $PM_1/PM_{2.5}$, $PM_{2.5}/PM_{10}$ and PM_{10}/TSP, respectively. All these ratios are high, showing high contribution of small particles.

3.3 Bioaerosols

Table 3 presents the average concentrations of bacterial and fungal aerosols collected in the indoor and outdoor air of rural nursery schools during the analysed winter and spring seasons. Bacteria levels (751-2588 CFU/m³) were higher than fungi levels (156-1549

CFU/m^3). The total concentrations of bacterial aerosols obtained in indoor air during the spring season were of a comparable level to those of the previous study conducted in urban nursery schools, which found levels of between 2545 and 2890 CFU/m^3. However, the concentrations recorded inside the urban classrooms during the winter season were in some cases almost five times higher [12] than the results obtained in the rural nursery school. Other studies [24] concerning IAQ in two nursery schools in Bydgoszcz, conducted between April and February 2014, reported higher average concentrations of bacteria in the indoor air (3697 CFU/m^3, range: 1520-7780 CFU/m^3) and lower average concentrations in the outdoor air (137 CFU/m^3, range: 100-180 CFU/m^3). Research performed in Ankara, Turkey, underlined that, among indoor urban environments, the highest concentrations of total bacteria aerosols were observed in kindergartens, at 649 and 1462 CFU/m^3 in the winter and summer seasons, respectively [25].

Table 3. Average concentrations of total bacterial and fungal colony-forming units per cubic metre in winter and spring season, as well as indoor/outdoor ratios.

Season		**Winter**			**Spring**		
Aerosol	**Location**	**Average concentration**	**SD**	**I/O ratio**	**Average concentration**	**SD**	**I/O ratio**
Bacterial aerosol, CFU/m^3	OUT	751	223.4	-	1428	138.7	-
	(I)	1990	301.9	2.65	2588	85.8	1.81
	(II)	1596	288.8	2.12	2223	291	1.56
Fungal aerosol, CFU/m^3	OUT	156	28.6	-	1549	222.4	-
	(I)	172	34.7	1.1	670	93.7	0.43
	(II)	241	21.5	1.55	707	119.5	0.46

The calculated average I/O ratios of bacterial and fungal aerosol levels according to groups of children for both winter and spring seasons, representing parallel indoor and outdoor samples collected for each sampling day, are given in Table 3. As can be seen from Table 3, the results indicate between two and four times higher indoor concentrations of bacterial aerosols than outdoor samples. The average I/O ratio calculated for all indoor and outdoor bacteria concentrations for the winter season was >1 so it can be concluded that the major sources of these bioaerosols are likely internal, such as building occupants (in this case children and their activities) as well as building materials that host microbiological growth (especially carpets).

According to Mann–Whitney U test results, a non-significant relationship was found between sample types (i.e., indoor (I) and (II) as well as outdoor bacteria levels) measured in both seasons ($p>0.05$).

3.4 Carbon dioxide

Based on the general guidelines concerning the quality of air inside non-residential buildings [5], the increase in CO_2 concentrations in relation to CO_2 concentration in outdoor air (ΔCO_2) was measured during the children's occupation of both classrooms of each of the studied building. Figure 3 depicts the classification of IAQ in each nursery school building during the children's occupation during compulsory care/teaching hours (8:00–13:00). The indoor concentrations of CO_2 revealed inadequate classroom air exchange rates. Most worryingly, during compulsory care/teaching hours, the air in the classrooms was often of low quality (IDA4). The highest contribution of IDA4 (60.0% of compulsory care/teaching time) prevailed in the younger children's classroom (II).

Our research confirmed the enumerated conditions, moreover, our results point out differences between older (I) and younger (II) children's classrooms, which can become more significant if we link inadequate ventilation with the various patterns of children's activities.

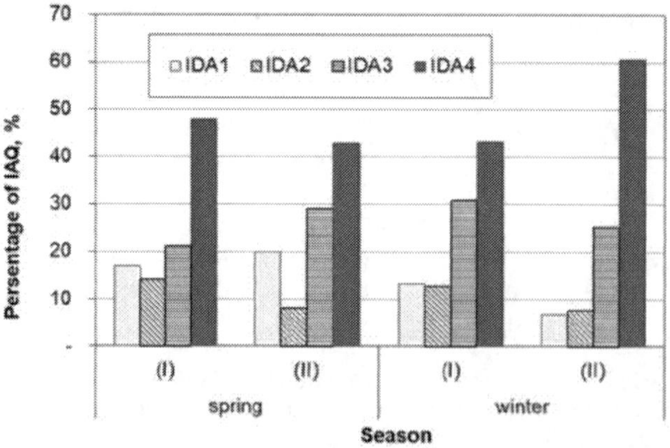

Fig. 3. Classification of IAQ (IDA) in nursery school building according to growth of CO_2 concentration (ΔCO_2).

4 Conclusions

Based on the results from this study, we draw some general conclusions here. The mean values of indoor VOC, PM, bacterial and fungal aerosol samples, as well as CO_2 concentrations, were higher than in outdoor samples. The conclusion can therefore be drawn that indoor sources are the main contributors of IAQ in nursery schools. The results clearly indicate the problem of high $PM_{2.5}$ and PM_{10} concentrations inside classrooms, exceeding the WHO short-term guidance values. Frequently monitored high levels of CO_2 exceeding 1000 ppm in relation to outdoor air (IDA4) also confirmed low indoor air quality inside classrooms, which is concerning in terms of the potential exposure effects on children's health.

The relationship between IAQ in older and younger children's classrooms was the most significant in PM (PM_1, $PM_{2.5}$, PM_{10} and TSP) and CO_2 concentrations. Compared to younger children, older children are more physically active, thus increasing PM re-suspension. The highest CO_2 concentration was observed in the classroom of younger children, who slept during the afternoon. In addition to the highest contribution of low IAQ – during 60% of teaching hours – the role of the afternoon nap seems significant.

This work was supported by the Faculty of Power and Environmental Engineering, Silesian University of Technology (statutory research). All the results used in this paper have received funding from the Polish-Norwegian Research Programme operated by the National Centre for Research and Development under the Norwegian Financial Mechanism 2009-2014 in the frame of Project Contract No Pol Nor/210247/20/2013.

References

1. EU Commission, *Progress towards the Common European Objectives in Education and Training (2010/2011)* (2011)
2. M. R. Ashmore, C. Dimitroulopoulou, Atmos. Environ. **43**, 128 (2009)
3. T. Salthammer, E. Uhde, T. Schripp, A. Schieweck, L. Morawska, M. Mazaheri, S. Clifford, C. He, G. Buonanno, X. Querol, M. Viana, P. Kumar, Environ. Int. **94**, 196 (2016)

4. P. T. B. S. Branco, M. C. M. Alvim-Ferraz, F. G. Martins, S. I. V Sousa, Atmos. Environ. **84**, 133 (2014)
5. PN-EN13779, *Ventilation for Non-residential Buildings. Performance Requirements for Ventilation and Room- conditioning Systems* (2008)
6. M. Griffiths, M. Eftekhari, Energy Build. **40**, 556 (2008)
7. ASHRAE, *Standard 62.1-2013 Ventilation for Acceptable Indoor Air Quality* (2013)
8. A. Mainka, E. Brągoszewska, B. Kozielska, J. S. Pastuszka, E. Zajusz-Zubek, Atmos. Pollut. Res. **6**, 1098 (2015)
9. SKC, *VOC Method Update SKC Appendices to EPA Method TO-17 Determination of Volatile Organic Compounds (VOCs) in Ambient or Indoor Air* (2012)
10. PN-EN12341, *Ambient Air. Standard Gravimetric Measurement Method for the Determination of the PM10 or PM2,5 Mass Concentration of Suspended Particulate Matter* (2014)
11. A. Mainka, B. Kozielska, AIMS Environ. Sci. **3**, 858 (2016)
12. E. Brągoszewska, A. Mainka, J. S. Pastuszka, Aerobiologia (Bologna). **32**, 469 (2016)
13. A. Mainka, E. Zajusz-Zubek, K. Kaczmarek, Int. J. Environ. Res. Public Health **12**, 7990 (2015)
14. K. H. Jung, F. Artigas, J. Y. Shin, Environ. Monit. Assess. **173**, 555 (2011)
15. Agency for Toxic Substances and Disease Registry ATSDR, 1 (2015)
16. Dz.U. Polish Journal of Laws, (2012)
17. WHO, *Air Quality Guidelines for Particulate Matter, Ozone, Nitrogen Dioxide and Sulfur Dioxide* (2006)
18. B. P. de Castro, G. de Souza Machado, G. F. Bauerfeldt, J. D. Nunes Fortes, E. M. Martins, Atmos. Environ. **104**, 22 (2015)
19. A. Zwoździak, I. Sówka, M. Fortuna, Annu. Set Environ. Prot./Rocz. Ochr. Środowiska **15**, 2022 (2013)
20. J. S. Pastuszka, W. Rogula-Kozłowska, E. Zajusz-Zubek, Environ. Monit. Assess. **168**, 613 (2010)
21. K. Ćwiklak, J. S. Pastuszka, W. Rogula-Kozłowska, Polish J. Environ. Stud. **18**, 579 (2009)
22. W. Rogula-Kozłowska, B. Błaszczak, S. Szopa, K. Klejnowski, I. Sówka, A. Zwoździak, M. Jabłońska, B. Mathews, Environ. Monit. Assess. **185**, 581 (2013)
23. W. Rogula-Kozłowska, K. Klejnowski, P. Rogula-Kopiec, L. Ośródka, E. Krajny, B. Błaszczak, B. Mathews, Air Qual. Atmos. Heal. **7**, 41 (2014)
24. Ł. Kubera, J. Studzińska, W. Dokładna, M. Małecka-Adamowicz, W. Donderski, Med. Pr. **66**, 49 (2015)
25. S. Mentese, A. Y. Rad, M. Arisoy, G. Gullu, Indoor Built Environ. **21**, 797 (2012)

E3S Web of Conferences **28**, 01023 (2018) https://doi.org/10.1051/e3sconf/20182801023
Air Protection in Theory and Practice

Analysis of the application of selected physico-chemical methods in eliminating odor nuisance of municipal facilities

Urszula Miller[1,*], *Agnieszka* Grzelka[1] , *Elżbieta* Romanik[1] and *Magdalena* Kuriata[2]

[1]Unit of Engineering and Protection of Atmosphere, Faculty of Environmental Engineering, Wroclaw University of Science and Technology, Pl. Grunwaldzki 9, 50- 377 Wroclaw
[2] Faculty of Environmental Engineering, Wroclaw University of Science and Technology, Pl. Grunwaldzki 9, 50- 377 Wroclaw

Abstract. Operation of municipal management facilities is inseparable from the problem of malodorous compounds emissions to the atmospheric air. In that case odor nuisance is related to the chemical composition of waste, sewage and sludge as well as to the activity of microorganisms whose products of life processes can be those odorous compounds. Significant reduction of odorant emission from many sources can be achieved by optimizing parameters and conditions of processes. However, it is not always possible to limit the formation of odorants. In such cases it is best to use appropriate deodorizing methods. The choice of the appropriate method is based on in terms of physical parameters, emission intensity of polluted gases and their composition, if it is possible to determine. Among the solutions used in municipal economy, there can be distinguished physico-chemical methods such as sorption and oxidation. In cases where the source of the emission is not encapsulated, odor masking techniques are used, which consists of spraying preparations that neutralize unpleasant odors. The paper presents the characteristics of selected methods of eliminating odor nuisance and evaluation of their applicability in municipal management facilities.

1 Introduction

In order to reduce the emission of odor nuisance, odorants formation can be prevented by the choice of technology and the proper conduct of the processes, or by using different methods of deodorizing the waste gases. Among the most commonly used methods, a group of physico-chemical methods can be distinguished, including sorption, oxidation and neutralization methods. The world-wide efforts aim to refine the methods and increase their potential to reduce odor nuisance of many objects. The variety of gases emitted and the variability of pollutant concentrations contained therein influences the efficiency of the methods and determines the choice of the particular case. Selection of methods and design of deodorization equipment should be subordinated to the following quantities: stream and

* Corresponding author: urszula.miller@pwr.edu.pl

composition of the contaminated gas, process efficiency and the amount of investment and operating costs. Choosing an effective method is often based on many years of engineering experience. The paper presents advantages and disadvantages of selected physico-chemical methods of deodorization together with the assessment of their applicability in municipal facilities.

2 Deodorization methods

2.1 Sorption methods

Sorption techniques (absorption and adsorption) for the reduction of odor nuisance are included in physical deodorization methods.

Absorption is one of the more effective methods, especially in the case of large gas streams and concentrations of the dominant odorants (e.g. H_2S, NH_3). Absorption is usually combined with a chemical reaction that results in the separation of impurities from the gas phase to thc liquid phase. The nature of the gas flow through the column (absorber) is determined by the designer at the selection stage and is closely related to the nature of the removed impurities. The absorption rate is influenced by the rate of penetration of the mass, which depends on the solubility of the gaseous pollutants. The disadvantage of this technique is the inconvenient wastewater with an aggressive nature, which also needs to be processed.[1, 2]
Sorption methods are highly effective against organic odorants, including volatile organic compounds. They can also be used to limit emissions of inorganic compounds such as ammonia [2]. The choice of method of contacting the gas with the liquid depends on the physicochemical properties of the absorption system
Most often during the deodorization process, air from the composting plant contaminated with fragrances is subjected to an absorption process in a crossflow scrubber. The air flows horizontally through a series of columns in which it contacts with the liquid containing neutralizing compounds. The system is characterized by a low pressure drops of spraying liquid and low operating costs. This solution allows multi-stage absorption of impurities without the need for additional connections between individual devices. The absorption liquid is sprayed on the construction elements of the filling, thus increasing the contact surface. By application of such construction, the contaminants of different physico-chemical structures can be removed at the same time. For example, one column removes ammonia (NH3) by spraying the contaminated air with sulfuric acid, and in further stages, (columns II and III) hydrogen sulphide (H2S) or mercaptans can be removed by distillation of hydrogen peroxide (H_2O_2) [1]. The choice and use of chemicals are dependent on the concentration, the type of contaminants and the required process efficiency. The treated gases coming out of thc scrubber arc then sent to thc biofilter.
Examples of two-stage systems for removing odors are [1]:

- Oława composting plant with a counterflow scrubber GSW-230 with a flow rate of 18,000 m^3/h - and biofilter with total area of 154 m^2
- Tychy - two counterflow scrubbers GSW-410-P with a flow rate of 2 × 65,000 m^3/h,
- Ostrołęka - two scrubbers with a flow of 2 × 20 000 m^3/h and a three-chamber biofilter with a total area of 400 m^2,
- Stalowa Wola - a counterflow scrubber GSW-300-P with a flow rate of the 48,000 m^3/h and a 2 × 192 m^2 biofilter.

Adsorption on activated carbon is based on the accumulation of molecules on the surface of the adsorbent due to the presence of intermolecular forces or the formation of chemical

bonds. The speed of this process is determined by the diffusion rate of the adsorbate to the surface. The adsorption efficiency is influenced by a number of factors, both surface and activated coal properties (pore distribution, particle size) and solution properties (pH, temperature).
The efficiency of the adsorption process increases with [3]:

- molecular weight drop of the removed impurities,
- temperature drop (with the temperature rise, the process is reversed - desorption),
- solubility drop of pollutants,
- increase contact time in the bed.

A comparison of the chemical absorption and adsorption on activated carbon methods is presented in table 1.

Table 1. Comparison of deodorizing methods [2, 4, 5, 6]

Method	Advantages	Disadvantages	Field of application
Chemical absorption	• Flexibility for variable inlet conditions • Process stability • High efficiency at high concentrations of pollutants • No adaptive time • Full automation of the process • Low pressure drop of sprayed liquid • Simple construction	• High investment costs • Consumption of chemicals • Operating costs • The presence of disruptive sewage, • Need for costly, corrosion-resistant materials • The presence of aggressive and dangerous reagents	- fat industry - tannery - tobacco fermentation - production of pharmaceuticals - production of fatty acids - waste water treatment - production of dyes, amines, phenolic resins, methyl methacrylate
Adsorption on activated carbon	• Low investment costs • Small surface • High performance at the beginning of the process	• High operating costs • Consumption and regeneration of activated carbon • Narrow application range • No ammonia removal, ammonia and odorant removal • High sensitivity to changes in temperature and humidity	- sewage treatment plants - fat industry - food industry - production of paints and varnishes - production of pesticides - thought the paper

Most of the described deodorizing installations are characterized by a multistage process. In the first stage, acidic sorption solutions (removal of alkaline impurities such as ammonia) are used, and in subsequent steps - e.g. NaOH solutions and sodium hypochlorite (removal of sulfur compounds) are used. The danger of emissions of organic chlorine compounds is avoided by ozone usage as an oxidizing agent.
These methods are used primarily in the deodorization of gases from wastewater treatment processes as well as in waste management, either alone or as a stage of the purification system. It is common practice to use spray scrubbers in combination with biological methods to moisturize and cool gases, which then go to the biofilter (bed drying prevention) and also reduce the acidification of the bed.

2.2 Oxidation methods

Methods of oxidation of odorous pollutants include: thermal oxidation, catalytic oxidation, and plasma methods [7]. Thermal oxidation is a process of oxidation of combustible gases at a high temperature in the presence of oxygen. Depending on the type and amount of pollutants, the oxidation process is carried out directly in flame or thermally. The efficiency of the thermal combustion is high, almost 100% [8]. During catalytic oxidation, the oxidation of gases takes place on the surface of the catalyst. This gives the same level of combustion at lower temperatures, and, therefore with lower fuel consumption [9]. The main parameters for designing of thermal or catalytic oxidation systems are: residence time, temperature and mixing. The average residence time for the oxidation of aroma compounds is from 0.25 to 0.6 s [10]. Non-thermal oxidation processes are: ozone, UV, non-thermal plasma. The non-thermal plasma technology uses strong alternating electrical currents or microwave radiation. So, produced plasma results in the decomposition of gaseous pollutants [11]. Comparison of oxidation methods is shown in Table 2.

Table 2. Comparison of oxidation methods [5]

Method	Parameters	Advantages	Disadvantages	Field of application
Thermal oxidation	- Temperature: 923-1173 K - Concentration of pollutants: high - Continuous measurement of carbon dioxide, oxygen, carbon monoxide and nitrogen oxides in waste gases	- Broad scope of application - Heat recovery	- High fuel consumption - Pre-treatment of gases - humidity reduction, dust removal, removal of solid and liquid particles - soot formation	- waste incineration plants - treatment of animal waste - coffee roasters - paintshop - production of synthetic rubber - refining of oil
Catalytic oxidation	- Temperature: 623-673 K - The need to use a catalyst - Concentration of pollutants: high	- High efficiency - Heat recovery - The gases may be slightly dusty or slightly damp - For the purification of hot gases	- Limited flow of purified air - Necessity to remove solids - Sensitive catalyst - Formation of sulfur dioxide and hydrogen chloride - The need for additional purification; - The need to ensure a constant concentration of pollution	- treatment of animal waste - plastics processing - coffee roasters - production of synthetic rubber - refining of oil - paintshop
Non-thermal plasma	- Reactive radicals and ions react with odorous compound - Electric power supply to the high voltage;	- For all gaseous pollutants - No waste	- The formation of excess ozone	- sterilization and disinfection - paintshop - combustion of medical waste processes - chemical processes - food industry

Studies on the use of photocatalytic oxidation were carried out using ammonia as an odorant [12]. Typical catalyst reactivity was compared with catalyst activity with additional UV radiation. The results are shown in Table 3.

Table 3. Results of photocatalytic oxidation of ammonia [12]

Catalyst composition	Efficiency	Reaction temperature without UV	Reaction temperature with UV
5%Cu/5%TiO_2/SiO_2	100 %	500 °C	350 °C
5%Cu/10%TiO_2/SiO_2	90 %	410 °C	300 °C

In the study of the use of plasma techniques in the reduction of odors [13] reverse discharge reactor (I-7), electric barrier-discharge plasma with a grid electrode (W-1) and electric barrier-discharge plasma with a spiral electrode (IEL-4) were compared. The results obtained for the reactors are summarized in table 3 and showed that for the IEL-4 reactor with decreasing propane concentration the deodorant efficiency decreased. Better results were obtained for the Sz-1 reactors, but they were characterized by smaller gas flows and greater energy input.

Table 4. Results of research on the use of plasma techniques [13]

Reactor	Pollutant	Gas flow	Ionization energy	Efficiency
I-7	acetone ethanol toluene xylene	215 l/h	600 J/l	> 50 %
Sz-1	cyclopentane pentane cyclohexane	370 l/h	10000 J/l	> 50 %
		36 l/h	36000 J/l	90 %
IEL-4	propane	44000 l/h	20 J/l	75 %

For the IEL-4 reactor with decreasing propane concentration, the deodorant efficiency decreased. Better results were obtained for the Sz-1 reactors, but they were characterized by smaller gas flows and greater energy input.

In municipal economy oxidation methods are used when the concentration of pollutants is very high, eg the combustion of fermentation gas or the catalytic oxidation of gases from waste incineration. Plasma methods, which allow simultaneous disinfection and sterilization, are also of great potential.

Ozone has excellent oxidizing properties, but it is characterized by high toxicity. Therefore, the main limitation of the use of ozone in the gas phase in the deodorization of gases is due to the need to remove large quantities of unreacted ozone. Better results are obtained with the use of liquid phase ozonation, what results in higher costs [14]. This method is recommended when gases contain non-biodegradable pollutants and it is not possible to use biological methods [15]. The ozone performance (at 50 ppm ozone concentration) is shown in Table 5.

Table 5. Results of photocatalytic oxidation of ammonia [15]

Compound	Efficiency, %
trichloroethene	35,5

tetrachloroethene	94,8
benzene	99,2
cyclohexane	97,5
toluene	99,1
chlorobenzene	99,8
1,2-dichlorobenzene	99,0
1,3,5-trichlorobenzene	99,9

2.3 Neutralization methods

In neutralization methods (modification of the aroma), there can be distinguished such methods as fogging, wet barrier and odor masking. Each of them relies on the introduction of volatile compounds, that changes the nature of the odor or the perceived olfactory sensation [5, 16]. Fogging is a method consisting of spraying of appropriate chemicals in the gas stream. It is used for surface sources, especially when it is not possible to capture gases in organized emission [17]. The technology has been created to reduce odor nuisance, above all, in open space (diffuse emission) for example: sludge fields, landfills of wastes, and space closed (organized emission) for example transshipment halls and waste sorting plants.

Table 6 summarizes the main advantages and disadvantages of odor neutralizing methods and examples of their applicability.

Table 6. Advantages, disadvantages and Field of application of odor masking [5]

Advantages	Disadvantages	Field of application
- moderate costs - ease of use and mobility - small storage area	- sensitivity to changes in an odor concentration - possibility of secondary odor formation - dependence on the distance from the odor source	- modernization of the installation - as a temporary solution - for a short period of time / seasonally - when it is possible to apply directly on the odor source

Table 7 shows the sample results of laboratory tests on the masking of unpleasant odors using essential oils. In the presented studies [18, 19, 20], pine, lemon, and orange flavoring oils were used in two different concentrations of sewage sludge from municipal sewage treatment plants. The results indicate that, as the dose of the oil increases, the odor retention time increases. Relatively the longest retention times were found in pine oil.

Table 7. Results of research on the use of essential oils form odor masking [18, 19, 20]

Type of essential oil	Odor retention time for essential oil dose [min]	
	0,06 cm^3/100 cm^3 sludge	0,18 cm^3/100 cm^3 sludge
pine	14	30
lemon	7	24
orange	5	25

Masking odors is not a deodorizing method. Sprayed fragrances cover the emitted odors, but they do not remove odorogenic compounds, and, by reacting with impurities, may cause secondary odor nuisances. Masking odors should therefore only be a temporary solution, eg during maintenance or failure of the gas purification system.A particular application may also be found in waste management in composting processes in closed pits when the odor emission is temporary (during throwing pits).

3 Summary

It is best to prevent odor emissions by preventing them from the formation. In the case of communal facilities it is not always possible, so the solution to the problem of odorant pollution is different methods of deodorization. The adsorption and absorption methods used for gases that are highly humid are commonly used, but the operating costs of these types of installations are relatively high. A very effective method is to combust contaminated gases, but for economic reasons, this technology is mostly used only for high temperature gases. There are also a number of indirect solutions (masking, fogging). However, they should only be used in the form of supplements to control non-toxic odorants which are characterized by low concentrations.

This work was supported by a statutory research No. 0401/0003/17

References

1. T. Pfliegensdörfe, M. Szołtys, Komunalny PLUS, **4**, (2015)
2. J. Kwaśny, W. Balcerzak, Archives of Waste Management and Environmental Protection, vol. **16**, 4, (2014)
3. M. Szołtys, Metody usuwania złowonnych zanieczyszczeń gazowych – porównanie technologii, http://tholander.pl/artykuly/metody-usuwania/ [available 10.11.2017]
4. I. Sówka, P. Zwoździak, A. Zwoździak, J. Zwoździak [in :]: Ekotoksykologia w ochronie środowiska. Polskie Zrzeszenie Inżynierów i Techników Sanitarnych. Oddział Dolnośląski, (2008)
5. *Department of Air and Climate Protection, Code of anti-odor nuisance, Warsaw 2016*
6. L. Wang, J. R. Taricska, Y. Hung, J. Eldridge, K. Hung Li, Air Pollut. Control Eng., **1**, 2004
7. M. Słomińska , S. Król, J. Namieśnik, Crit Rev Env Sci Tec, **43,** (2013)
8. M.I. Szynkowska, E. Wojciechowska, A. Węglińska, T. Paryjczak, Przem Chem, **88**, 6 (2009)
9. G. Busca, C. Pistarino, J Loss Prevent Proc, **16** (2003)
10. D. Gabriel, X. Gamisans, R. Muñoz and J. Lafuente, [in:] J.M Lema; S. Suárez, Innovative wastewater treatment & resource recovery technologies : impacts on energy, economy and environment. IWA Publishing, London, UK (2017)
11. M. Schlegelmilch, J. Streese, R. Stegmann, Waste Manage, **25**, (2005)
12. M.I. Szynkowska, E. Leśniewska, E. Maćkiewicz, A. Węglińska, Z. Gorzka, J. Rogowski, J, Góralski, Ł. Kowalczyk, W. Maniukiewicz, A. Żarczyński, T. Paryjczak, [in:] M.I. Szynkowska, J. Zwoździak, Współczesna problematyka odorów, (Wyd. Naukowo-Techniczne, Warszawa 2010)
13. W. Mielcarek, R. Kacprzyk, J. Subocz, J. Rutkowski, G. Gryglewicz, S. Gryglewicz, K. Prociów, J. Warycha, T. Czapka, [in:] M.I. Szynkowska, J. Zwoździak, Współczesna problematyka odorów, (Wyd. Naukowo-Techniczne, Warszawa (2010)

14. J. Waluś, P. Tatoj, T. Kaczyńska, M. Palica, K. Chmiel, Annual Set the Environment, (2001)
15. J. Ozonek [in:] M.I. Szynkowska, J. Zwoździak, Współczesna problematyka odorów. Wyd. Naukowo-Techniczne, Warszawa (2010)
16. Description of bioArcus technology, www.bioarcus.pl [available 15.11.2017]
17. I. Sówka, J. Zwoździak, M. Szklarczyk, A. Nych, M. Skrętowicz [in:] M.I. Szynkowska, J. Zwoździak, Współczesna problematyka odorów. Wyd. Naukowo-Techniczne, Warszawa (2010)
18. A. Kowalczyk, J. Kutryn, T. Piecuch, Annual Set the Environment, **12**, (2010)
19. A. Kowalczyk, T. Piecuch, Inżynieria ekologiczna, **25**, (2011)
20. A. Kowalczyk, T. Piecuch, L. Andriyevska, Annual Set the Environment, **15**, (2013)

E3S Web of Conferences **28**, 01024 (2018) https://doi.org/10.1051/e3sconf/20182801024
Air Protection in Theory and Practice

Seasonal variations in chromium concentration in urban atmospheric aerosol in the city of Radom

Artur Molik[1]*, *Marzena* Trojanowska[1], *Monika* Łożyńska[2] and *Ryszard* Świetlik[1]

[1]Kazimierz Pulaski University of Technology and Humanities, Department of Environmental Protection, 26-600 Radom, B. Chrobrego 27, Poland
[2]Warsaw University of Technology, Faculty of Building Services, Hydro and Environmental Engineering, 00-653 Warsaw, Nowowiejska 20, Poland

Abstract. This paper presents seasonal variations in the occurrence of chromium in urban particulate matter (PM). Radom is an interesting area for this type of research due to its typical antiquated heating infrastructure. PM samples of two particulate fractions PM2.5 (particulates ≤ 2.5 μm) and PM10 (particulates ≤ 10 μm) were collected using a Copley cascade impactor Hi-Flow MOUDI Model 130. Chromium was determined in the samples using the GF-AAS technique. The concentration of chromium in PM2.5 was in the range of 0.38 to 3.2 ng/m^3, whereas in PM10 it ranged from 0.43 to 4.1 ng/m^3.

1 Introduction

Atmospheric particulates are a dominant form of urban air pollution. They originate from both natural and anthropogenic sources. It is assumed that the coarse fraction (2.5 - 10 μm) is of natural origin whereas anthropogenic processes are the source of the fine fraction (≤ 2.5 μm). The distribution of particle sizes and chemical compositions determine the impact of the atmospheric aerosol on the environment. The composition of particulate matter is complex and heterogeneous [1, 2]. Particulate matter with particle diameters of below 2.5 μm is very dangerous for living organisms. It penetrates deep into the alveoli from where it can enter the circulatory system [2, 3]. Various toxic compounds are transported with the particles. Heavy metals are the group of pollutants to which particular attention is paid [2-4].

Heavy metals can be found primarily in the fine fraction [3-5]. They are generally regarded as toxic for living organisms. Several of them, e.g. Cr, Cd and Ni, are carcinogenic. Therefore, they are permanently monitored in the environment [3, 4]. The characteristic feature of heavy metals is their stability – they do not undergo biodegradation. They usually remain in the environment in the form in which they were emitted. Their bioaccumulation increases their danger to living organisms [6, 7].

* Corresponding author: artur.molik@uthrad.pl

Chromium is one of the heavy metals whose presence in excessive amounts is a threat to the biotic component of the environment. Anthropogenic sources are responsible for 60% to 70% of chromium emission to the atmosphere while natural sources emit 30% to 40% [8]. With a share of about 42%, non-industrial combustion processes play a dominant role in the chromium emission in Poland. Combustion processes in industry, manufacturing processes and combustion processes in the energy production and transformation, 19%, 17% and 13% respectively, are considerably less important. The share of road transport has been increasing steadily in recent years with about 9% in 2015 [9].

There are many publications on the occurrence and speciation of heavy metals, including chromium [2-7]. There has been an increasing interest in seasonal variations in the concentration of heavy metals in atmospheric aerosol and its probable effect on the estimation of inhalation hazard for city dwellers [11-18, 21, 24]. There is little interest in chromium in this respect, particularly in domestic literature [19-20, 22-23].

This work presents the results of the study of seasonal fluctuations in the occurrence of chromium in urban atmospheric aerosol in the city of Radom. The aim of the study was to diagnose main chromium emission sources and to create a basis for prediction of air pollution in Radom.

2 Materials and methods

2.1 Study area and sampling collection

Radom is a medium-sized city (pop. 217,000) located in the southern part of the Mazowsze province. It covers an area of 111.8 km^2. The city is an interesting site for this type of study due to its outdated heating system which is typical of Poland.

The aerosol sampling site was located on the grounds of the Faculty of Materials Science and Design of the University of Technology and Humanities in Radom (Fig. 1). Aerosol samples were collected in four successive seasons (2015/2016) using a Hi-Flow MOUDI Model 130 cascade impactor (Copley Scientific). Using the device it is possible to separate particle fractions PM10 and PM2.5.

The particulates were deposited on cellulose filters produced by POCh. Before and after collection of aerosol samples the filters were weighed three times using a Mettler Toledo MX5 microbalance and a mean mass of the filter was determined on this basis. The rate of air flow through the impactor was 100 ± 2.5 L/min. Each sample was collected during seven days, the sampler operated for 12 hours per day – each working hour was followed by a one-hour break for cooling of the pump which sucked in the air.

2.2 Analytical procedures

A Milestone MLS 1200 Mega microwave digestion unit was used to extract chromium from the collected samples of atmospheric particulates. One fourth of the filter was placed in a Teflon vessel containing 5 mL 65% HNO_3 and 1 mL 30% H_2O_2. The samples were digested under standard conditions. After digestion the solutions were filtered into 25 mL volumetric flasks which were then made up to the mark with deionized water.

Fig. 1. The location of the sampling site in the Radom city.

The determination of the chromium concentration in the digests from individual particle fractions of the particulates was carried out by means of a Perkin-Elmer AAS 3100 spectrometer equipped with a graphite furnace HGF 600. Standard solution of chromium (J. T. Baker Inc.) with concentration of 1000 μg/mL was used to calibrate the device. LOD (instrumental) was found to be 0.2 μg/L of Cr or 0.3 ng/m^3 as the concentration of Cr in air.

3 Results and discussion

The concentrations of PM10, PM2.5 and chromium are given in Table 1.

Table 1. Concentrations of PM10, PM2.5 and chromium.

Sample number	Sampling time	PM10		PM2.5	
		PM [μg/m^3]	Cr [ng/m^3]	PM [μg/m^3]	Cr [ng/m^3]
P.1	Spring	18	1.1	14	1.0
P.2		21	1.5	17	1.4
P.3		9.7	0.78	6.3	0.66
P.4	Summer	32	0.82	25	0.64
P.5		27	2.3	22	2.2
P.6		24	1.3	17	1.1
P.7		43	2.2	33	1.8
P.8		23	1.8	19	1.4
P.9		22	1.1	17	0.75

Table 1. Concentrations of PM10, PM2.5 and chromium (continued).

Sample	Sampling	PM10	PM2.5

number	time	PM [μg/m³]	Cr [ng/m³]	PM [μg/m³]	Cr [ng/m³]
P.10	Autumn	31	0.88	24	0.65
P.11		50	4.1	38	3.2
P.12		51	1.7	41	1.2
P.13		57	0.75	47	0.72
P.14		68	0.56	59	0.41
P.15		49	0.43	40	0.38
P.16		54	1.5	47	0.88
P.17		61	1.5	54	1.3
P.18		59	1.2	49	0.92
P.19	Winter	56	2.4	52	2.0
P.20		46	1.6	42	1.4
P.21		41	1.7	34	1.4
P.22		39	2.3	32	1.7
P.23		60	2.0	50	1.5
P.24		61	2.2	52	1.6
P.25		34	2.3	30	2.1

The concentration of PM10 ranged from 9.7 to 68 μg/m³. Its highest value was observed in autumn whereas the lowest one was found in spring. Distinct seasonal variations can be observed in the occurrence of PM10. Mean concentrations of this fraction decreased in the order: autumn (53.3 μg/m³) > winter (48.1 μg/m³) > summer (28.5 μg/m³) > spring (16.2 μg/m³) (Fig. 2). The concentration of PM2.5 ranged from 6.3 to 59 μg/m³. The highest concentration of this fraction was found in autumn and the lowest one was observed in spring. The concentrations of PM2.5 also underwent seasonal changes. Its mean concentrations decreased in the order: autumn (44.3 μg/m³) > winter (41.7 μg/m³) > summer (22.2 μg/m³) > spring (12.4 μg/m³). PM2.5 concentrations were lower than the corresponding PM10 concentrations but the differences were not statistically valid.

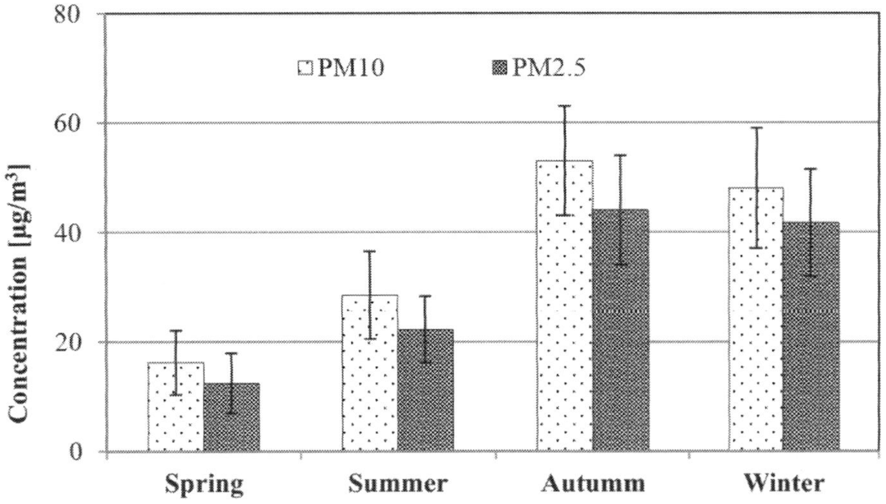

Fig. 2. Concentrations of PM2.5 and PM10 in urban air in Radom.

Seasonality also had an effect on the content of PM2.5 in PM10. In the spring-summer period the ratio of PM2.5/PM10 was 76% and 78%, respectively. In the autumn-winter period the ratio was higher: 83% and 87%, respectively.

The determined PM10 and PM2.5 concentrations are typical of urban areas in various parts of the world [19-24]. Table 2 shows a comparison of the mean values obtained in this study with the data from other publications.

Table 2. Comparison of mean concentrations of PM10, PM2.5 and Cr in various cities in the world.

City (Country)	Sampling time	PM10		PM2.5		References
		PM [μg/m³]	Cr [ng/m³]	PM [μg/m³]	Cr [ng/m³]	
Radom (Poland)	Spring	16.2	1.13	12.4	1.02	This study
	Summer	28.5	1.59	22.2	1.32	
	Autumn	53.3	1.40	44.3	1.07	
	Winter	48.1	2.07	41.7	1.67	
Zabrze (Poland)	Winter	42	4.98	-	-	[19]
Kraków (Poland)	Winter	-	-	50	7.1	[20]
	Summer	-	-	16	<LOD	
Kraków (Poland)		-	-	22	2.5	[22]
Zabrze (Poland)	Winter	-	-	228	12-54	[23]
Vienna (Austria)		30.4	0.8	21.5	0.4	[24]
Oporto (Portugal)		53.2	6.08	38.8	3.05	[21]
Lecce (Italy)	Warm season	24.8	-	15.2	-	[12]
	Cold season	34.7	-	22.5	-	
Wuhan (China)	Winter	-	-	142.9	8.97	[11]
	Spring	-	-	132.7	9.04	
	Autumn	-	-	101.8	8.20	
	Summer	-	-	48.6	5.41	
Changsha (China)	Winter	170	26.5	-	-	[15]
Hong Kong (China)	Winter	-	-	-	4.1	[18]
	Summer	-	-	-	6.8	
	Summer			16	<LOD	

The concentration of chromium in PM10 ranged from 0.43 to 4.1 ng/m³ (Table 1). Its highest value was found in winter while its lowest value was observed in spring. Seasonal changes can also be observed in the occurrence of chromium in this fraction. Its average concentrations decreased in the order: winter – 2.07 ng/m³, summer – 1.59 ng/m³, autumn 1.40 ng/m³ and spring - 1.13 ng/m³ (Fig. 3).

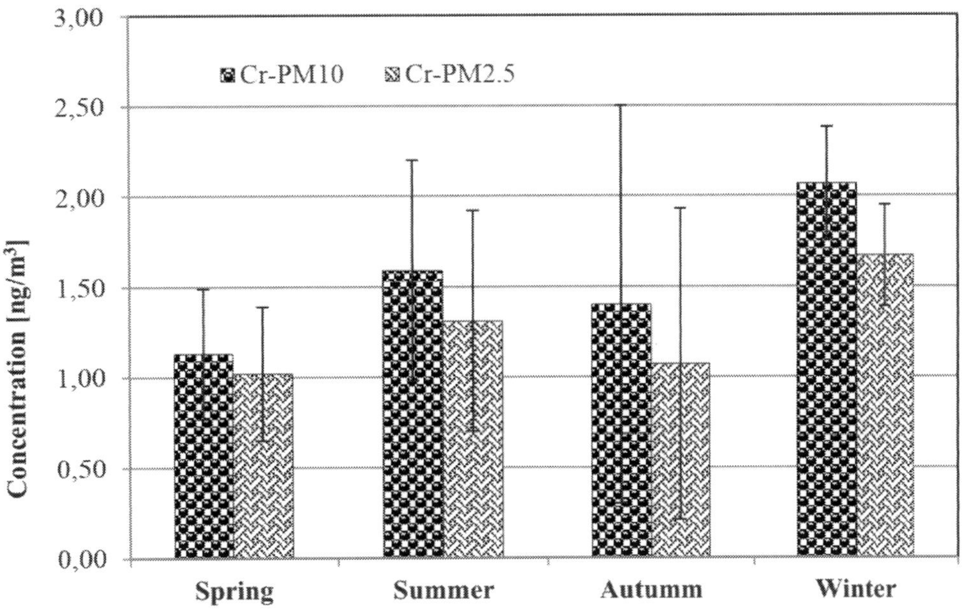

Fig. 3. Concentration of chromium in PM2.5 and PM10 fractions.

The concentration of chromium in PM2.5 varied in the narrow range of 0.38 to 3.2 ng/m^3 (Table 1). Its highest concentration occurred in winter and the lowest in spring. In the case of chromium present in PM2.5 it is also possible to observe the effect of seasonality on its average concentration: winter – 1.67 ng/m^3 > summer – 1.32 ng/m^3 > autumn – 1.07 ng/m^3 > spring – 1.02 ng/m^3. It is, however, worth noting that statistically important differences refer only to mean chromium concentrations in the winter-spring period. The characteristic feature of the presence of chromium in the urban air in Radom is also a dominant share of the respirable fraction (76% - 90%) (Fig. 4). This is a very unfavourable situation for Radom inhabitants, as it increases health risk connected with inhalation hazard.

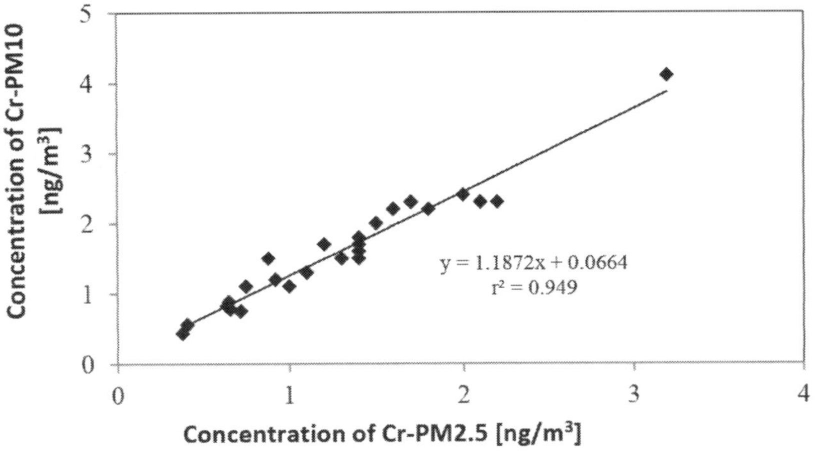

Fig. 4. Interdependence of concentrations of Cr-PM2.5 and Cr-PM10 fractions in urban air in Radom in a one-year period.

The concentrations of chromium in PM10 and PM2.5 are comparable to those quoted by other authors for urban areas [19-22, 24].

The obtained results of PM2.5 and PM10 concentrations confirm the opinion of Regional Inspectorate for Environmental Protection that the combustion processes in the communal and living sector are the main source of PM in Radom.

It is probable that the observed increase in the concentration of chromium in the winter period results from the same cause. However, without additional analyses, it is difficult to unequivocally conclude whether the main source is combustion in heating plants or burning of fuels in household stoves. The fact that the numbers of samples collected in each of the seasons varied does not facilitate drawing unequivocal conclusions.

4 Conclusion

- The results obtained confirmed seasonal changes in chromium concentration in Radom urban air.
- The highest values of chromium concentrations were observed in aerosol samples collected in winter.
- Chromium occurring in PM2.5 equals 76% to 90% of chromium present in PM10.

We gratefully acknowledge the financial support from the National Science Center (Poland) grant No. 2011/01/B/ST10/06757, which allowed us to carry out this study.

References

1. M. Malandrino, M. Casazza, O. Abollino, C. Minero, V. Maurino, Chemosphere **147**, (2016)
2. C. Samara, D. Voutsa, Chemosphere **59**, (2005)
3. S. M. Tagliani, M. Carnevale, G. Armiento, M. R. Montereali, E. Nardi, M. Inglessis, F. Sacco, S. Palleschi, B. Rossi, L. Silvestroni, A. Gianfagna, Atmos. Environ., **153**, (2017)
4. D. Jandacka, D. Durcanska, M. Bujdos, Transport. Res. D-Tr. E., **50**, 397-408, (2017)
5. M. J. Gatari, J. Boman, A. Wagner, S. Janhäll, J. Isakson, Sci. Total Environ., **368**, 675-685 (2006)
6. Ch. Hsu, H. Chiang, S. Lin, M. Chen, T. Lin, Y. Chen, Sci. Total Environ., **541**, (2016)
7. S.M. Talebi, Environ. Res., **92**, (2003)
8. Ch. Seigneur, E. Constantinou, Environ. Sci. Technol., **29**, (1995)
9. B. Dębski, A. Olecka, K. Bebkiewicz, Z. Chłopek, I. Kargulewicz, K. Kubica, J. Rutkowski, D. Zasina, M. Zimakowska - Laskowska, M. Żaczek, *Krajowy bilans emisji SO_2, NO_X, CO, NH_3, NMLZO, pyłów, metali ciężkich i TZO za lata 2014 - 2015 w układzie klasyfikacji SNAP* (Warszawa, 2017)
10. K.L. Mandiwana, N. Panichev, T. Resane, J. Hazard. Mater., **B136**, 379-382 (2006)
11. F. Zhang, Z. Wang, H. Cheng, X. Lv, W. Gong, X. Wang, G. Zhang, Sci. Total Environ., **518-519**, (2015)
12. D. Cesari, G. E. De Benedetto, P. Bonasoni, M. Busetto, A. Dinoi, E. Merico, D. Chirizzi, P. Cristofanelli, A. Donateo, F. M. Grasso, A. Marinoni, A. Pennetta, D. Contini, Sci. Total Environ., **612**, (2018)
13. M. Arhami, V. Hosseini, M. Z. Shahne, M. Bigdeli, A. Lai, J. J. Schauer, Atmos. Environ., **153**, (2017)
14. B. Ambade, Urban Climate, **9**, (2014)

15. K. Zhang, F. Chai, Z. Zheng, Q. Yang, J. Li, J. Wang, Y. Zhang, J. Environ. Sci-China., **26**, (2014)
16. A. Saffari, N. Daher, M. M. Shafer, J. J. Schauer, C. Sioutas, Environ. Pollut., **181**, (2013)
17. J. Gao, H. Tian, K. Cheng, L. Lu, Y. Wang, Y. Wu, Ch. Zhu, K. Liu, J. Yhou, X. Liu, J. Chen, J. Hao, Atmos. Environ., **99**, (2014)
18. S. Y. Jiang, D. S. Kaul, F. Yang, L. Sun, Z. Ning, Sci. Total Environ., **533**, (2015)
19. W. Rogula-Kozłowska, B. Kozielska, K. Klejnowski, S. Szopa, Arch. Environ. Prot., **39**, (2013)
20. L. Samek, A. Gdowik, J. Ogarek, L. Furman, Environment Protection Engineering, **42**, (2016)
21. K. Slezakova, M. C. Pereira, M. A. Reis, M. C. Alvim-Ferraz, J. Atmos. Chem., **58**, (2007)
22. L. Samek, L. Furman, M. Mikrut, A. Regiel-Futyra, W. Macyk, G. Stochel, R. van Eldik, Chemosphere **187**, (2007)
23. J. S. Pastuszka, W. Rogula-Kozłowska, E. Zajusz-Zubek, Environ. Monit. Assess., **168**, (2010)
24. H. Puxbaum, B. Gomiscek, M. Kalina, H. Bauer, A. Salam, S. Stopper, O. Preining, H. Hauck, Atmos. Environ., **38**, (2004)

Regions of pollution with particulate matter in Poland

Kacper Rawicki, *Małgorzata* Czarnecka, and *Jadwiga* Nidzgorska-Lencewicz

Department of Meteorology and Landscape Architecture, West Pomeranian University of Technology in Szczecin, ul. Papieża Pawła VI 3, 71-459 Szczecin, Poland

Abstract. The study presents the temporal and spatial variability of particulate matter concentration in Poland in the calendar winter season (December-February). The basis for the study were the hourly and daily values of particulate matter PM_{10} concentration from the period 2005/06 – 2014/15, obtained from 33 air pollution monitoring stations. In Poland, the obligation to monitor the concentration of the finer fraction of particles smaller than 2.5μm in aerodynamic diameter was introduced only in 2010. Consequently, data on $PM_{2.5}$ concentration refer to a shorter period, i.e. 2009/10 – 2014/15, and were obtained from 23 stations. Using the cluster analysis (k-means method), three regions of comparable variability of particulate matter concentration were delineated. The largest region, i.e. Region I, comprises the northern and eastern central area of Poland, and its southern boundary is along the line Gorzów Wlkp-Bydgoszcz-Konin-Łódź-Kielce-Lublin. Markedly smaller Region II is located to the south of Region I. By far the smallest area was designated to Region III which covers the south west area of Poland. The delineated regions show a marked variability in terms of mean concentration of both PM fractions in winter (PM_{10}: region I - 33 $\mu g{\cdot}m^{-3}$, region II - 55 $\mu g{\cdot}m^{-3}$, region III - 83 $\mu g{\cdot}m^{-3}$; $PM_{2.5}$: region I - 35 $\mu g{\cdot}m^{-3}$, region II - 50 $\mu g{\cdot}m^{-3}$, region III - 60 $\mu g{\cdot}m^{-3}$) and, in the case of PM_{10}, the frequency of excessive daily limit value.

1 Introduction

Almost all winter seasons in Poland, as in other countries in Central and East-Central Europe, are characterised by very high concentrations of particulate matter PM_{10} and $PM_{2.5}$. The main source of this phenomenon is increased emission of particulate pollutants generated by combustion processes in stationary sources, predominantly the municipal and household sector. Data provided by IOŚ-PIB [1] show that combustion of fuels for energy generation, mainly coal in stationary sources [2], accounts for approximately 75% of PM_{10} national emission and almost 100% of SO_2 national emission. Aerosanitary situation is determined by pollution emission values as well as by weather conditions which, on the one hand, result in intensification of heating process and, on the other, may contribute to effective dispersion or accumulation of pollutants leading to smog situations [3-9].

Particulate matter (PM), also known as particulates, is a complex mixture of different chemical components including water soluble ions, trace metals and organic compounds which originate from a wide range of natural and anthropogenic sources [10-12]. The negative effect of particulates on human health is widely documented in literature on the subject [13-16]. The current available data indicate that atmospheric PM pollution is the 6th (out of 43 taken into consideration) leading risk factor contributing to more than 3 million deaths worldwide every year [17]. The International Agency for Research on Cancer has classified air pollution in general, as well as PM as a separate component of air pollution mixtures, as carcinogenic [18]. The report by European Environmental Agency [19] states that in 2015, a total of 19 % of the EU-28 urban population was exposed to PM_{10} levels above the daily limit value (50 $\mu g \cdot m^{-3}$). Additionally, the report indicates that 7% of the EU-28 urban population was exposed to $PM_{2.5}$ levels above the EU limit value. Taking into account the stricter guidelines of WHO, these percentages increase accordingly for PM_{10} - up to 53%, for $PM_{2.5}$ - up to 82%.The latest report by WHO [20] on air quality in cities is even more alarming. The assessment is based on mean annual $PM_{2.5}$ concentration in cities and, according to the report, 33 out of 50 most polluted cities in Europe are located in Poland. Among the EU-28 member states, Poland is ranked as being the 3rd country with the highest estimated number of premature deaths (46 020 a year!) attributed to $PM_{2.5}$ [19].

The aim of the present paper is to determine the regions of Poland which are characterised by similar variability of PM concentration in the winter season.

2 Materials and methods

The study was based on hourly and daily values of particulate matter PM_{10} concentration in the calendar winter season (December-February) in the period 2005/06 – 2014/15, obtained from 33 air pollution monitoring stations. In Poland, the obligation to monitor the concentration of the finer fraction of particles smaller than 2.5μm in aerodynamic diameter was introduced only in 2010. Consequently, data on $PM_{2.5}$ concentration refer to a shorter period, i.e. 2009/10 – 2014/15, and were obtained from only 23 stations with complete measurement series. The values of hourly and daily concentration were characterised per various time steps (month, season, multiannual period) with the use of mean and maximum values, standard deviation, upper and lower quartile, coefficient of variation, and the number of days on which the applicable EU limit value for PM_{10} was exceeded. The aforementioned characteristics of PM_{10} which were obtained from a longer measurement series and additionally refer to the finer fraction of particulate matter (2.5 μm) were analysed using the cluster analysis method. The present study employs *k*-means (Euclidean distance) which belongs to non-hierarchical cluster analysis methods. This method allows determination of *k* of the clusters characterised by the possibly greatest difference. The analysis was made using Statistica 12 software which includes implementation of a v-fold cross-validation test allowing identification of the optimum number of clusters. Various combinations (sets) were analysed, taking into consideration all as well as some of the aforementioned PM_{10} characteristics, according to hourly and daily values. In the end, three regions and their areas were determined on the basis of compilation of four characteristics: lower and upper quartile, variation coefficient of hourly concentration values, and the number of days on which the daily limit value was exceeded. The fact that identical results were obtained in the analysis which additionally included other characteristics of PM_{10} concentration, i.e. means and standard deviation (also for daily means), only confirms the obtained delineation of the regions.

3 Results and discussion

The area of the three regions of pollution with particulate matter in the calendar winter season in presented in Fig. 1 Region I was found to be the largest in area, covering the Pomeranian and Masurian Lakeland, and the Masovian Lowland. Region II was generally classified as covering the Greater Poland Lowland, Silesian Lowland, and the Silesian Upland, Lesser Poland and Lublin Upland. Region III was found to be by far the smallest in area, and comprised 5 cities located in the south of Poland in poorly ventilated basins and well developed river valleys.

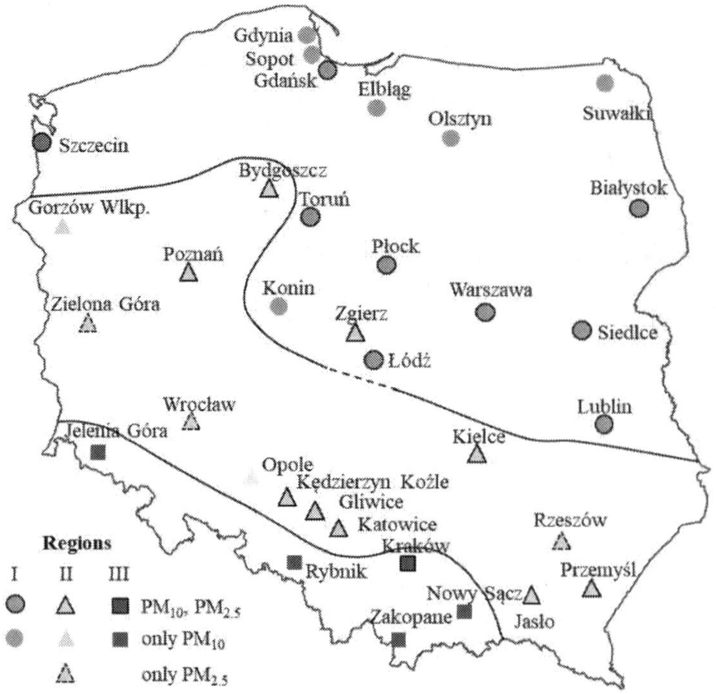

Fig. 1. Regions of pollution with particulate matter during calendar winter (Dec-Feb).

Variation in terms of pollution with PM between the delineated regions is most clearly demonstrated by PM_{10} concentrations in winter. In region I, the concentration is approximately 33 $\mu g \cdot m^{-3}$. By comparison, in region II the concentration is approximately 1.5

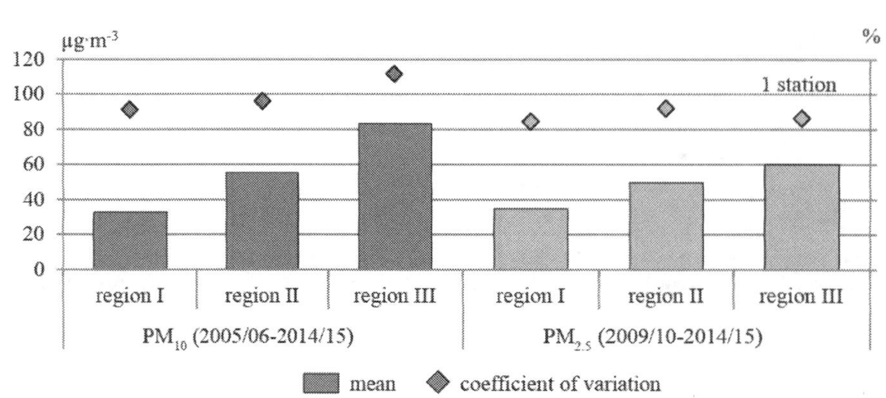

Fig. 2. Mean ($\mu g \cdot m^{-3}$) and coefficient of variation (%) of hourly concentrations of particulate matter in designated regions.

times higher and in region III it is as much as 2.5 times higher. The differences between mean $PM_{2.5}$ concentrations are significantly smaller. This can result from the fact that the data series was shorter and did not include the four winter seasons from 2005/06 to 2008/09, which were characterised by high variability in PM_{10} concentration (Tab. 1, Fig. 2).

The most polluted region III is distinguished by not only higher average particulate matter concentration but also by much higher variability of PM_{10} concentration. In this region, variation coefficient of hourly PM_{10} concentrations between the winter seasons amounts to approximately 112%, whereas in regions I and II it does not exceed 100%. In the latter regions, greater difference in terms of variability of particulate matter concentration are found for $PM_{2.5}$ fraction rather than PM_{10}.

Table 1. Mean and maximum hourly concentrations of PM_{10} ($\mu g{\cdot}m^{-3}$) in designated regions during winter (Dec-Feb) from 2005/06 to 2014/15.

Winter (XII-II)	Region I		Region II		Region III	
	mean	max	mean	max	mean	max
2005/06	41.2	599.8	74.0	769.0	105.9	1052.0
2006/07	21.6	252.3	32.5	354.0	51.6	455.0
2007/08	25.8	460.9	43.8	400.4	79.4	671.0
2008/09	34.0	375.8	62.2	865.4	77.3	618.0
2009/10	40.4	487.5	69.0	732.0	101.5	771.0
2010/11	33.6	344.3	66.5	687.0	105.4	763.0
2011/12	32.0	705.0	52.6	639.0	78.0	842.0
2012/13	34.9	483.6	56.2	552.0	86.2	1258.0
2013/14	31.5	328.7	48.7	456.7	74.4	847.0
2014/15	30.5	433.6	46.4	425.3	70.9	985.0

In all regions, on average, the highest PM_{10} concentration was recorded in the winter season of 2005/06, and the lowest (by almost two times) in the subsequent winter season of 2006/07 which, according to thermal conditions classification by Lorenc [21], was classified as very warm (Tab. 1). In this particular season, across the country, also the absolute maximum hourly concentration values were by far the lowest in the 10-year long period under analysis. The maximum hourly concentrations were recorded in all of the regions in different winter seasons, yet in each case the values exceeded the mean seasonal values by more than 10 times. In the analysed period, the differences between the mean seasonal PM_{10} concentration did not exceed 20 $\mu g/m^3$ in region I, in region II the difference was twice as high, and in region II – approximately 54 $\mu g/m^3$.

The regions also demonstrate substantial variation in terms of the frequency of the daily PM_{10} limit value being exceeded (Fig. 3). In the most polluted region III, excessive daily concentration is recorded on more than half of the days of the calendar winter, in region II – approximately 38%, and in region I – only on 16% of days. The first of the analysed periods, i.e. 2005/06, was characterised by on average the highest PM_{10} concentration values and frequent exceeding of the daily limit value. However, in other winter seasons

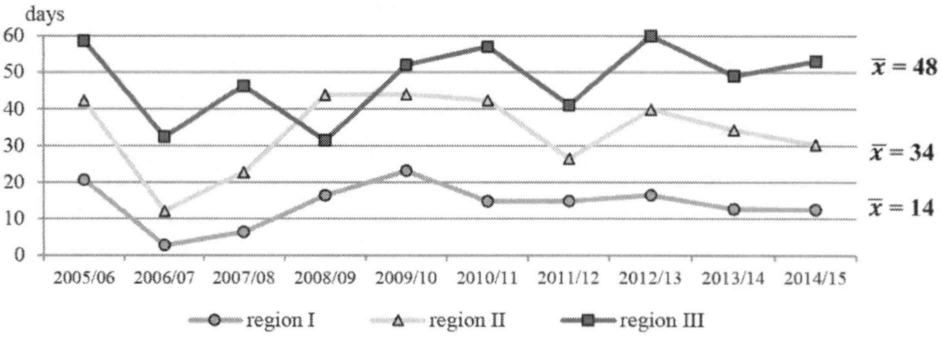

Fig. 3. Number of days exceeding the daily limit value of PM_{10} in designated regions during the winter (Dec-Feb) from 2005/06 to 2014/15.

under analysis, the UE limit value was exceeded with comparable and at times even higher frequency, depending on the region. In region III the daily limit value was exceeded with similar frequency (on approximately 2/3 of the days) also in 2012/13. In region II EU limit value was exceeded on more than 40 days in subsequent winter seasons, i.e. 2008/09, 2009/10 and 2010/11, and in region I in the winter of 2009/10.

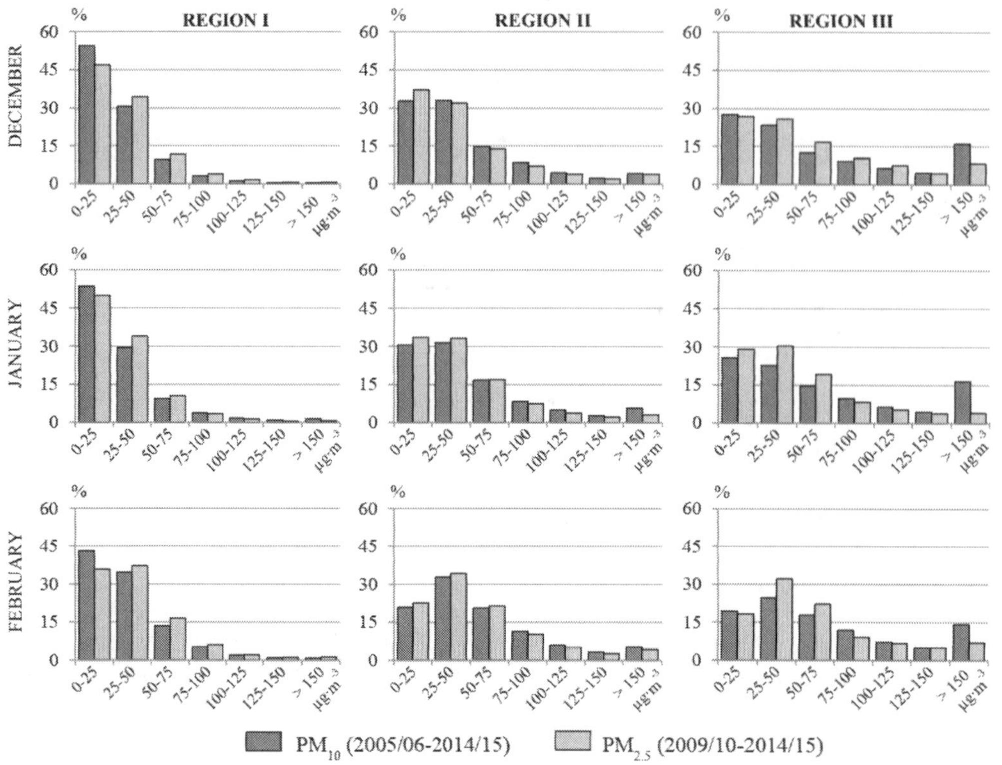

Fig. 4. Frequency distribution of hourly concentrations of particulate matter in designated regions by month.

The characteristics of both fractions of particulate matter is supplemented by frequency histograms of the adopted hourly concentration values according to the three months of the calendar winter season (Fig. 4). In all regions, distribution of the adopted concentration range in December and January shows some similarity, yet in February the distribution is slightly different. In region I, the prevalent concentration values do not exceed 25 μg/m^3 – recorded on more than half of the days in December and January. In regions II and III, in both aforementioned months, the concentration values of 25-50 μg/m^3 were recorded only slightly less frequently. However, in February, such concentration values are predominant particularly in region II. In region I, hourly particulate matter concentration of more than 100 μg/m^3 accounted for less than 4% of the cases in all winter months.

4 Summary

The extent of the regions marked by specific values of pollution with particulate matter in the calendar winter season is generally consistent with the increasing from the north to the south emission values and worsening conditions of pollutants dispersion and deposition. The increase in emission results from high concentration of industrial, as well as household and traffic sources due to high population density [22, 23]. The degree of dispersion is dependent on atmospheric conditions and topographic factors [2, 3, 5-9, 24, 25]. Poor conditions for natural ventilation in the south of Poland are associated with low wind speed, greater frequency and duration of atmospheric calm, less frequently occurring unstable equilibrium, and more frequent surface inversions [4, 24, 25], including orographic inversions.

The three regions delineated on the basis on the results of the 10-year long measurement series of PM_{10} particulate matter concentration are predominantly characterised by differences in mean seasonal concentration values and frequency of exceeding the daily limit value. The results are representative as the analysed winter seasons include not only the winter seasons marked by average thermal conditions, but the winter seasons classified as very warm as well as cold. Undoubtedly, extending the measurement series and including a greater number of stations monitoring $PM_{2.5}$ fraction would allow to revise the delineation of the regions determined on the basis of the results for the calendar winter seasons in the period 2005-2015.

References

1. Instytut Ochrony Środowiska – Państwowy Instytut Badawczy (IOŚ-PIB), Warszawa, *Krajowy bilans emisji SO_2, NO_X, CO, NH_3, NMLZO, pyłów, metali ciężkich i TZO w układzie klasyfikacji SNAP i NFR. Raport podstawowy* (2015)
2. M. Reizer, K. Juda-Rezler, Air Qual Atmos Health, **9**, 5, (2016)
3. M. Czarnecka, J. Nidzgorska – Lencewicz, Annual Set The Environment Protection, **17**, (2015)
4. J. Nidzgorska – Lencewicz, M. Czarnecka, Theor Appl Climat, **119**, 3-4, (2015)
5. M. Czarnecka, J. Nidzgorska – Lencewicz, Env. Prot. Eng. **44**, 2, (2017)
6. T. Olszowski, Ecol Chem Eng S. **22**, 3, (2015)
7. J. Jędruszkiewicz, B. Czernecki, M. Marosz, Inter. J Environ. Health Res., (2017)
8. B.Czernecki,M. Połrolniczak, L. Kolendowicz, M. Marosz, S. Kendzierski, N. Pilguj, J Atmos Chem, **74**, (2017)
9. R. Oleniacz, M. Bogacki, M. Rzeszutek, A. Kot, Ochrona powietrza w teorii i praktyce (red. J. Konieczyński) **2**, (2014)
10. W. Rogula-Kozłowska, G. Majewski, P.O. Czechowski, Environ Monit Assess, **187**, (2015)
11. G. Majewski, W. Rogula-Kozłowska, Theor Appl Climatol **125**, (2016)

12. A. Zwozdziak, M.I. Gini, L. Samek, W. Rogula-Kozlowska, I. Sowka, K. Eleftheriadis, J Aerosol Sci, **103**, 38–52 (2017)
13. K. Widziewicz, W. Rogula-Kozłowska, K. Loska, Atmos Pollut Res, **7**, 884-894 (2016)
14. M. Tainio, J. Kukkonen, Z. Nahorski, Arch Environ Prot, **36**,1, 29-39 (2010)
15. L. Shi, A. Zanobetti, I. Kloog, B. A. Coull, P. Koutrakis, S. J. Melly, J.D. Schwart, Environ Health Perspect, **124**,1, 46–52 (2015)
16. F. Dominici, R.D. Peng, M.L. Bell, L. Pham, A. McDermott, S.L. Zeger et al., JAMA, **295**, 10, 1127–1134 (2006)
17. S.S. Lim, T. Vos, A.D. Flaxman, G. Danaei et al., Lancet, **380**, 2224–2260, (2012)
18. International Agency for Research on Cancer (IARC), *Outdoor air pollution a leading environmental cause of cancer deaths*, Press Release **221** (2013), (http:// www.iarc.fr/en/media-centre/iarcnews/pdf/pr221_E. pdf) accessed 31 October 2017.
19. European Environment Agency (EEA), *Air quality in Europe - 2017 report*, **13** (2017)
20. World Health Organization (WHO), *WHO global urban ambient air pollution database (update 2016). Ambient (outdoor) air pollution database, by country and city*
(http://www.who.int/phe/health_topics/outdoorair/databases/cities/en/) accessed 31 October 2017
21. H. Lorenc, *Termiczno-opadowa ocena klimatycznych sezonów roku w Polsce oraz tendencje czasowo-przestrzenne*. Projekt badawczy M-9, IMGW, Maszynopis, (2000)
22. Central Statistical Office (GUS), *Environment*. Statistical Information and Elaborations, Warsaw (2015)
23. Państwowy Monitoring Środowiska - Inspekcja Ochrony Środowiska, *Ocena jakości powietrza w strefach w Polsce za rok 2016*. Warszawa (2017)
24. A. Palarz, Prace Geograficzne, **138**, 29-43 (2014)
25. A. Bokwa, Prace Geograficzne, **126**, 41-51 (2011)

Research on chromium and arsenic speciation in atmospheric particulate matter: short review

Katarzyna Nocoń[1] *, *Wioletta* Rogula-Kozłowska[1], and *Kamila* Widziewicz[1]

[1]Institute of Environmental Engineering Polish Academy of Sciences, 34 M. Skłodowskiej-Curie St., 41-819 Zabrze, Poland

Abstract. Atmospheric particulate matter (PM) plays an important role in the distribution of elements in the environment. The PM-bound elements penetrates into the other elements of the environment, in two basic forms - those dissolved in the atmospheric precipitation and those permanently bound to PM particles. Those forms differs greatly in their mobility, thus posing a potential threat to living organisms. They can also be an immediate threat, while being inhaled. Chromium (Cr) and arsenic (As) belong to the group of elements whose certain chemical states exhibit toxic properties, that is Cr(VI) and As(III). Thus, recognition of the actual threat posed by Cr and As in the environment, including those present in PM, is possible only through the in depth speciation analysis. Research on the Cr and As speciation in PM, more than the analogous studies of their presence in other compartments of the environment, have been undertaken quite rarely. Hence the knowledge on the speciation of PM-bound As and Cr is still limited. The state of knowledge in the field of PM-bound Cr and As is presented in the paper. The issues related to the characterization and occurrence of Cr and As species in PM, the share of Cr and As species mass in different PM size fractions, and in PM of different origin is also summarized. The analytical techniques used in the speciation analysis of PM-bound Cr and As are also discussed. In the existing literature there is no data on the physical characteristics of Cr and As (bound to a different PM size fractions), and thus it still lack of data needed for a comprehensive assessment of the actual environmental and health threat posed by airborne Cr and As.

1 Introduction

Chromium (Cr) and arsenic (As) are two elements widely common in environment. They occur in different chemical forms, which in case of As are more differentiated. The circulation of chromium and arsenic in the environment mostly involve their biologically active forms, present in an ionic forms, and by definition bioavailable for living organisms. The impact of chromium and arsenic on human's health depends mostly on their oxidation state. It is documented that Cr(III) takes a part in the processes of the formation and transformation of lipids, glucose and proteins, while Cr(VI) form is a known human

* Corresponding author: katarzyna.nocon@ipis.zabrze.pl

carcinogen, characterized by a high toxicity [1]. After inhalation Cr(VI) migrate easily into respiratory track, where it cross the cell membranes and is further transported into the bloodstream. Similarly to Cr, also the toxicity of As is related to the different chemical forms and oxidation states in which it can be found. It is recognized, that the inorganic arsenic species are more toxic than the organic ones (except trivalent organoarsenicals), and that As(III) is more toxic than As(V). At the cellular level, As(III) interacts with proteins and enzymes, causing oxidative stress and ablation of DNA by its methylation [2].

Considering the fact of As and Cr toxicity it can be stated that understanding their speciation is more important, than knowing its total contents in the exact elements of the environment [3]. All over the world the quantitative assessment of As and Cr species in all elements of the environment such as water, soil, sediments, wastes, plants, particulate matter etc. is a huge challenge. While looking into scientific literature it appears that particulate matter is probably the least recognized among above mentioned.

Atmospheric particulate matter (PM), also referred as aerosol, represents mixture of solid, liquid, semi-volatile and volatile compounds. PM is very complex, regarding particle size and composition, as it may contain metals, metalloids, inorganic ions, organic compounds, black carbon and biogenic material (e.g. algae, pollen, bacteria or fungi) [4]. Airborne chromium (i.e. PM-bound Cr) exists in two valence states Cr(III) and Cr(VI). Similarly arsenic exist in PM in two valence states As(III) and As(V), in inorganic and/or organometalloid forms. Five different species of As have been detected in PM so far: the two inorganic species arsenite (As(III)) and arsenate (As(V)) and the organoarsenic methylated forms of As(V), that is methylarsonate (MA), dimethylarsinate (DMA) and trimethylarsine oxide (TMAO) [5].

The occurrence of different oxidation states of PM-bound As and Cr is dependent on their origin. It has been estimated that approximately two thirds of the Cr and As emissions to the atmosphere is related to anthropogenic sources. The main industrial Cr sources are: ferrochrome metallurgy, brick manufacturing, plating, fuels combustion, waste incineration, the production of Cr-containing chemicals, mainly chromates and dichromates, pigments, Cr trioxide and Cr salts. Arsenic sources include mainly metal smelting, fossil fuels combustion and pesticide use, therefore 60% of the global anthropogenically-generated atmospheric arsenic is attributable to copper smelting and coal combustion. Natural sources of Cr and As emission include volcanic eruptions and erosion of soils and rocks. As can also be released from low temperature volatilization of organoarsenic compounds, forest fires and sea spray [6-8].

The prevailing compounds of considered elements in PM are inorganic compounds of As(V) and Cr(III). Cr(VI) contribute from 3 to 8% of the total chromium emission into atmosphere and it's lifetime is relatively short 0.7-4.8 days [6,9]. The compounds of Cr and As in PM as well as the shares of its particular species are shaped by emission sources.
X-ray analysis of PM collected in industrial area showed that arsenic is present as: $Ca_3Sr_2(AsO_4)2.5(PO_4)0.5(OH)$, As_2O_3 and As_2O_5 [10] and chromium is present in the chemical forms: Cr_2O_3, Cr–Fe spinel, $Cr(OH)_3$, $Cr(NO_3)_3$, $Cr_2(SO_4)_3$, and $CrCl_3$. Most of them are not soluble in water in contrary to Cr(VI) compounds, which are water-soluble unless existing as $PbCrO_4$, $BaCrO_4$ and $ZnCrO_4$ (emitted during utilization of chromate pigments and dyes). Soluble Cr(VI) is reported to be dominant in mobile and fuel combustion emission [11].

The source of PM emission also affects the mass shares of particles in specific size ranges (size fractions) and in the total PM mass. While most mining operations generate coarse PM, high temperature processes, such as smelting and coal combustion, are typically associated with fine PM. Coarse PM (particles with aerodynamic diameter ≥ 2.5 μm) are produced usually by mechanical processes such as crushing and grinding of ore and may be resuspended via wind erosion and mechanical disturbance. Fine PM (particles with aerodynamic diameter ≤2.5 μm) are produced during smelting and combustion through the condensation of high temperature vapors, diffusion and coagulation [12]. It is known that PM generated by anthropogenic sources consists mostly of fine particles, with the domination of particles below 1 μm [13, 14]. Generally, it can be stated, that fine particles, mainly due to the chemical composition, are characterized by much greater surface area than coarse one. Many studies confirm connection between the size of the surface area, and the ability of PM particles to cumulate metals and metalloids. In the Polish urban areas, approximately 70-90% of total PM-bound metal mass is bound to submicron particles (PM_1) [15-17].

The aim of the paper is to synthetize the state of knowledge on the presence of chromium and arsenic species in atmospheric particulate matter available in the national literature including such information as: (i) characterization and occurrence of Cr and As species in PM, (ii) the mass share of Cr and As species in different PM size fractions and in PM of different origin, (iii) the analytical techniques used in the speciation analysis of PM-bound Cr and As.

2 Application of Chromium and Arsenic Speciation Analyses in PM Samples

Among PM-bound elements in urban air, Cr and As are those, which can affect human health. The negative health effect, after exposure by inhalation of PM-bound Cr and As, depends mainly on their oxidation states, solubility of their compounds and their mass distribution respect to inhaled particles [1, 12]. It is established that the smaller is the PM diameter, the higher its capacity to penetrate into the respiratory track: particles with aerodynamic diameter equal to 10 μm can reach the larynx, 2.5 μm the bronchi, and 1 μm the alveoli [18]. For this reason interest in species-selective analysis especially of fine PM samples has increased in recent decades.

Research on Cr and As and their species show that the concentrations of their species in PM are spatially dependent and are mainly shaped by local emission. Higher concentration of PM-bound Cr and As and their species (Cr(VI) and As(V)) were determined in the regions where local emission is mainly industrial, in comparison to areas, where the influence of industry is not prevailing, or where the area is located far from the direct impact of the industrial source emission [19-24]. For example, in urban areas, close to chromium emission sources in particular, Cr(VI) concentration can achieve values, e.g. 20 ng/m^3 close to a foundry [19] and 20-70 ng/m^3 around a ferrochrome smelter [20]. Talebi [21] reported 5.4–8.2 ng/m^3 ambient concentration of PM-bound Cr(VI) in Isfahan (Iran), where two large iron and steel factories were located nearby. The PM_{10}-bound Cr(VI) concentrations in Belgium were 5.2 ng/m^3 close to stainless steel factory and 1.2 ng/m^3 in location further away [22]. Świetlik et al. [23] determined Cr species in PM (TSP) collected in 4 locations characterized by domestic and vehicular emission and in one with an industrial emission (co-fired plants and metalworking plants) in Radom (Poland). The PM from industrial zone turned out to be almost fivefold richer in Cr(III). The level of Cr(VI) in air over the whole city was similar and the mean concentration of PM-bound Cr(VI) was 6.0 ng/m^3. In non-polluted areas mean concentration of Cr(VI) was lower than 0.5 ng/m^3 [24].

While examining the influence of industry on PM –bound As levels Oliveira et al. [25] found that close to the cooper smelter the contents of As species in TSP were 1.2ng m^{-3} As(III) and 10.4 ng m^{-3}As(V). In Taiwan in China the average concentrations of TSP-bound arsenic species at industrial area (thermal power plant) were 1.06 ng/m^3 and 2.49 ng/m^3 for As(III) and As(V), respectively; while in residential area they were 0.83 ng/m^3

for As(III) and 1.96 ng/m^3 for As(V) [26]. More examples of As species concentration from various mining operations are presented in review paper by Martin et al. [12].

Most of studies on Cr and As speciation in PM has been performed on total PM samples (so called total suspended particles, TSP) or either PM_{10} and $PM_{2.5}$. Speciation of PM-bound Cr and As in size-segregated samples has been studied only to a little extent. Inorganic arsenic species were determined for example in $PM_{2.5-10}$ and $PM_{2.5}$ in Tapada and Sines in Portugal [27], in PM_{10} and $PM_{2.5}$ in Huelva in Spain [28], in Antwerp in Belgium [10] in $PM_{2.5}$ and PM_1 in Zabrze, Poland [29]. These studies indicate the presence of As(V) as the dominating specie in the PM samples, followed by As(III); only Tirez and co-workers [10] showed that the dominant specie in $PM_{2.5}$ was As(III).

Chromium species was determined in $PM_{2.5-10}$ and $PM_{2.5}$ in Wielgegund, Southern Africa [30], in PM_{10} and $PM_{2.5}$ in New Jersey, USA [31, 11], and in $PM_{2.5}$ and PM_1 in Zabrze, Poland [29]. In Africa the majority of Cr(VI) was found in $PM_{2.5}$ [30]. This observation was explained by the nature of PM-bound Cr(VI) - emitted by pyrometallurgical plants in the studied region and the atmospheric lifetimes of different PM size fractions. Yu et al. and Huang et al. [11, 31] found that at the MERI site (~700 m from New Jersey Turnpike) concentrations of PM_{10}-bound Cr(VI) were higher than concentrations of $PM_{2.5}$-bound Cr(VI) both in summer and in winter. Widziewicz et al. [29] noticed higher concentrations of $PM_{2.5}$-bound Cr(VI) than PM_1-bound Cr(VI) only in summer. Generally, the higher concentration of Cr(VI) and higher ratio Cr(VI)/Cr(III) were observed in summer season compared to the winter season. Authors explained their findings by the fact, that the airborne Cr(III) can be oxidized to Cr(VI) by reactions with reactive oxygen species (ROS), characterized by strong oxidative capacity. Photoreactions are expected to be more intensive in the summer, also as a result of the presence of manganese in the ambient air [6]. Multiple linear regression analysis confirmed that the ambient concentrations of Cr(VI) were significantly affected by meteorological factors (i.e. temperature and humidity) and reactive gases particles (i.e. O_3, Fe and Mn) [31]. The mass distribution of Cr(VI) in respect to the particle size (8 PM fraction) was determined at the William site (~50 m from New Jersey Turnpike) [31]. Authors found that greater mass of Cr(VI) was allocated in the size range below 2.5 μm and the maximum of mass size distribution of Cr(VI) were found around the particle size of 2.5 μm. The results of Cr species masses distribution among 10 PM fractions from an industrial site in Northern Italy indicated that both mass concentration of Cr(VI) and Cr(VI)/Cr(III) ratio increase with the decrease in particle size [32]. The mass size distribution of Cr(VI) was unimodal and their maximum was in the range 0.56-1.0 μm. Mass size distribution of Cr(III), instead, was a bi-modal with maxima in the ranges 1.0-1.8 and 5.6-10 μm [32]. The mass size distribution of Cr(III) is consistent with the distribution of total chromium in size-fractionated PM (13 PM fractions) collected in Zabrze (Poland) during winter [16, 17]. In Zabrze, Cr were accumulated mostly in fine particles and had bimodal mass size distribution with maxima in the ranges 1-1.6 and 6.8-10 μm. Arsenic was accumulated mostly in submicron particles and had tetramodal distribution with maxima in the ranges 0.108-0.17; 0.4-0.65; 2.5-4.4 and 6.8-10 μm. As speciation in size-segregated PM samples has not been found in literature. Therefore it seems reasonable to undertake further research on chromium and arsenic speciation in size-segregated PM.

3 Sample preparation and analytical techniques for PM-bound Cr and As speciation

Speciation of PM-bound Cr and As as an analytical process involves a few step, that are typically sampling, sample storage, sample pretreatment (extraction), separation of species and detection. Each of these steps of the procedural chain can influence the accuracy of determination of every specie of particular element, and each has to be optimized with respect to the other.

According to regulations referring to sampling of atmospheric air, PM can be collected on filters made of glass or quartz fiber filters, and cellulose or polycarbonate or PTFE (Teflon) membranes. The filters/membranes differ among each other for instance by flow resistant and so called "blank values" for determined element. Usually blank filters are initially prepared in order to minimize the analyte background ("blank values" in clean filters/membranes), i.e. minimize of positive artifacts in chemical analysis. The selection of the material of the filters is an important matter, because both filters and PM (deposited on filters) have to be digested or extracted with a suitable solution for posterior analysis of PM-bound Cr and As present. The desired extraction method must be therefore compatible with the analytical detection methods used and ideally require a small amount of PM sample. Since nineteen's a few methods using the alkaline digestion, for Cr(VI) determination in solid matrices, including atmospheric PM have been developed [32]. Among mentioned, standard methods dedicated for PM (ASTM, NIOSH and OSHA) differ on sample pretreatment level - filter material (PVC, quartz, Teflon), except ASTM, and/or extraction conditions (NaOH, Na_2CO_3, $NaHCO_3$) but all are fallowed by Ion Chromatography and UV detection after DPC derivatization (Table 1). US EPA 3060A alkaline digestion method for determination of Cr(VI) in solid matrices was also used for PM samples [23, 29, 30]; its modified version was used by Huang et al. [11]. Chromium in extracts has been analyzed also by such techniques as: catalytic cathodic stripping voltammetry with adsorption of Cr(III)-DTPA complexes (CCSV-DTPA), graphite furnace atomic absorption spectroscopy (GF-AAS), inductively coupled plasma optical emission spectrometry (ICP-OES) (Table 1).

Over the past few years, two review papers have been published on the methodology used to determine the inorganic and organic species of arsenic in PM samples [34, 35]. Therefore in the presented work the authors focused only on the latest publications (Table1). As shown in Table 1 PM samples for arsenic species determinations were prepared by acid extraction/digestion with using solution such as: HNO_3, H_3PO_4, $NH_2OH \cdot HCl$, ascorbic acid or with citrate buffer. The latest trends in detection element species concern the so-called hyphenated techniques [36]. Determination of As species in PM has been performed with a very sensitive HPLC methodology coupled with hydride generation and atomic fluorescence spectrometry (HPLC-HG-AFS) or with inductively coupled plasma mass spectrometry (HPLC-ICP-MS).

The environmental conditions in which the extraction of analytical species is carried out are conducive to its transformation. Especially Cr(VI) is very reactive, and inter-conversion between Cr species can easily occur during sampling and sample processing. So very important in the speciation analyses is the optimization of the extraction procedure (time and conditions of ambient air), followed by compatible detection method, in order to minimize changes of analytes oxidation state. Because there is no CRM for Cr and As species in PM, accuracy of the method (including possible determined element species interconversion) can be evaluated mainly by recovery of known concentration of proper species standards added to PM sample. In case of Cr recovery of Cr (VI) both by external standard method from soil CRM and by speciated isotope-dilution mass spectrometry (SIDMS) method (EPA Method 680) from isotope Cr (VI) and Cr (III) enriched PM sample were studied [33, 38, 39]. SIDMS method occurred to be useful in correction of possible Cr species interconversion for determination of their soluble forms [11, 39].

Table 1. Comparison of methodologies in speciation of PM-bound Cr and As

Location	PM fraction Cr/As species (chemical analysis) Cr/As species (calculations)	Type of filter Sample extraction/digestion	Technique of chemical analysis(*)	Type of location	References
Radom (Poland)	TSP Cr(VI) and Cr_{total} Cr(III)= Cr_{total}- Cr(VI)	Quartz EPA 3060A: Na_2CO_3+NaOH	CCSV-DTPA	Five sites with different municipal structure	Świetlik et al. (2011)
Wielgegund (Southern Africa)	$PM_{2.5\text{-}10}$/$PM_{2.5}$ Cr(VI)	Teflon Na_2CO_3+NaOH	IC-UV	Industrial – pyrometallurgical smelter, coal-fired power station	Venter et al. (2016)
Belgium	PM_{10} Cr(VI)	Cellulose $NaHCO_3$	IC-UV	Industrial-stainless steel factory	Tirez et al. (2011)
Katowice (Poland)	$PM_{2.5}$/PM_1 Cr(VI) and Cr_{total} Cr(III)=Cr_{total}- Cr(VI)	Quartz EPA 3060A: Na_2CO_3+NaOH	GF-AAS	Urban traffic site	Widziewicz et al. (2016)
Tunis (North Italy)	PM_{10} PM0.18-PM18 Cr(VI) and Cr (III)	Cellulose, teflon, quartz Na_2CO_3+$NaHCO_3$	CAdSV-DTPA ICP-OES	Industrial, peri-urban site	Catrambone et al. (2013)
New Jersey (USA)	$PM_{2.5}$ $PM_{0.18\text{-}18}$ Cr(VI) soluble	Cellulose filter cleaned with HNO_3, pretreated with $NaHCO_3$ Extraction in 4% HNO_3	IC-ICP-MS	Urban traffic site	Yu et al. (2014)
New Jersey (USA)	PM_{10} Cr(VI) PM_{10} Cr(VI) soluble	Teflon Na_2CO_3+NaOH Cellulose filter cleaned with HNO_3, pretreated with $NaHCO_3$, samples extracted in 4% HNO_3	IC-ICP-MS IC-UV	Urban traffic site, suburban	Huang et al. (2014)
Huelva (Spain)	PM_{10}/$PM_{2.5}$ As(V) and As(III)	Quartz $NH_2OH\cdot HCl$	HPLC-HG-AFS	Industrial-copper smelter	Sanchez-Rodas et al. (2012)
Antwerp (Belgium)	PM_{10}/$PM_{2.5}$ As(V) and As(III)	Cellulose Ascorbic acid+ H_3PO_4	HPLC- ICP-MS	Industrial –metallurgical plant	Tirez et al. (2015)
Katowice (Poland)	$PM_{2.5}$/PM_1 As(III)and As_{total} As(V)=As_{total}- As(III)	Quartz Citric acid+sodium citrate	HG-AAS	Urban traffic site	Widziewicz et al. (2016)
Juan (China)	$PM_{2.5}$ As(V) and As(III)	Glass fiber membrane 1% HNO_3	HPLC- ICP-MS	Industrial - coal-fired power station	Yin et al. (2017)

(*) Abbreviations: IC-UV Ion Chromatography–UV detection; GF-AAS Graphite Furnace Atomic Absorption Spectroscopy; HG-AAS Hydride Generation Atomic Absorption Spectrometry; CCSV-DTPA Catalytic Cathodic Stripping Voltammetry with adsorption of DTPA complexes; ICP-OES Inductively Coupled Plasma Optical Emission Spectrometry; IC-ICP-MS Ion Chromatography-Inductively Coupled Plasma–Mass Spectrometry; HPLC-ICP-MS High Performance; Liquid Chromatography–Inductively Coupled Plasma-Mass Spectrometry; HPLC-HG-AFS High Performance Liquid Chromatography–Hydride Generation-Atomic Fluorescence Spectrometry

4 Overall summary and conclusions

Chemical speciation of toxic elements, such chromium and arsenic, in an environmental samples gives a crucial information on their amounts present in different oxidation form, that are characterized by different toxicity, availability, and reactivity. The share of toxic forms of airborne As and Cr in their total PM-bound mass is of great importance in environmental and human health effects. Most interesting seems to be the information on the contents of toxic forms of Cr and As in fine particulate matter, characterized by long residence time in atmosphere and free migration to the respiratory tract.

Till now the most research works on Cr and As speciation concern their presence in $PM_{2.5}$, $PM_{2.5-10}$ or PM_{10-40}. There is a scarce research on chromium and arsenic speciation in ultrafine PM fraction and on the mass size distribution of their species in PM. Published results of research on mass size distribution of Cr in atmospheric air (two studies) indicate that their toxic specie - Cr(VI) tend to cumulate in fine $PM_{2.5}$. Published results on As species in coarse and fine fraction (collected during the same research campaign) indicated that toxic As(III) was rather tend to cumulate in finer one.

It can be stated that in the present state of knowledge there is not enough data on PM-bound Cr and As distribution within size-fractionated particles for a comprehensive assessment of the actual threat posed by atmospheric Cr and As. In connection with the above, it is still a challenge to describe the nature of Cr and As species distribution of in the function of the PM aerodynamic diameter in relation to all factors shaping the distribution.

The work was financed in the frame of project „Studies on the speciation of chromium and arsenic in size-fractionated urban aerosol" (UMO-2016/23/N/ST10/03032) by National Science Center, Poland (NCN)

References

1. IARC, IARC Monogr. Eval. Carcinog. Risks Hum. **49** (IARC Press, LYON, 2012)
2. K.T. Kitchin, K. Wallace J. Inorg. Biochem. **102**, 8 (2008)
3. N. Unceta, F. Séby, J. Malherbe, O.F. Donard, Anal. Bioanal. Chem. **397**, 15 (2010)
4. WHO, (Chapter 7.3 Copenhagen, Denmark, 2000b)
5. A. Jakob, K. Roth, K. Haas, E. M. Krupp, A. Raab, P. Smichowski, D. Gomez and J. Feldmann, J. Environ. Monit, **12** 8 (2010)
6. C. Seigneur, E. Constantinou, Environ. Sci. Technol. **29**, 10 (1995)
7. D. Kimbrough, Y. Cohen, A. Winer, L. Creelman, C. Mabuni, Crit. Rev. Environ Sci. Tech. **29**, 47 (1999)
8. J. Matschullat, Sci. Total Environ. **249**, 16 (2000)
9. WHO, (Chapter 6.1 Copenhagen, Denmark, 2000)
10. K. Tirez, C. Vanhoof, J. Peters, L. Geerts, N. Bleux, E. Adriaenssens, E. Roekens, S. Smolek, A. Maderitsch, R. Steininger, J. Goettlicher, F. Meirer, C. Streli, P. Berghmans, J. Anal. At. Spectrom. **30**, 15 (2015)
11. L. Huang, C.H. Yu, P.K. Hopke, P.J. Lioy, B.T. Buckley, J.Y. Shin, Z.T. Fan Aerosol Air Qual Res. **14**, 11 (2014)
12. H. R. Martin, K. Dowling, D. Pearce, J. Sillitoe, F. Singarayer, Geosci. **4**, 48 (2014).
13. W. Birmili, A. G. Allen, F. Bary, R. M. Harrison, Environ. Sci. Technol. **40**, 10 (2016)

14. B. Chen, A. F. Stein, P. G. Maldonado, A. M. Sanchez de la Campa, M. Gonzalez-Castanedo, N. Castell, N. J. de la Rosa, Atmos. Environ. **71**, 11 (2013)
15. W. Rogula - Kozlowska, B. Błaszczak, K. Klejnowski, Arch. of Environ. Prot. **37**, 2, 9 (2011)
16. W. Rogula-Kozłowska, G. Majewski, P. Czechowski, Environ Monit Assess. **187**, 240 (2015)
17. W. Rogula-Kozłowska, Air Qual. Atmos. Health. **9**, 18 (2016)
18. J. S. Brown, T. Gordon, O. Price, B. Asgharian, Part. Fibre Toxicol. **10**, 12 (2013)
19. P. Krystek, R. Ritsema, Int. J. Mass Spectrom. **265**, 7 (2007)
20. K. L. Mandiwana, N. Panichev· T. Resane, J. Hazard. Mater. **136**, 4 (2006)
21. S.M. Talebi, Environ. Res. **92**,3 (2003)
22. K. Tirez, G. Silversmit, N. Bleux, E. Adriaensens, E. Roekens, K. Servaes, C. Vanhoof, L. Vincze, P. Berghmans, Atmos. Environ. **45,** 10 (2011)
23. R Świetlik, A. Molik, M. Molenda, M. Trojanowska, J. Siwiec, Atmos. Environ. **45**, 5 (2011)
24. J.H Seinfeld, S.N Pandis, *Atmospheric Chemistry and Physics - From Air Pollution to Climate Change.* (John Wiley & Sons, 2006).
25. V. Oliveira, J.L. Gómez-Ariza, D. Sánchez-Rodas, Anal Bioanal Chem, **382**, 6 (2005)
26. G.Ch. Fang, Ch.S. Huang, Ch.Y. Chang, J.H. Huang, Ch.K. Liu, Y. J. Zhuang, Y.F. Xiao, Y.Ch. Kuo, Ch. Y. Huang, K.H. Tsai, Environ. Earth Sci. 75, **124** (2016)
27. M.M. Farinha, Z. Slejkovec, J.T. van Elteren, H.T. Wolterbeck, M.C. Freitas, J. Atmos. Chem. **49** 11 (2004)
28. D. Sanchez-Rodas, A.S. de la Campa, V. Oliveira, J. de la Rosa, Atmos. Environ. **42**(26) 11 (2012)
29. K. Widziewicz, W. Rogula - Kozlowska, K. Loska, Atmos. Pollut. Res. **7**, 11 (2016)
30. A. D Venter, P. Beukes, P. Gideon van Zyl, M. Josipovic, K. Jaars, V. Vakkari, Atmos. Pollut. Res. **7,** 6 (2016)
31. C.H. Yu, L. Huang, J.Y. Shin, F. Artigas, Z.T. Fan, Atmos. Environ. **94,** 8 (2014)
32. M. Catrambone, S. Canepari, C. Perrino, E3S Web of Conferences 1 (2013), 10.1051/e3sconf/20130107005
33. R.E. Wolf, S.A. Wilson, Open-File Report 2010–1243 (U.S. Geological Survey, Reston, Virginia 2010)
34. A.S. Lewis , K. R. Reid, M. C. Pollock, S. L. Campleman, J Air Waste Manag Assoc. **62**,1, 16 (2012)
35. D. Sanchez-Rodas, A.M. de la Campa, l. Alsioufi , Anal. Chim. Acta. **898**, 18 (2015)
36. R. Michalski, S. Szopa, M. Jabłońska, A. Łyko, Sci. World J. **2012**, 17 (2012)
37. X. Yin, L. Wang, X. Yu, S. Du, H. Zhang, Z. Zhang, Pol. J. Environ. **26**, 6 (2017)
38. Q. Meng, Z. Fan, B. Buckley, L. Lin, L. Huang, C-H. Yu, R. Stiles, L. Bonanno, Atmosph. Environ. **45**, 7 (2011)
39. L. Huang, C-H. Yu, P. K. Hopke, J. Y. Shin, Z. Fan, J. Air Waste Manag. Assoc. **64,** 7 (2014)

Application of the Data Mining Methods to Assess the Impact of Meteorological Conditions on the Episodes of High Concentrations of PM10 along the Polish – Czech Border

Leszek Ośródka*, *Ewa* Krajny, and *Marek* Wojtylak

IMWM-NRI, Department Modelling of Air Pollution, 10 Bratków Str., 40-045 Poland

Abstract. The paper presents an attempt to use selected data mining methods to determine the influence of a complex of meteorological conditions on the concentrations of PM10 (PM2.5) proffering the example of the regions of Silesia and Northern Moravia. The collection of standard meteorological data has been supplemented by increments and derivatives of measurable weather elements such as vertical pseudo-gradient of air temperature. The main objective was to develop a universal methodology for the assessment of these impacts, i.e. one that would be independent of the analysed pollution. The probability of occurrence (at a given location) of the assumed concentration level as exceeding the value of the specified distributional quintile was adopted as the discriminant of the incidence. As a result of the analyses conducted, incidences of elevated concentrations of air pollution particulate matter PM10 have been identified and the types of weather responsible for the emergence of such situations have also been determined.

1 Introduction and Aim of the Study

In spite of continuous measures undertaken by the European Union (EU) in order to improve air quality (overarching strategies: "*EU Thematic Strategy on Air Pollution*", "*EU Clean Air Policy Package*") and despite the implementation of many legal instruments (such as the "*Air Quality Directive*", or the "*National Emission Ceilings Directive*"), air pollution remains one of the most poignant problems in Central Europe [1]. Scientific research is proffering increasingly copious evidence of the negative impact of air pollution on the health of the population, both as a direct exposure and as an indirect result of exposure to pollutants accumulating in plants, soil or water [2]. The quality of atmospheric air in view of the presence of pollutants such as lead (Pb), sulphur dioxide (SO_2) or benzene (C_6H_6) has been systematically improving. On the other hand, a significant percentage of the population, especially in highly urbanised areas, is still affected by harmful levels of pollutants such as suspended particulate matter PM in the planetary boundary layer (O_3), nitrogen dioxide (NO_2) or polycyclic aromatic hydrocarbons such as

* Corresponding author: leszek.osrodka@imgw.pl

benzo(a)pyrene (BaP). The cause of poor air quality adversely impacting on health is the emission of substances from transport, industry, agriculture or households. The reduction of air pollution with substances such as particulate matter requires actions to reduce the natural or antropogenic emissions from different sources (point, area,, linear).

It is generally known that meteorological conditions determine the spreading of pollutants from their sources and their dispersion in the atmosphere. Constituting the most important and decisive meteorological factor here are the direction and speed of the wind, as well as the temperature of the air and precipitation. The wind affects the horizontal and vertical spreading of air pollution. The influence of air temperature on the level of pollution is both indirect and direct. Altitude-specific changes in the air temperature, or the so-called vertical thermal structure, shape the stability of the atmosphere. The more stable the thermal structure is – or as the temperature lowers, does not alter (isothermy) or even increases (inversion) with the altitude - the more unfavourable are the conditions for horizontal scattering of pollutants. Additionally, In wintertime, the air temperature "controls" the amount of pollutant emissions, thus indirectly influencing the concentration volumes. In turn, atmospheric precipitation causes diluting, reducing the level of the substance in the air. Insolation is also strongly impactful on photochemical pollution.

The aim of the study is to comprehensively assess the influence of the meteorological factors on the formation of episodes of high levels of particulate matter in areas with a high degree of anthropopressure.

A review of the reference literature identifies many works related to the study of the influence of meteorological conditions on the formation of high concentrations of air pollutants; however, most of them either constitute a description of such cases [3] or an undertaking to statistically analyse the influence of particular meteorological elements on air quality [4]. Nonetheless, considering the fact that the episodes of high concentrations of pollutants are not linearly dependent on particular meteorological elements but are the result of many factors, such assessments must be deemed as incomplete. Hence, in this paper an attempt has been made to use data mining methods that allow a more comprehensive assessment of the influence of many meteorological elements on air quality.

This article presents the results of quantification of the above-mentioned relations between suspended particulate matter PM10 in the planetary boundary layer and the weather in the cross-border area of Silesia (PL) and Moravia (CZ). A collection of standard meteorological data has been supplemented with increments and derivatives of measurable weather elements such as the vertical pseudogradient of air temperature.

1.1 Data and the Scope of the Study

The analysis of weather conditions responsible for episodes of high concentrations of air pollutants: suspended particulate matter and ground ozone was carried out for the area of the Polish - Czech border area in the Silesian and Moravian regions. On the Czech side, this region included the Moravian-Silesian region comprising the Ostrava-Karviná agglomeration as well as the Bruntál, Frydek-Mistek, Nový Jičin and Opava counties. On the Polish side, it was a region of the Silesian Voivodeship encompassing the Upper Silesian agglomeration, and the rural counties of Bielsko, Cieszyn, Racibórz, Rybnik, Wodzisław, Żywiec, followed by the town counties of Bielsko-Biala, Jastrzębie-Zdrój, Rybnik and Żory. The total study area amounted to 10 370 km^2, of which 52.5% was the Czech territory inhabited by 38.0% of the population in relation to the total number of persons for the entire area. In turn, the population density of 648.9 persons /km^2 was twice the size on the Polish side of the border [5, 6]. The described regions of the Silesian and Moravian-Silesian voivodships are among the most urbanised and industrialised in Poland and in the Czech Republic. This area is very diverse in terms of physical, geographical and socio-economic conditions. It is located in four provinces: the Central European Lowlands, the Polish Highlands, the Czech Massif and the Western Carpathians [7, 8]. In terms of its climate, it is located in a temperate zone with a characteristic annual course of four seasons. As part of the study, a collection of measurement data covering a period from October 2006 to March 2011 was analysed. The meteorological data was obtained from the measurement network of national meteorological services, four of which belong to the Czech Hydrometeorological Institute (ČHMÚ): Ostrava-Mošnov, Ostrava-Poruba, Lučina, Červená, Lysá hora and three to the Polish Institute of Meteorology and Water Management National Research Institute (IMGW-PIB): Bielsko-Biała, Katowice, Racibórz. The table 1 below shows the air quality monitoring stations included in the analyses.

Table 1. Air quality PM10 on monitoring stations used in the analysis.

Station name		Type of station / area	Localization Latitude N / Longitude E *degrees (°), minutes ('), seconds (") /* Elevation above sea level *(m a.s.l.)*
CZ - ČHMÚ	Bohumín	B/S	49°54'15" / 18°20'50" / 200
	Český Těšín	B/U	49°44'56" / 18°36'35" / 285
	Frýdek-Místek	B/S	49°40'18" / 18°21'04" / 290
	Havířov	B/U	49°47'28" / 18°24'25" / 260
	Karviná	B/S	49°51'50" / 18°33'05" / 238
	Opava - Kateřinky	B/U	49°56'42" / 17°54'34" / 255
	Orlová[*)]	B/U	49°52'32" / 18°26'01" / 266
	Ostrava - Českobratrská (hot spot)[*)]	T/U	49°50'23" / 18°17'24" / 215
	Ostrava - Fifejdy	B/U	49°50'21" / 18°15'49" / 220
	Ostrava – Poruba (ČHMÚ)[*)]	B/S	49°49'31" / 18°09'33" / 242
	Ostrava - Přívoz[*)]	I/U	49°51'23" / 18°16'11" / 207
	Ostrava - Radvanice (ZÚ)[*)]	I/S	49°48'25" / 18°20'21" / 263
	Ostrava - Zábřeh	B/U	49°47'46" / 18°14'50" / 236
	Studénka	B/R	49°43'15" / 18°05'22" / 231
	Třinec - Kosmos	B/U	49°40'05" / 18°40'40" / 320
	Věřňovice[*)]	B/R	49°55'29" / 18°25'22" / 203
PL - Silesian Voivodship Inspector of Environmental Protection	Bielsko-Biała	B/U	49°48'36" / 19°01'37" / 365
	Cieszyn	B/U	49°44'00" / 18°38'20" / 353
	Dąbrowa Górnicza	B/U	50°19'36" / 19°13'52" / 293
	Gliwice	B/U	50°16'36" / 18°39'18" / 236
	Godów[*)]	B/U	49°55'19" / 18°28'17" / 205
	Katowice-Muchowiec	B/U	50°15'36" / 18°58'30" / 273
	Rybnik	B/U	50°06'36" / 18°30'58" / 245
	Tychy	B/U	50°06'00" / 18°59'24" / 252
	Wodzisław Śląski	B/U	50°00'36" / 18°27'18" / 271
	Zabrze	B/U	50°18'00" / 18°47'53" / 257
	Żywiec	B/U	49°41'36" / 19°12'22" / 352

type of station: T – traffic, I – industrial, B – background; type of area station: U – urban, S – suburban, R – rural; [*)] - complementary station; ZÚ station (Public Health Institute Ostrava).

1.2 Methodological Principles of Data Analysis

Although meteorological conditions determine the formation of high concentrations of particulate matter, however, the existing dependencies are not linear. Taking this into consideration, data mining methods were used to evaluate the influence of meteorological factors on the quality air [9, 10]. The main aim was to elaborate a universal methodology, i.e. one that would be independent of the analysed pollution. The probability of occurrence of a given concentration level in excess of the value of a particular quantile distribution was assumed as the discriminant of the episode.

1.2.1 Identification of Episodes High Concentrations of PM10

First, the measurement data were unified, replacing the absolute values of PM10 with the probability of occurrence of a given concentration at the station. For data sequences $x_1,\dots,x_n$ representing average daily PM10 concentrations the probabilities were calculated from the moving averages, according to the following formula:

$$p_i = \frac{\overline{\overline{\{x \in \{x_i,\dots,x_n\}; x \geq x_i\}}}}{n} \qquad (1)$$

where: symbol $\overline{\overline{}}$ denotes the cardinality (numerical amount) $i=1,\dots,n$.
Thus, the concentrations were characterised by the frequency of their occurrence together with the higher values. Subsequently, average probability for each day was calculated for all stations actively measuring on that day, resulting in an averaged (area average) sanitary air quality. The lower probability corresponded to the higher levels of the analysed pollution:

$$\Pr(day, month, year) = \frac{\sum_{k=1}^{N} p_k}{L_s} \qquad (2)$$

where: L_s – the number of measuring stations operating on a given day.

1.2.2 Identification of Meteorological Situations Associated with High Concentrations of PM10

The characteristics of the daily meteorological situation were obtained by averaging of a number vector consisting of 59 hourly elements and the sum of the daily precipitation. The weather for each day was represented by the p vector, as per the following formula:

$$p^{(i)} = [\tilde{p}_1^{(i)},\dots,\tilde{p}_{59}^{(i)}] \qquad i=1,\dots,N. \qquad (3)$$

Thus, for the period from October 2006 – March 2011 (N) for PM, a collection of 2008 cases was obtained. In order to standardise the data, a transformation was performed for each coordinate of the meteorological vectors:

$$p_j^{(i)} = \frac{\tilde{p}_j^{(i)} - \tilde{p}_j^{\min}}{\tilde{p}_j^{\max} - \tilde{p}_j^{\min}} \qquad where$$

$$\tilde{p}_j^{\min} = \min_{i=1,..,N} \tilde{p}_j^{(i)} \qquad \tilde{p}_j^{\max} = \max_{i=1,..,N} \tilde{p}_j^{(i)} \tag{4}$$

In this way, data was prepared in the form of vectors $p = [p_1, ..., p_{59}]$ with coordinates in the [0,1] interval, which were used for further analyses.

1.2.3 Determining the Weather Patterns Responsible for the Occurrence of Smog Episodes

The $P := \{p^{(1)},\ldots,p^{(N)}\}$ sequence of daily probabilities of occurrence of a given pollution level was sorted in an ascending order. It was then assigned a sequence of weather vectors corresponding to the given sanitary air quality. In the next stage, grouping of set elements was performed. In order to allow clustering of the most similar meteorological situations, a limit distance $r > 0$ was determined. The value of the r parameter was determined based on the clustering index referred to as the Davis-Bouldin index. For the start, the vector of the weather for which the worst sanitary air quality occurred was assumed as the pattern. And this weather was the first one to be included in the kth group. Examining the P sequence starting from the second position, the kth group was subsequently enlarged by those $p^{(l)}$ whose Manhattan distance from $p^{(1)}$ was lower than the set r value; afterwards, the $p^{(l)}$ vector was removed from the P sequence.

$$\sum_{j=1}^{59} \left| p_j^{(1)} - p_j^{(l)} \right| < r \quad l=2,..,M. \tag{5}$$

Thus, groups of several of numerical amount from several to dozens of vectors were designated. However, there have been cases - especially for smog situations with the highest concentration of pollutants (unusual weather) - in which the resultant group was one-element. On the other hand, a certain imperfection of the Davis-Bouldin index is constituted by the fact that it achieves a minimum of zero for one-element groups. Therefore, the proposed modification of this index for the purposes of these analyses entailed an insertion of the $2/\overline{\overline{S_i}}$ component where S_i represents the numerical amount of the ith group:

$$DB_{\text{mod}} = \frac{1}{n}\sum_{i=1}^{n} \frac{2}{\overline{\overline{S_i}}} + \max_{j,j\neq i} \frac{\sigma_i + \sigma_j}{d(c_i, c_j)} \quad c_x - \text{centroid of the } x^{\text{th}} \text{ cluster} \tag{6}$$

where: σ_x – an average distance of the elements of the xth cluster from its centroid, $x=1,\ldots,n$ (the number of all clusters), and $d(c_i,c_j)$ is the centroid distance. Grouping was performed many times for different r values, assuming as appropriate such r_{min} for which DB_{mod} (r_{min}) reached the lowest value.

Due to the large variability of meteorological conditions, which resulted in the small number of groups containing meteorological situations responsible for the highest concentration of pollutants, it was decided to cluster similar groups into larger aggregates. For this purpose, a c centroid was calculated for each group, or, in other words, a vector composed of 59 coordinates formed by averaging the same coordinates of the weather vectors belonging to the given group. Then proceeding as described above, the resulting clusters were combined into groups, which were referred to as weather patterns. As a result for each pollutant, three weather patterns (types) were assigned for each PM level.

2 Results of the Study and Discussion

The result of the conducted research was a clustering of common features (meteorological elements) with high mean area concentrations of PM pollutants, followed by the determination of the types of weather responsible for such situations.

2.1 Particulate Matter PM

Table 2 shows for selected weather stations the characteristics of the basic meteorological elements that distinguish separate types of weather responsible for particulate matter PM episodes.

Table 2. Meteorological characteristic of the types of weather responsible for the situations of high PM10 concentrations.

Meteo.element (24h mean values)	Meteorogical station	Cool half of the year (Oct – Mar)	Winter season (Nov – Feb)	Episode PM10					
				Type I		Type II		Type III	
		Probability		Size of data set / Probability					
		0.412	0.403	16	0.020	78	0.142	35	0.111
Mean area concentration 24-hr daily PM10		60.1	64.5	258.0		105.2		115.5	
Air temperature [°C]	Ostrava-Mošnov	2.7	0.8	-13,0		-0.2		-3.1	
	Červená	0.1	-1.8	-8.6		-1.7		-3.1	
	Lysá hora	-2.3	-3.8	-8.2		-3.5		-4.1	
	Bielsko-Biała	2.8	0.9	-10.9		0.3		-3,0	
	Katowice	2.4	0.5	-12.2		-0.4		-2.3	
	Racibórz	2.8	0.9	-13.8		-0.4		-2.7	
Wind speed [m·s⁻¹]	Ostrava-Mošnov	3.9	4.1	1.9		2,0		2,0	
	Červená	4.4	4.5	3.6		3.4		3.2	
	Lysá hora	8,0	8.4	6.7		8.5		6.8	
	Bielsko-Biała	3.4	3.5	1.4		2.3		1.9	
	Katowice	2.8	2.9	1.5		2,0		2,0	
	Racibórz	3.6	3.7	1.7		2,0		1.8	
Calm daily [%]	Ostrava-Mošnov	7.5	6.5	25.5		15.2		18.8	
	Červená	0.1	0.1	0.3		0.2		0.1	
	Lysá hora	0	0	0		0		0	
	Bielsko-Biała	4.8	5,0	19,0		6.8		6,0	
	Katowice	7.5	6,0	17.4		11.5		12.4	
	Racibórz	3.8	3.6	9.4		7.5		8.3	
Wind direction [degrees]	Ostrava-Mošnov	237	236	220		23		39	
	Červená	241	240	184		149		76	
	Lysá hora	248	247	118		164		113	
	Bielsko-Biała	207	210	131		80		88	
	Katowice	235	220	135		85		90	
	Racibórz	220	215	133		86		111	
Height of the cloud base [km]	Ostrava-Mošnov	5.6	5.3	6.4		5.1		4.6	
	Katowice	5.6	5.4	8.2		5.4		5.8	

 https://doi.org/10.1051/e3sconf/20182801027

Meteo.element (24h mean values)	Meteorological station	Cool half of the year (Oct – Mar)	Winter season (Nov – Feb)	Episode PM10					
				Type I		Type II		Type III	
		Probability		Size of data set / Probability					
		0.412	**0.403**	**16**	**0.020**	**78**	**0.142**	**35**	**0.111**
Relative air humitity [%]	Ostrava-Mošnov	84,0	86,0	77.9		89.6		89.2	
	Katowice	83.7	86,0	78.7		86.6		86.3	
Atmospheric pressure a.s.l. [hPa]	Ostrava-Mošnov	1017.7	1017.9	1036.0		1010.3		1029.2	
	Katowice	1017.5	1017.6	1035.8		1010.8		1029.3	
Effective snnshine duration [hr]	Ostrava-Mošnov	25.2	17.8	47.3		12.1		29.7	
	Katowice	24.9	17.4	44.5		17.9		29.4	
Cloud cover [octant]	Ostrava-Mošnov	5.9	6.3	2.8		6.6		4.7	
	Katowice	5.8	6.1	2.1		6.4		4.9	
Vertical pseudogradient of temperature [°C· 100 m^{-1}]	Ostrava-Mošnov - Červená	0.5	0.5	-0.9		0.3		0	
	Ostrava-Mošnov - Lysá hora	0.5	0.4	-0.4		0.3		0.1	
Atmos. precipitation [mm]	Ostrava-Mošnov	1.1	1.1	0		1.1		0	
	Katowice	1.7	1.8	0.1		1.5		0	

An analysis of the results (Table 2) showed that the highest concentrations of particulate matter PM10 were observed in the weather type no. I, which is characterised by the lowest probability of occurrence and occurs at very low mean daily air temperatures (at all stations < -10 °C). Such low temperatures are accompanied by a very low wind speed (except for plateaus) and a long windless period. In weak wind conditions, there is also a dominant inflow from the southern sector. Among other weather patterns, this one is characterised by the lowest relative air humidity, the highest insolation, no precipitation or precipitation of a small daily sum thereof, as well as 24-hour inversion of temperature. The weather type no. II episodes that account for the highest number of dust episodes, but with the lowest mean regional concentration values occur at slightly negative air temperatures with relatively high wind velocities considering the dust episode conditions and the lowest number of silences among all the differentiated weather patterns. Dominant here is a clear influx of air masses from the north and northeast, with high humidity and low insolation. A slightly positive vertical pseudogradient of temperature indicates a slight inversion of temperature during these episodes. Episodes of this type feature atmospheric precipitation with relative frequency. The weather type no. III, which accounts for less than 30% of the dust episodes, is characterised by a mean daily air temperature of about -4.0°C, the highest wind speeds among all other types of weather (excluding the plateaus), the influx of air masses from the east, and a relatively high total radiation with no precipitation. Its characteristic feature is isothermy. As can be seen from the results shown in Table 1, with the weather type no. I, the probability of occurrence of an episode of an at least regional scale is almost total. In type no. II, the probability of occurrence of an episode amounts to about 60%. In turn, in type no. III, the probability is estimated at 75%.

3 Summary

Areas which have been anthropogenically strongly transformed and thus those which are marked with dominance of industrial and urban infrastructure are particularly vulnerable to the occurrence of high concentrations of particulate matter base. The identification and subsequent testing of the development phases of such episodes had been hitherto performed using classical synoptic analysis methods or descriptive statistics. In contrast, this paper proposes a comprehensive approach to the subject - the use of data mining methods, which, according to the authors, allows for a more complete indication of a group of meteorological factors which influence high concentrations of particulate matter. The applied methodology helped delineate several groups of meteorological situations responsible for the formation of episodes of high concentrations of PM10. It seems that this methodology - despite the fact that it does not simply refer to the origin of such situations - does allow for an objective identification (i.e. such that is independent of the researcher's intuition) of a set of meteorological factors that are conducive to their engendering. This method may therefore constitute an important auxiliary tool in both short-term air quality forecasting, and - what is perhaps even more important – it may create an element of an air quality management system in areas of particular vulnerability.

The scientific study was elaborated within the framework of the project entitled "Air Quality Information System in the Polish - Czech Border Area in the Region in Silesia and Moravia", which was founded by the Operational Program for Cross - Border Cooperation between the Czech Republic and the Republic of Poland (POWT RCz – RP 2007-2013) (registration number CZ.3.22/1.2.00/09.01610) and co-financed by the ERDF (European Regional Development Fund).

References

1. EEA Report, Air quality in Europe –2017 report. No 13/2017
2. WHO, Ambient air pollution: A global assessment of exposure and burden of disease. 2016
3. J. He, S. Gong, Y. Yu, L. Yu, H. Mao, C. Song, S. Zhao, H. Liu, X. Li, R. Li, *Air pollution characteristics and their relation to meteorological conditions during 2014–2015 in major Chinese cities*. Environ. Pollut., **223**, (2017)
4. K. Juda-Rezler, M. Reizer, M. and J.P Oudinet, *Determination and analysis of PM10 source apportionment during episodes of air pollution in Central Eastern European urban areas: The case of wintertime 2006*. Atmos. Environ. **45,** 36, (2011)
5. Central Statistical Office (GUS), Area and Population in the Territorial Profile in 2017. Statistical Information and Elaborations. Warszawa 2017, http://stat.gov.pl/en
6. Czech Statistical Office (CZSO), https://www.czso.cz (accessed 15.11.2017)
7. J. Demek, P. Mackovčin P. (eds.) et. al., Hory a nížiny - zeměpisný lexikon ČR. Agentura ochrany přírody a krajiny ČR - AOPK ČR (2006)
8. J. Kondracki, *Geografia regionalna Polski*. WN PWN Wyd.3, Warszawa 2011
9. A. Łachwa, *Rozmyty świat zbiorów, liczb, relacji, faktów, reguł i decyzji*. AOW Exit, Warszawa 2001
10. Osowski, *Sieci neuronowe*. WNT, Warszawa 1997

E3S Web of Conferences **28**, 01028 (2018) https://doi.org/10.1051/e3sconf/20182801028
Air Protection in Theory and Practice

Preliminary comparative assessment and elements of equivalence of air pollution measurement results of portable monitoring stations with using stochastic models

Tomasz Owczarek[1]*, *Mariusz* Rogulski[2] and *Artur* Badyda[2]
[1]Akademia Morska w Gdyni, Wydział Przedsiębiorczości i Towaroznawstwa, ul. Morska 83, 81-225 Gdynia, Polska
[2]Warsaw University of Technology, Faculty of Building Services, Hydro and Environmental Engineering, Nowowiejska 20, 00-653 Warsaw, Poland

Abstract. This article is an attempt to assess the suitability of new non-reference mobile measuring devices for the analysis of PM_{10} concentrations. The aim is to compare the concentration daily values measured by these devices with the concentrations obtained by the reference device. It also examines the possibility of building models that correct these values to equivalent reference values. The analysis allows to conclude that the properties of the obtained measurements indicate a good chance of constructing an effective model of correcting the concentration values. Comparison of frequency, time variation and correlation of concentrations indicates the ability to demonstrate the equivalence of mobile devices with the reference method.

1 Introduction

The quality of the air and its impact on the environment, in particular on human health, is an issue that is driving more and more widespread groups of society. Therefore, the problem of measuring concentrations of pollutants in a fast, cheap and, above all, accurate and effective way is a problem for scientists and companies that produce measuring devices.

Among the air pollutions, particulate matters occupy a special place. They are a mixture of particles of different structure, origin and size. They may be of natural origin (arising from erosion, evaporation or natural physicochemical transformations) or anthropogenic (fuel combustion, communication emissions, industrial emissions) [1, 2]. The chemical composition of the particulate matter can vary greatly and depends largely on where it is located [3]. It may contain heavy metals (lead arsenic, nickel, cadmium), polycyclic aromatic hydrocarbons (especially benzo(a)pyrene) and anions (eg. SO_4^{2-}, NO_3, Cl^-) and cations (eg. Na+, K^+, Ca^{2+}, Mg^{2+}, NH_4^+) (for example [2, 4-6]).

* Corresponding author: t.owczarek@wpit.am.gdynia.pl

There are three factions of particulate matter depending on the dust particle size: PM_{10} (dust containing particles up to 10 μm in diameter), $PM_{2.5}$ (up to 2.5 μm in diameter) and PM_1 (up to 1 μm in diameter). Studies have shown that the presence of dust in the air has a negative impact on human health [7]. This is especially true for smaller particulate matters diameter. Thicker fractions of dust can settle in the lungs and entire upper respiratory tract, while smaller particles can penetrate the lungs into the bloodstream and enter the brain, heart and other organs. They can cause upper respiratory tract diseases (asthma, COPD, dyspnea), but also cause heart disease, nervous system, neurosis and many others [8, 9].

The concentration of pollutants in the air is examined with using methods described in detail in the legislation. For the concentration of particulate matter, the gravimetric method is used as the reference method. This method is considered as a reference method and as such is considered to be the most accurate method [10, 11]. Unfortunately, due to the method of data collection, the results for this method are obtained once a day. In addition, the research is relatively expensive, there is a limited number of measuring devices (currently in Poland is about 200), and the measurements take place in very carefully selected places. As an alternative to measuring by reference method are measurements made by using automatic measuring devices.

Different measurement methods are used for their construction: with using a tapered element oscillating microbalance (TEOM), β -radiation attenuation or light scattering (optical method). The advantage of these devices is their relatively low cost and high flexibility of measurements. The disadvantage is unknown level of measurement accuracy. Therefore, in order to be able to use the measurement results from these devices, it is necessary to demonstrate their equivalence with the results obtained by the reference method. Equivalence procedures include a comparative study of the results obtained by both methods and the risk assessment of the results of severely deviated reference results [1, 12, 13].

The aim of the study is to compare the daily values of PM_{10} concentrations measured with using five new, mobile devices based on optical sensors with concentration values obtained by a measuring device owned by the Voivodeship Environmental Protection Inspectorate (VIEP), treated as reference values.

2 Methodology

The comparison of PM_{10} concentrations from the reference device and mobile devices was made both in the frequency domain and in the time domain. The purpose was to determine whether the values obtained by the test devices and the reference device are correlated, whether they have a similar frequency structure and whether the changes in time are similar. Confirmation of these facts will increase the likelihood of test measurements with the reference measurements and will allow further research to confirm the equivalence of new devices with the reference method, after the application of equations or correction factors.

The PM_{10} concentration measurements were conducted in Nowy Sącz. The VIEP reference device (Ref) and two mobile devices (U2 and U4) were located in the city center at a relatively short distance from each other. The other three units were located in the suburbs of the city, one to three kilometers from the center, in urbanized locations (U1 on the north, U3 on the south-west and U5 on the south-east of the center). The location of the equipment was dependent on the decision of the local authorities, which co-organized this tests. Because the devices were not close to each other or in the immediate vicinity of the reference device, it was not possible to compare directly the concentration values. For comparison time series analysis was used to describe the behavior of the data and to detect its similarities.

The study used data from measurements from September 2016 to September 2017. Devices provide data on PM_{10} concentrations hourly, but for comparison with the reference device and later with the reference method, the measurements were aggregated to the daily values by calculating the 24 hourly averages. 376 observations were thus obtained, which were used in further analysis.

The analysis of the measurement series used descriptive statistics methods such as graphs, means and standard deviations as well as Pearson correlation coefficients. The Principal Components Analysis (PCA) was used to verify the similarity of the concentration factors obtained on all devices. PCA's goal is to rotate a multidimensional coordinate system that describes factors affecting the magnitude of the concentration to maximize the variability (variance) of the particular coordinates. The results of this analysis allow us to assess whether each of the series describing the concentrations affects the same factors or whether they are formed by a different set of external factors.

One of the analyzed elements of the obtained PM_{10} concentrations is their comparison with the frequency function. The similarity of periodic variations of the obtained series may indicate the convergence of the concentrations behavior obtained on the mobile devices with the measurement results of the reference device. In such a situation, to obtain convergent values with a device equivalent to the reference, only the correction using a linear model is sufficient.

To compare the time series in the frequency domain, used model based on Fourier series. A transform is used to divide a periodic function (a series of measurements) into a finite sum of trigonometric functions. This allows for a very precise description of the behavior and consequently a comparison of the observed PM_{10} concentrations. Fourier series is a sum of trigonometric functions whose coefficients are determined for each frequency separately. The periodogram is the sum of the squares of the coefficients calculated for specified frequency:

$$P_k = a_k^2 + b_k^2 \tag{1}$$

Periodogram can be interpreted as a variance corresponding to fluctuations of a particular frequency or period. A detailed description of the Fourier series is in [14].

One of the important part of the analysis of the PM_{10} concentration series is the question of whether the series is sufficiently informative to be able to describe it in terms of mathematical models and possible predictions of future values. The tool used for this purpose is an autocorrelation analysis, it means analysis of the relationship of series values with the same series delayed for a certain time.

In the study is also used the analysis of total and partial autocorrelation. Total autocorrelation of the order t is the occurrence of the correlation between the values of the series and the values of the series delayed by t periods. In the case of autocorrelation of the order t, the occurrence of autocorrelation is investigated, after the effect of all autocorrelation of rows 1 to t-1 is removed. Demonstration of the autocorrelation allows us to determine the relationships that occur in a series that can be interpreted as a charge of information or a memory of a series. This information can be used to model or predict a series value. Detailed information on the statistical methods about described statistical methods can be found in [15].

3 Results

In the analysis of the series of observed concentrations of PM_{10}, the behavior of the concentration values in the whole examined period was compared first. The observed concentrations are shown in figure 1.

Based on figure 1, it can be stated that in the cold months PM_{10} concentrations were high, reaching up to 200-300 μg/m^3 in January and February, with high variability of the obtained concentrations. It is also characteristic for this period of considerable divergence in the measurements obtained on different devices.

Lowest concentrations were obtained in warm months. The PM_{10} concentrations in June and July were between 5 and 25 μg/m^3. In this period, they also had the lowest variability and the highest similarity of values.

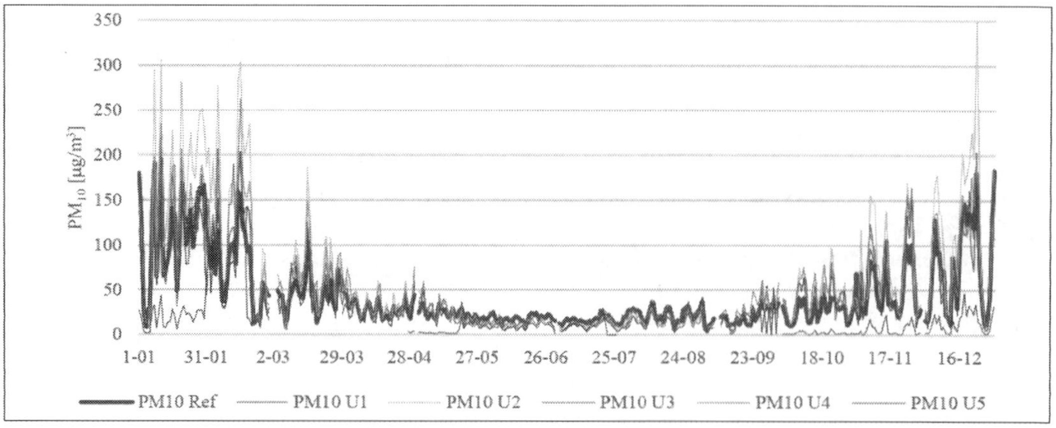

Figure 1. PM_{10} measurement results in Nowy Sącz (source – own study based on own research results)

A very low PM_{10} concentration is noted on the device number 5. This device is located in a part of the city, which is very few buildings and away from the busy roads. The location of the device was most likely to cause such low concentrations.

In order to eliminate influences in the same way on the tested devices as for the reference device, analysis was performed on the difference in concentrations. The comparison of the obtained differences is shown in figure 2.

It is noticeable high value of differences in the cold months disappearing in the warm months. In the case of three devices, however, these differences do not take up large values. In the case of two devices, especially in colder months, there are large differences of the systematic nature. In the case of the U2 device, the differences in most cases are negative (the values obtained on the device are higher than the value from the reference device), while in the case of the U5 meter the positive values are dominant (the device indicates significantly lower values than the reference device).

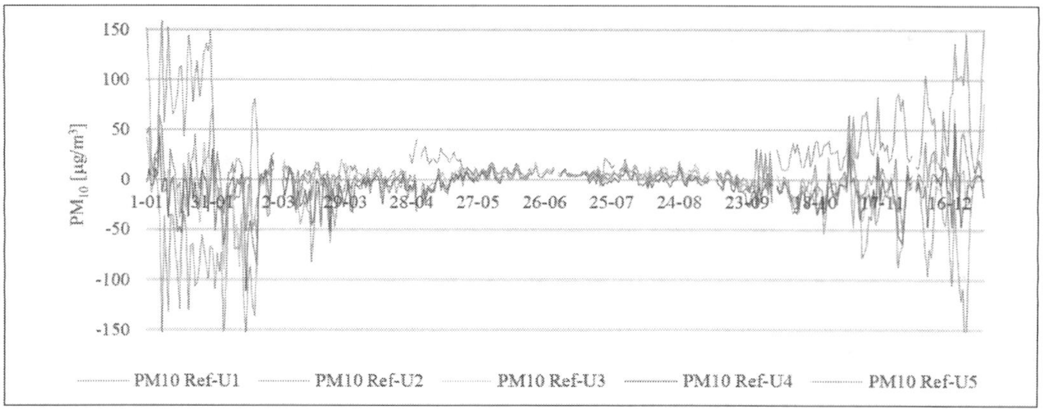

Figure 2. Differences in PM_{10} concentration in VIEP station and tested devices (source – own study based on own research results)

 https://doi.org/10.1051/e3sconf/20182801028

Hypothesis about the similarity of PM_{10} concentration changes obtained on mobile devices and reference device can be confirmed by estimating and comparing the values of linear correlation coefficients between the measured values of these devices. Correlation coefficients are shown in figure 3.

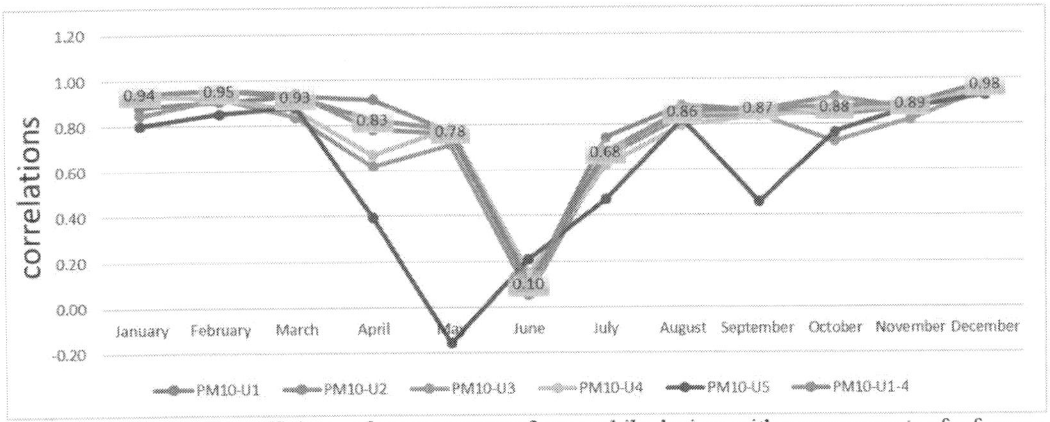

Figure 3. Correlation coefficients of measurements from mobile devices with measurements of reference device (source – own study based on own research results)

The values of correlation coefficients in the vast majority of the studied months are high, ranging from 0.8 to close 1. This is consistent with changes in PM_{10} values measured on the reference device and on mobile devices. The correlation coefficients for the U5 is different (r=0.1). Due to the specific location, they were removed and average correlation coefficients for other devices (U1-U4) were calculated. Their values (placed in frames in Figure 3) indicate a very high compliance of observations from mobile devices and reference device.

The exception to the above conclusions is June, in which the values of all correlation coefficients have practically approached zero. Such an abnormal situation may be explained either by a very low volatility of data from this month (standard deviations from 3.21 to 4.24).

Further part of the analysis was devoted to the identification of factors influencing the measured values of PM_{10} concentrations. The results of Principal Components Analysis is shown in figure 4 and table 1.

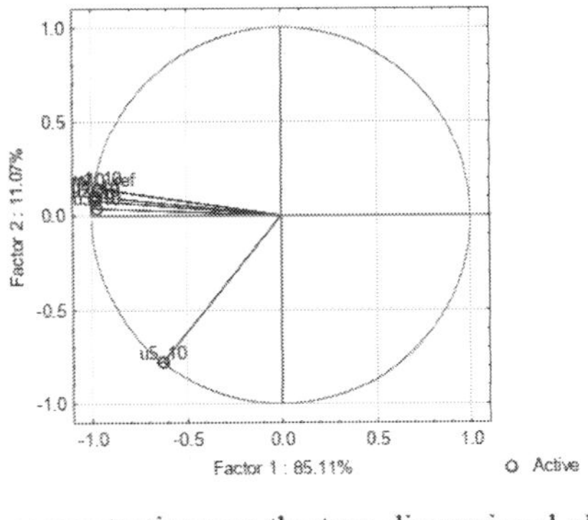

Figure 4. Projection of PM_{10} concentrations on the two-dimensional plane of the factors in the principal components analysis (source – own study based on own research results).

The results of the main components analysis indicate that for all the mobile devices except the U5 device had the same impacts as for the reference device. The compatibility of the factors in the analysis is confirmed in table 1.

Principal component model results for all devices, except U5, have a similar factors (PC1 from -0.98 to -0.96, PC2 form 0.04 to 0.14), indicating that the same concentration factor values were used for the measurements of the concentrations from these devices. Even if the concentration values had different values, they were affected by the same factors and in the same way.

Table 1. Main component loads for PM_{10} concentrations.

Model PCA	*Ref*	*U1*	*U2*	*U3*	*U4*	*U5*	*Own values*	*Percentage of the total*	*Total*
PC1	-0.96	-0.97	-0.98	-0.97	-0.98	-0.63	5.11	85.1%	96.2%
PC2	0.14	0.14	0.08	0.04	0.10	-0.78	0.66	11.1%	

Source. Own study based on own research results.

The obtained model of the main components effectively describes the variability of the analyzed feature. Two components with the highest charges (own values for PC1=5,11, and for PC2=0,66) explain the variation of concentrations in over 96%. However, it is not possible to give their interpretation, in other words – identified factors.

To eliminate the main factors affecting PM_{10} concentrations during the investigated period, the main components were analyzed for the result of subtraction of the concentrations obtained on mobile devices and reference device. The results of PCA analysis are shown in table 2.

Table 2. Principal component loads for subtractions of PM_{10} concentrations.

Model PCA	*Ref-U1*	*Ref-U2*	*Ref-U3*	*Ref-U4*	*Ref-U5*	*Own values*	*Percentage of the total*	*Total*
PC1	-0.72	-0.69	-0.82	-0.92	0.17	2.54	50.9%	82.2%
PC2	-0.49	0.65	-0.28	-0.02	-0.91	1.57	31.3%	

Source. Own study based on own research results

The results of the analysis indicate a greater divergence of factors describing differences in PM_{10} concentrations. Two components PCA model (own values 2.54 and 1.57) describes variation in 82.2%. After eliminating the main aggregate at the concentrations obtained on the reference device, the remaining factors do not allow for clear conclusions. However, similar results of U1, U3 and U4 observations, as well as slightly different behavior of U2 and U5 devices, can be observed. In the whole study, the U5 behaves differently from the rest of the devices, while the behavior of the U2 is considerably more subtle, and only appears after the removal of the major concentration factors.

The periodogram values, obtained by Fourier spectral analysis, for PM_{10} concentrations observed on all devices, together with the correlation coefficients for mobile devices and reference device, show a high likelihood of the series in terms of frequency distributions. Pearson correlation coefficients for periodograms range from 0.88 to 0.98. The PM_{10} concentration components for all analyzed devices have the same structure and are very similar[†].

[†] Due to the limited size of the article, detailed Fourier analysis results are not presented

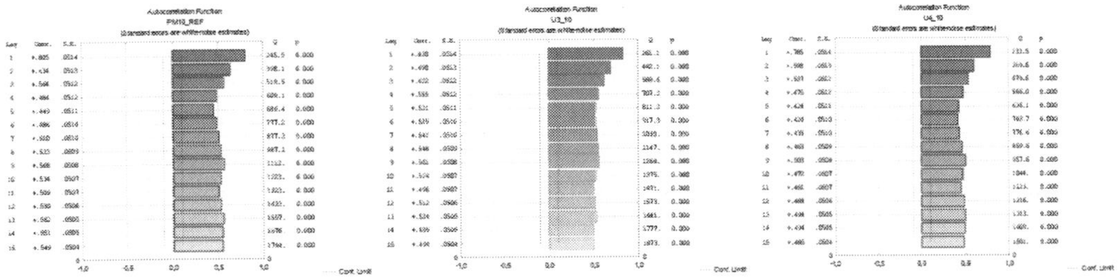

Figure 5. Selected autocorrelation charts of PM_{10} concentration from mobile devices (source – own study based on own research results).

Results of autocorrelation analysis of PM_{10} concentrations, for tree selected devices, are shown in figure 5.

As the result of the analysis observed strong correlations between the test series and the delayed series for at least 15 days. Values of one-day delay are from 0.75 to 0.83. With larger delays, autocorrelation values are smaller but decrease only to a certain level. With delays of 12 to 15 days, correlation coefficients take values close to 0.5.

The autocorrelation analysis is supplemented by partial autocorrelation analysis. Sample results of partial autocorrelation for reference device and devices U1 and U3 are shown in figure 6.

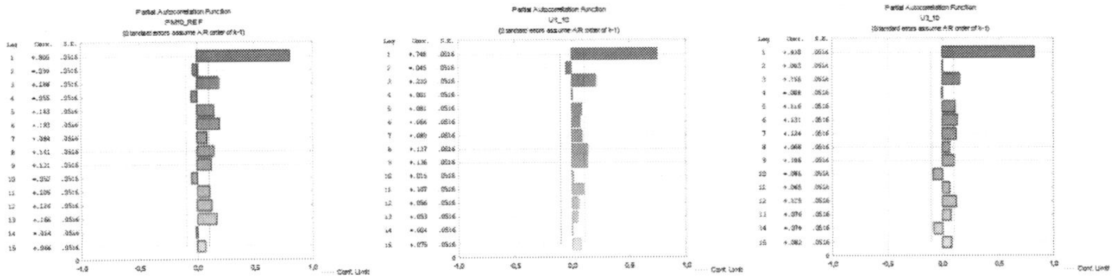

Graph 6. Partial autocorrelation selected charts of PM_{10} concentrations from mobile devices (source – own study based on own research results).

Partial autocorrelation analysis showed strong autocorrelation with one-day delayed sequences and statistically significant autocorrelation with delayed series of 3, 7 or 8 and 12 or 13 days. It is therefore possible to assume that the information contained in the series of PM_{10} concentrations can be used to construct a valid mathematical model describing these concentrations.

4 Conclusions

The PM_{10} concentrations obtained on mobile devices differ noticeably from the results obtained at the reference station in the cold months (average results in January of references station 113.21 μg/m^3, mobile devices – from 97.74 to 162.01 μg/m^3). In warm months these differences are small (in July: Ref – 16.85 μg/m^3, and mobile: from 9.45 to 17.10 μg/m^3). As the concentration decreases, their variability decreases, so that in the warm months it is at a very low level (standard deviation of results in January from 12.33 to 81.66 μg/m^3, in June – 3.51 to 4.24 μg/m^3). In the warmer months, there were also greater divergences between the concentration indications for different measuring devices (average standard deviation of results in January 45.17 μg/m^3, in June – 3.61 μg/m^3).

There is a clear relationship between the concentration measurements obtained by both methods. The correlation coefficients of PM_{10} concentrations obtained with using the reference method and mobile devices calculated for individual months, except for June, have high values, form 0.68 to 0.98. The values of these indicators show that although the concentration values obtained by the different devices are not the same, there is a strong correlation between them that causes the concentration values to change in a similar way.

These observations also confirm the results of the series analysis in terms of frequency of changes over time. The results of the series decomposition into the periodograms corresponding to the individual oscillation frequencies by Fourier analysis show very high convergence in this respect. Correlation coefficients calculated for periodograms (values form 0.88 to 0.98) show that the structure of fluctuations in PM_{10} concentrations obtained by different methods is very close to one another.

The main components analysis proves that the values obtained by the majority of devices are affected by the same or similar factors. The charge values of the components are very similar to each other, and the resulting model translates a decisive part of the variability of the observed concentrations.

The analysis of autocorrelation the series of concentrations shows that there is a large autocorrelation with series delayed by many periods (from 0.75-0.83 for 1-day delay, to 0.5 for 15-day delay). It can be concluded that, thanks to the charge of information on previous events, stored in series, it is possible to use these data to build an effective model describing the behavior of PM_{10} concentrations.

References

1. J. Gębicki, K. Szymańska, Atmos. Environ. **54,** (2012)
2. G. Dongarrà, E. Manno, D. Varrica a, M. Lombardo, M. Vultaggio, Atmos. Environ. **44**, (2010)
3. A. Zwozdziak, L. Samek, I. Sowka, L. Furman, M. Skretowicz, Sci. World. J. (2012)
4. G. Majewski, W. Rogula-Kozłowska, P.O. Czechowski, A. Badyda, A. Brandyk, Atmosphere, **6,** (2015)
5. W. Rogula-Kozłowska, G. Majewski, P.O. Czechowski, P. Rogula-Kopiec, EPE **43**, 1 (2017)
6. *Ustawa prawo ochrony środowiska*, 27 kwietnia 2001r. Dz.U. z 2017 poz. 519
7. *Health risk assessment of air pollution. General principles* (2016), http://www.euro.who.int /en/health-topics/environment-and-health/air-quality/publications/2016/health-risk-assessment-of-air-pollution.-general-principles-2016
8. A. Badyda, J. Grellier, P. Dąbrowiecki, Adv. Med. Exp. Bio. **944** (2016)
9. A. Badyda, P. Dąbrowiecki, P.O. Czechowski, G. Majewski, Respir. Physiol. Neurobiol. **209,** (2014)
10. *Powietrze atmosferyczne -- Standardowa grawimetryczna metoda pomiarowa do określania stężeń masowych frakcji PM10 lub PM2,5 pyłu zawieszonego*, PN-EN 12341:2014, 24 lipca 2014r.
11. *Dyrektywa Parlamentu Europejskiego i Rady w sprawie jakości powietrza i czystszego powietrza dla Europy*, 2008/50/WE, 21 maja 2008 r.
12. J. Gębicki, K. Szymańska, Pol. J. Environ. Stud. **20**, 6 (2011)
13. *Ambient Air – Automated measuring systems for the measurement of the concentration of particulate matter (PM_{10}; $PM_{2,5}$)*, CEN/TS 16450, European Committee for Standardization (2013)
14. G.M. Fichtenholz, Rachunek różniczkowy i całkowy, Tom 3 (Państwowe Wydawnictwo Naukowe, Warszawa, 1985)
15. A. D. Aczel, *Complete Business Statistics* (Richard D. Irwin, Inc., 1989)

Use of metallurgical dust for removal chromium ions from aqueous solutions

Magdalena Pająk[1,*], *Agnieszka* Dzieniszewska[1], *Joanna* Kyzioł-Komosińska[1] and *Michał* Chrobok[1]

[1]Institute of Environmental Engineering Polish Academy of Science, 41-819 Zabrze, 34 M. Skłodowskiej-Curie Street, Poland

Abstract. The aim of the study was to determine the potential for the application of dust from steel plant as an effective sorbent for removing Cr(III) and Cr(VI) in the form of simple and complex ions – Acid Blue 193 dye from aqueous solutions. Three isotherms models were used to interpret the experimental results namely: Langmuir, Freundlich, and Dubinin–Radushkevich. Estimated equations parameters allowed to determine the binding mechanism. Based on laboratory studies it was found that the dust was characterized by high sorption capacities for Cr ions and dye from the aqueous solution. The sorption capacity of the dust for Cr(III) and Cr(VI) ions depended on the degree of oxidation, pH of solution and kind of anion and changed in series: Cr(III)-Cl pH=5.0> Cr(III)-SO_4 pH=5.0> Cr(III)-Cl pH=3.0> Cr(III)-SO_4 pH=3.0> Cr(VI) pH=5.0> Cr(VI) pH=3.0. Dust was also characterized by a high maximum sorption capacity of dye at a range of 38.2 – 91.7 mg/g, depending on the dose of dust. Based on the study it was found that dust from a steel plant, containing iron oxides, can be used as low-cost and effective sorbent to remove pollutions containing chromium ions, especially from acidic wastewater.

1 Introduction

Among the different pollutants of an aquatic ecosystem, dyes are a large and important group of industrial chemicals. They are originated from dyeing process in: textile, cosmetic, paper and leather industries [1-3]. Metal-complex dyes, often used in the dyeing process, contain in their structure chromium ions which are bound in the form of complex ions. This dyes are water-soluble, have high light stability and wet properties. Unused dyes and auxiliary substances, such as hexavalent chromium ions, enter the environment along with wastewater.

Chromium species exist in water in two different oxidation states: Cr(VI) and Cr(III). Both forms of chromium differ considerably in chemical properties and environmental impact. Trivalent chromium is a nutrient and essential microelement at trace levels, on the other hand hexavalent forms of chromium are highly toxic and carcinogenic. Depending on the source (kind of industry), chromium in wastewater occurs as strongly hydrated cations $[Cr(H_2O)_6]^{3+}$, as anions $Cr_2O_7^{2-}$ or $HCrO_4^-$ (galvanic, textile wastes) or as complexed

* Corresponding author: magdalena.pajak@ipis.zabrze.pl

neutral, anionic or cationic compounds (textile dyeing wastes) [4]. According to current Polish legislative regulations of the Ministry of the Environment the concentrations of the Cr(VI) ions and total Cr in sewage introduced into water should be lower than 0.05 – 0.5 mg/L and 0.5 – 1.0 mg/L, respectively. Due to low concentration limit for chromium ions in wastewater, it is necessary to eliminate it thoroughly from contaminated effluents. Several treatment technologies have been developed to remove chromium from water and wastewater. The common methods include chemical precipitation, ion exchange, membrane separation, sedimentation, and adsorption [5, 6]. Adsorption onto activated carbon is effective in the removal of contaminants from wastewater at a wide range of them but its regeneration after the use is troublesome [2, 7-9]. Due to the inability to widely used, arises the need of searching the cost effective substitutes with high sorption capacity. Such sorbents are: clays, peat, brown coal, zeolites, composts and metallurgical wastes [10-16].

Residue such as dusts and sludges that originate from steel plants in course of steel production are only partially recycle. Its chemical composition, including high iron content that suggests possibility of application as sorbent for chromium ions [15, 16].

The aim of this study was to determine the sorption capacity of metallurgical dust for the Cr(III), Cr(VI) ions and metal-complex dye – Acid Blue 193, as well as determining the most efficient and economical (by minimizing the sorbent treatment) process conditions. Moreover, using the nonlinear sorption isotherms the values of parameters in the Freundlich, Langmuir and Dubinin-Radushkevich sorption equations were estimated.

2 Materials and methods

2.1 Research materials

Dust from de-dusting of sintering belt obtained from steel plant was used for laboratory studies. Dust contained high amount of iron compounds (Fe – 51.66%, FeO – 6.07%, Table 1). Presence of calcium oxide contribute to high pH in water suspension (pH=10.11) and high buffer capacity (plot not included) indicating potential to counteract pH changes, for example in reaction with acidic solutions.

Table 1. Chemical composition.

Fe	FeO	SiO_2	CaO	MgO	Al_2O_3	Mn	Na_2O	K_2O	Zn	S
51.66	6.07	7.02	7.72	1.45	1.45	0.34	0.14	0.15	0.07	0.18

Sorption studies were carried out using metal-complex dye – Acid Blue 193 (AB-193) and Cr(III) and Cr(VI) ions.

Figure 1 shows the characteristics of the Acid Blue 193 dye: structural formula of dye, CAS number, molecular formula, molecular weight, wave length at which concentrations were measured and pH value in water at 1 g/L.

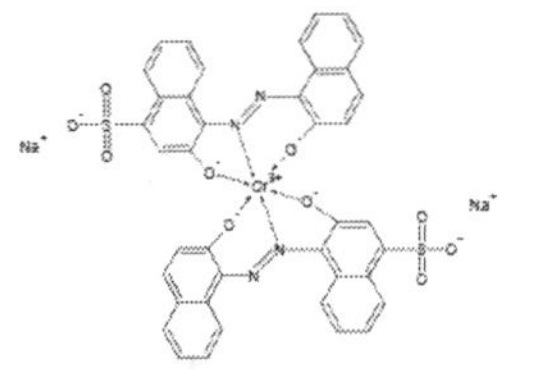

CAS: 12392-64-2

C40-H22-Cr-N4-O10-S2.2Na

M 880.7 g/mol

pH 6.70

Fig. 1. Characteristics of Acid Blue 193 dye.

Chromium (III) ions solution was prepared by dissolving $CrCl_3 \cdot 6H_2O$ or $KCr(SO_4)_2 \cdot 12H_2O$ in water. Solution of chromium (VI) ions was prepared by dissolving K_2CrO_4.

2.2 Methods

2.2.1 Sorption process

Laboratory experiments on sorption of metal-complex dye and Cr(III) and Cr(VI) were carried out under the batch method at room temperature, contact time of the solid phase (S) to the solution (L) – 24 hours, the initial concentrations of dye within the range 1 – 1000 mg/L, and of Cr ions 0.1 – 1000 mg/L, the solid phase to the solution ratio for AB-193 dye 1:50, 1:100, 1:200, for Cr ions 1:100, and the pH in the Cr(III) and Cr(VI) stock solutions 3.0 or 5.0 adjusted using 0.1M KOH and 0.1M H_2SO_4, respectively.

The initial (C_0) and equilibrium (C_{eq}) concentrations of the dye were determined using UV-vis spectrometry (Spectrometer Varian Cary 50 Scan UV-VIS) and Cr ions concentrations were determined by atomic absorption spectrophotometry (Spectra 200, Varian).

Amount of dye and Cr ions adsorbed (q) by the used sorbent was calculated from the formula:

$$q = (C_0 - C_{eq}) \cdot \frac{V}{m} \quad (1)$$

where: m – sorbent mass (g), V – solution volume (L).

Percentage pollutions removal (R) was calculated from the formula:

$$R = \frac{C_0 - C_{eq}}{C_0} \cdot 100\% \quad (2)$$

2.2.2 Determination of parameters in sorption isotherms

In order to estimate sorbent maximum sorption capacity and to determine the mechanism of the binding of the dye and Cr ions, three isotherm equations were used: Freundlich, Langmuir and Dubinin-Radushkevich.

Freundlich Isotherm of the formula [17]:

$$q = K_F \cdot C_{eq}^{\ 1/n} \quad (3)$$

where: K_F – Freundlich isotherm constant ((mg/g)·(L/mg)$^{1/n}$), 1/n – measure of surface heterogeneity (the higher the 1/n value, the more favorable the adsorption).

Langmuir Isotherm represented by the following equation [18]:

$$q = \frac{QK_L C_{eq}}{1 + K_L C_{eq}} \quad (4)$$

where: Q – maximum sorption capacity relative to dye/Cr ions (mg/g), K_L – Langmuir isotherm constant (L/mg).

Dubinin-Radushkevich Isotherm represented by the equation:

$$q = q_D \cdot exp(-\beta \varepsilon^2) \quad (5)$$

where: q_D – the sorption capacity of studied dust for dye/Cr ions (mmol/g), β – the constant (mol^2/J^2), ε is the Polanyi potential (J/mol), which is equal to:

$$\varepsilon = RT\, ln(1 + \frac{1}{C_{eq}}) \quad (6)$$

where: R – gas constant (8.314 J/mol·K), T – absolute temperature (K).
The β constant was used to determine the free energy (E) of the adsorption:

$$E = \frac{1}{(2\beta)^{1/2}} \tag{7}$$

The E value indicates that it is ion exchange adsorption (8<E<16 kJ/mol), or physical adsorption (E<8 kJ/mol) [19].

To determine the values of parameters in the sorption isotherms the nonlinear regression method was used (Statistica ver. 9.0). Apart from determination coefficient four different error functions of non-linear regression basin were employed in this study to find out the most suitable isotherm models to represent the experimental data [20]:

the sum of the errors square $$SSE = \Sigma(q_{cal} - q_{meas})^2 \tag{8}$$

nonlinear chi square test $$\chi^2 = \sum \frac{(q_{cal} - q_{meas})^2}{q_{meas}} \tag{9}$$

where: q_{cal} – calculated, q_{meas} – measured.

3 Study results and discussion

3.1 Sorption of Acid Blue 193 dye onto dust

Figure 2 shows experimental sorption isotherms for steel plant dust and the percentage removal Acid Blue 193 from aqueous solutions, depending on the ratio of solid phase (S) to solution (L).

The results showed that at a ratio of S:L=1:50 (dust dosage 20 g/L) AB-193 dye was completely bound, at the initial concentration ranging from 1 – 100 mg/L, sorption capacity changed in a range 0.0545 – 4.96 mg/g. At the ratio of S: L=1:100 (dust dosage of 10 g/L) and 1:200 (dust dosage of 5 g/L) dye was completely bound only at concentration equal 1 mg/L. As the initial concentration increased, the sorption capacity of the dust increased.

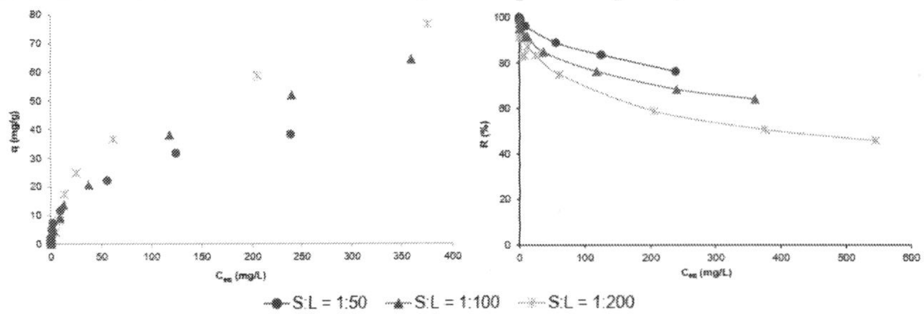

Fig. 2. Experimental sorption isotherms and the percentage removal from aqueous solutions of AB-193 dye on dust.

At the maximum initial concentration of 1000 mg/L and the ratio S:L=1:50 AB-193 dye was bound in an amount of 38.2 mg/g, in the ratio S:L=1:100 and 1:200 were bound in an amount 64.31 mg/g and 91.7 mg/g, respectively. With the increase of the dye initial concentration in the solution and the decrease the dosage of dust, it was observed a reduction of the degree its removal from the solution (removal efficiency). At the maximum dye initial concentration in solution, the removal rate was 76.17%, 64.11% and 45.71%, respectively. The trend indicates that the sorbent percentage was proportional to the sorbent mass and was in agreement with the formula:

$$R = \frac{q \cdot m}{C_0 \cdot V} \cdot 100\% \qquad (10)$$

obtained by transforming Eq. 1 and substituting it in Eq. 2.

Sorption process of the AB-193 dye was carried out at practically constant pH values in the range of 12.75 – 13.02.

Genc and Oguz [16] were carried out studies on the removal of metal-complex dyes containing chromium using metallurgical wastes. Scientists have found much worse research results (such as q = 3.2 mg/g for Acid Yellow 99, q = 0.6 mg/g for Acid Red 183) under the same conditions of sorption (such as S:L = 1:100, C_0 = 100 mg/L).

By comparing the sorption capacity and the removal rate of AB-193 dye depending on the ratio solid phase to the solution (S:L), it was found that the most favorable conditions of its binding are at S:L=1:100 (Fig. 3). Therefore, it was decided that the Cr ions sorption studies would be carried out only at the ratio S:L equal 1:100.

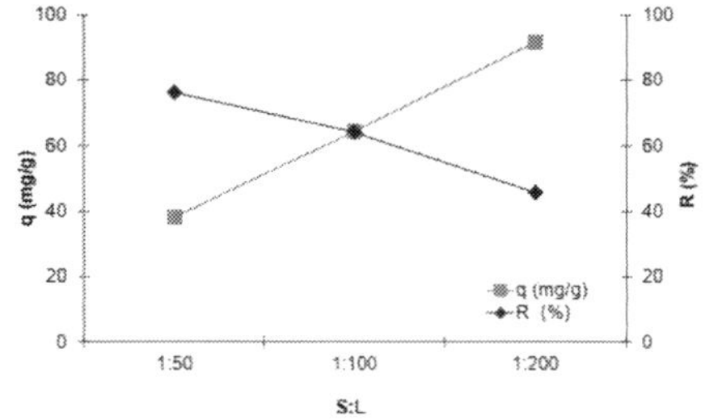

Fig. 3. Sorption (q) and removal efficiency (R) of AB-193 as a function dust dose.

3.2 Sorption of chromium (III) and (VI) onto dust

Figure 4 shows experimental sorption isotherms and the percentage removal from aqueous solutions of chromium (III) and (VI) on dust.

It was found that the sorption course and the sorption capacity depended on the degree of oxidation of Cr ions, pH of the solution, type of anion in solution, and decreases in following order: Cr(III)-Cl pH = 5.0> Cr(III)-SO_4 pH = 5.0> Cr(III)-Cl pH = 3.0> Cr(III)-SO_4 pH = 3.0> Cr(VI) pH = 5.0> Cr(VI) pH = 3.0.

The sorption capacity of Cr(III) ions from chloride solution at pH 5.0 and at initial ions concentration in range 0.1 – 990 mg/L increased from 0.0079 to 98.8 mg/g. The removal efficiency of Cr(III) ions in the studied range of solutions was steady at 99.90%. The sorption process was carried out at high pH value of the equilibrium solution in the range of 9.4 – 7.4. High solution pH indicate that Cr ions precipitate in form of $Cr(OH)_3$.

Cr(III) ions from sulfate solutions at pH 5.0 were bonded by dust in lower amounts than from chloride solutions. The sorption capacity varied from 0.018 to 62.0 mg/g. In the concentration range of 0.1 to 500 mg/L, the removal of Cr(III) ions from the solution was at a constant level of 99.99% and the equilibrium concentration the pH in the solution varied from 8.0 to 7.4, which also indicates the possibility of $Cr(OH)_3$ precipitation. The maximum sorption capacity of dust was 62.0 mg/g, and a removal efficiency – 69.66%.

At solution of initial pH=3.0 decrease in sorption capacity, for Cr(III) ions, occurs both in chloride and sulfate solution. Sorption from chloride solutions varied in the range of 0.0138 – 38.0 mg/g and from sulfate solutions – 0.0109 – 23.0 mg/g, while the degree of removal Cr(III) ions from the solutions in the range of 99.0 – 36.19% and 99.0 – 21.29%, respectively. Sorption process of Cr(III) was carried out at practically similar pH values in the range of 9.0 – 3.54 and 9.0 – 3.80, respectively. At the same time it was found that, in initial concentration range from 0.1 to 100 mg/L, ions from both solutions were sorbed completely and solution pH was above 7.5.

In the experiment $CrCl_3 \cdot 6H_2O$ and $KCr(SO_4)_2 \cdot 12H_2O$ were used and values of pH of the stock solutions were 2.76 and 3.05, respectively. For adjustment pH solution to values 5.0 and 3.0 were used 0.1 M KOH solution. K^+ ions were in competition for sorption centres with Cr(III) and H^+ ions mainly at higher concentration of Cr(III) ions. Presence of K^+ ions reduce of sorption capacity of dust for Cr(III) and maximum sorption capacity was considerably higher for chloride than sulfate solution.

Studied dust has high buffer capacity and suspension pH due to calcium oxide presence. Hence pH in equilibrium solutions was higher than in starting solutions. Moreover pH above 5.0 causes Cr precipitate in form of $Cr(OH)_3$.

Presence of Fe compounds in dust causes that Cr(III) ions can be bound by hydrated iron oxides in the reaction $3FeO(OH) + Cr^{3+} = (FeOO)_3Cr + 3H^+$.

Dust sorbed Cr(VI) ions in smaller quantities in corresponds to Cr(III) ions. From solutions of pH 5.0 they were sorbed in range 0.008 to 18.0 mg/g and from solutions of pH 3.0 from 0.012 to 16.0 mg/L, no influence of solution pH upon dust sorption capacity was found. Removal degree of ions from solution changes in following order 90.0 – 18.0% and 90.0 – 16.0%, respectively. Sorption of Cr(VI) ions occurred at similar pH 9.22 – 6.99.

Iron oxide belongs to compounds of high point of zero net charge (pH_{PZC}> 8.0) value, which is defined as the pH at which the charge of the colloidal particles is equal to zero. The minerals are positively charged due to the protonation of the surface when the solution pH is lower than their pH_{PZC} values and negatively charged due to the deprotonation of the surface when the solution pH is higher than the pH_{PZC}. Hence iron compounds at pH below 8.0 have positively charged surface and can bond anionic contaminants ($Cr_2O_7^{2-}$) or anionic dyes.

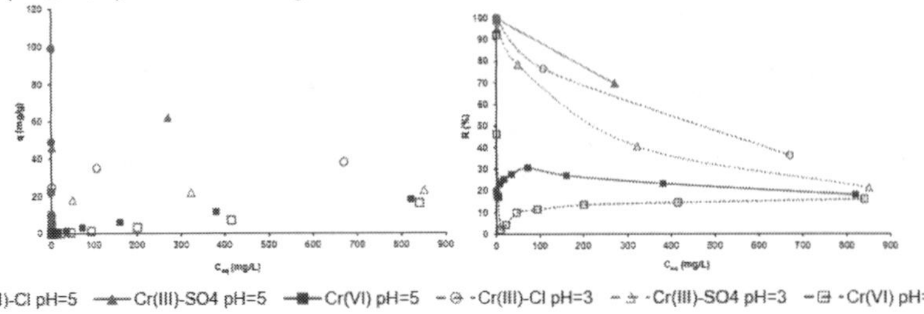

Fig. 4. Sorption isotherms and the percentage removal of Cr(III) and (VI) for studied dust.

3.3 Freundlich, Langmuir and Dubinin-Radushkevich adsorption isotherm models

The isotherm parameters for all the models studied along with correlation coefficient R^2 and values of error for functions of non-linear regression are listed in Table 2. The plots of the experimental data and the predicted data from Langmuir, Freundlich and Dubinin-Radushkevich (D-R) isotherm models of studied dust are shown in Figure 5.

The R^2 values suggest that the all isotherms provide a good models of the sorption of AB-193 on metallurgical dust. Values of 1/n less than 1 in Freundlich isotherm show the favorable nature of adsorption of AB-193 on dust for all S:L ratio. The saturated monolayer sorption capacity, q_{max} for dye, were higher than the experimental values. At the same time, the D-R isotherm model also well described the AB-193 sorption onto dust. The mean free energy values (E) for adsorption of dye onto dust were above 8.0 kJ/mol and indicate that the sorption process had an ion-exchange character. On the other hand Langmuir and D-R isotherm exhibited lower SSE and χ^2 values then Freundlich considered to be a better fit compared to the Langmuir and D-R isotherms.

Table 2. Parameters of sorption isotherms estimated on Freundlich, Langmuir and Dubinin-Radushkevich models.

	Cr(III)-Cl		**Cr(III)-SO4**		**Cr(VI)**		**Acid Blue 193**		
	pH=5	**pH=3**	**pH=5**	**pH=3**	**pH=5**	**pH=3**	**1:50**	**1:100**	**1:200**
				Freundlich isotherm					
1/n	2.40	0.2161	0.1774	0.1804	0.7133	0.8670	0.3564	0.4959	0.6628
K_F	10785	19.52	13.25	7.4055	0.1537	0.00617	5.478	3.470	1.440
R^2	0.9930	0.7556	0.8471	0.9046	0.9942	0.9998	0.9975	0.9984	0.9967
SSE	84.76	284.4	323	43.7	1.93	1.04	4.20	7.20	32.02
χ^2	44.52	331.8	742.6	124.62	1.95	0.51	0.84	1.29	1.61
				Langmuir isotherm					
$q_{exp.}$	98.99	38.00	62.00	23.00	18.00	16.00	38.20	64.30	91.70
Q	Not estimated	63.17	36.23	24.94	35.22	45.34	40.22	78.18	171.6
K_L		1.1523	4.033	12.81	0.00128	0.00045	0.0332	0.0962	0.0021
R^2		0.8844	0.9683	0.9277	0.9996	0.8872	0.9542	0.9822	0.9934
SSE		46.79	507.6	256.4	2.15	13.87	79.22	84.96	64.15
χ^2		7.30	65.77	108.1	2.25	4.51	12.49	7.98	7.94
				Dubinin-Radushkevich isotherm					
q_D	322.9	1.743	0.933	0.534	1.139	2.545	0.135	0.3556	0.896
β	0.0107	0.0019	0.0016	0.0016	0.0113	0.0204	0.0028	0.0043	0.0065
E	6.83	17.90	16.90	17.71	6.66	4.95	13.31	10.72	8.78
R^2	0.9639	0,8191	0.9011	0.9282	0.9995	0.9988	0.9943	0.9979	0.9955
SSE	197.9	186.5	828.5	25.92	0.14	0.28	9.86	9.93	43.99
χ^2	167.4	121.9	498.4	90.17	0.11	0.30	3.82	1.55	4.96

K_F ((mg/g)·(L/mg)$^{1/n}$), q (mg/g), Q (mg/g), K_L (L/mg), q_D (mol/g), β (mol^2/kJ^2), E (kJ/mol)

It was determined that Cr(III) ions sorption, from chloride solution of pH 5.0, cannot be modeled by Langmuir isotherm. However Freundlich and D-R equations well described the sorption process. Values of 1/n parameter, in Freundlich equation, above 1 indicate precipitation of Cr(III) ions in form of $Cr(OH)_3$. The E parameter in D-R isotherm was below 8.0 kJ/mol what indicates physical character of sorption.

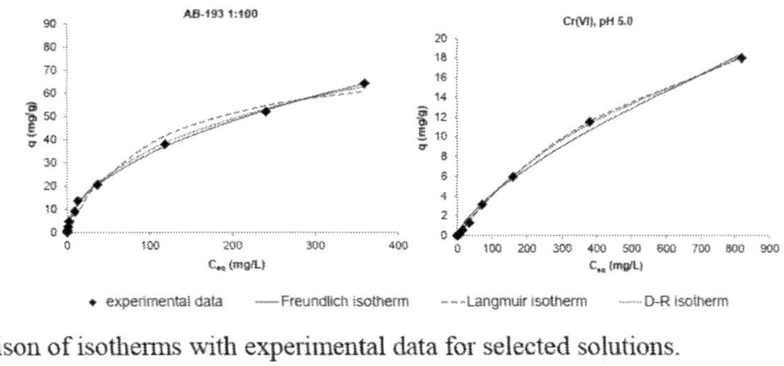

Fig. 5. Comparison of isotherms with experimental data for selected solutions.

From applied models, Langmuir isotherm well described sorption of Cr(III) ions from chlorides, at pH 3.0 as well as from sulfate solutions at pH 3.0 and 5.0. Correlation coefficient R^2 was in range 0.88 – 0.97.

All studied isotherm well described the sorption of Cr(VI) at pH 5.0 and 3.0. Values of 1/n below 1 show the favorable nature of adsorption and the values of E parameter below

8.0 kJ/mol indicate physical character of sorption. Langmuir the saturated monolayer sorption capacity values, Q parameter indicates high sorption properties of dust for Cr(VI) ions. Value of Q for sorption at pH 3.0 was higher than the Q for sorption at pH 5.0.

4 Summary

Conducted studies of sorption capacity of steel plant dust for Cr ions indicate that sorption capacity depends upon oxidation state, solution pH, as well as anion kind and changes from 99.0 – 16.0 mg/g in following order Cr(III)-Cl pH = 5.0> Cr(III)-SO_4 pH = 5.0> Cr(III)-Cl pH = 3.0> Cr(III)-SO_4 pH = 3.0> Cr(VI) pH = 5.0> Cr(VI) pH = 3.0. Metallurgical dust possess high maximal sorption capacity towards AB-193 dye on level 38.2 – 91.7 mg/g depending upon dust dosage.

Studies upon possibility of application of residual dust from steel plants indicate that mentioned dust may be used as effective sorbent for removal of Cr(III) and Cr(VI) ions from aqueous solutions including acidic solutions.

The research was conducted at the Institute of Environmental Engineering, Polish Academy of Sciences in Zabrze, as part of the research project UMO-2012/05/N/ST8/03149.

References

1. M. Wawrzkiewicz, M. Wiśniewska, A. Wołowicz, V. M.Gun'ko, V.I. Zarko, Micropor. Mesopor. Mat. **250**, 128 (2017)
2. D. Kołodyńska, J. Krukowska, P. Thomas, Chem. Eng. J. **307**, 353 (2017)
3. H.S. Rai, M.S. Bhattacharyya, J. Singh, T.K. Bansal, P. Vats, U.C. Banerjee, Crit. Rev. Env. Sci. Tec. **35**, 219 (2005)
4. J. Kyzioł-Komosińska, C. Rosik-Dulewska, A. Dzieniszewska, M. Pająk, I. Krzyżewska, Environ. Prot. Eng. **40**, 1, 5 (2014)
5. S. Babel, T.A. Kurniawan, Chemosphere **54**, 7, 951 (2004)
6. G. Crini, Bioresource Technol. **97**, 1061 (2006)
7. M.T. Yagub, T.K. Sen, S. Afroze, H.M Ang, Adv. Colloid Interface Sci. **209**, 172 (2014)
8. V.K. Gupta, Suhas, J. Environ. Manage. **90**, 2313 (2009)
9. N. Barka, A. Assabbane, Y. Aît Ichou, A. Nounah, J. Appl. Sci. **6**, 3, 692 (2006)
10. I. Chaari, B. Moussi, F. Jamoussi, J. Alloy. Compd. **647**, 720 (2015)
11. R. Elmoubarki, F.Z. Mahjoubi, H. Tounsadi, J. Moustadraf, M. Abdennouri, A. Zouhri, A.E. Albani, N. Barka, Water Resources and Industry **9**, 16 (2015)
12. Y.S. Ho, G. McKay, Chem. Eng. J. **70**, 2, 115 (1998)
13. Y.S. Ho, G. McKay, Process Biochem. **38,** 7, 1047 (2003)
14. N. Mirzaei, H.R. Ghaffari, K. Sharafi, A. Velayati, G. Hoseindoost, S. Rezaei, A.H. Mahvi, A. Azari, K. Dindarloo, J. Environ. Chem. Eng. **5**, 4, 3151 (2017)
15. C.N.R. Amaral, F.N. Feiteira, R.C. Cruz, V.O. Cravo, R.J. Cassella, W.F. Pacheco, J. Environ. Chem. Eng. **4**, 4, 4184 (2016)
16. A. Genc, A. Oguz, Desalination **264**, 78 (2010)
17. C. Ng, J.N. Losso, W.E. Marshall, R.M. Rao, Bioresource Technol. **85**, 2, 131 (2002)
18. A.O. Dada, A.P. Olalekan, A.M. Olatunya, O. Dada, IOSR-JAC **3**, 1, 38 (2012)
19. A. Özcan, E.M. Öncü, A.S. Özcan, Colloid. Surface. A **227**, 90 (2006)
20. K.Y. Foo, B.H. Hameed, Chem. Eng. J. **156**, 2 (2010)

E3S Web of Conferences **28**, 01030 (2018) https://doi.org/10.1051/e3sconf/20182801030
Air Protection in Theory and Practice

Distribution of gaseous and particle-bound Hg concentrations at the sites representative for urban and non-urban zones of Silesia Province

Halina Pyta[1,*], *Krzysztof* Słaby[1], *Krzysztof* Rajczykowski[1], and *Tadeusz* Sadowski[2]

[1]Institute of Environmental Engineering of the Polish Academy of Sciences in Zabrze, Poland
[2] Regional Inspectorate for Environmental Protection in Katowice, Poland

Abstract. The basic features of the distribution of total gaseous (TGM) and particle-bound mercury (PBM) concentrations were determined for a five locations representative for urban (Bielsko-Biała, Lubliniec, Zabrze) and rural areas (Godów, Złoty Potok) of Silesia Province. Gaseous mercury concentrations were measured (1) continuously - the automatic 1h TGM measurements in Zloty Potok and Zabrze and (2) non-continuously manual 24h TGM measurements with a pre-concentration of the Hg on gold traps (Bielsko-Biała, Lubliniec, Godów). The PBM concentrations were measured non-continuously by taking PM2.5 samples. The Hg content was determined by using a CVAAS method. The highest average concentration of TGM was recorded in Zabrze (2.8ng/m^3), significantly lower (2.0ng/m^3) in Bielsko-Biała and in the non-urban station in Godów, the lowest concentration (<2.0 ng/m^3) was observed in Lubliniec and at the regional background station in Zloty Potok. The results obtained for TGM concentrations exceeded the European average level of 1.5 ng/m^3 (AirBase, 2014). The highest average PBM concentration, associated with PM2.5, was obtained in Zabrze (70pg/m^3), more than 20% lower results were obtained in Bielsko-Biała and Godów, finally, the lowest one (lower by about 40% in comparison with Zabrze) were obtained in Lubliniec and Złoty Potok. Moreover, an enrichment of Hg concentration in PM was observed with the increasing of the PM content, during the heating season.

1 Introduction

Mercury (Hg) is a naturally occurring element found in the environment in metallic form (elemental mercury Hg^0 in liquid or gaseous state) or in various inorganic (mercuric chloride $HgCl_2$, mercuric sulphide HgS, mercurous chloride HgCl) and organic compounds (methylmercury MeHg, ethylmercury EtHg). Mercury is a persistent substance that comes from natural and anthropogenic sources. It is a toxic pollutant, causing adverse health effects on humans and wildlife. Released to the atmosphere it can be transported over long distances and deposited on the Earth's surface, leading to contamination of aquatic and terrestrial ecosystems. Atmosphere is the main pathway for Hg distribution on a global and

* Corresponding author: halina.pyta@ipis.zabrze.pl

 https://doi.org/10.1051/e3sconf/20182801030

regional scale. Atmospheric Hg is dominated (95-99%) [1-2] by the gaseous elemental mercury (GEM, Hg^0) with the background concentrations in the Northern Hemisphere from 1.3 to 1.7 ng/m^3 [3]. Other forms of atmospheric Hg are operationally defined as reactive gaseous mercury (RGM) and particulate bound mercury (PBM). Their concentrations are much lower (expressed in pg/m^3). Despite this, RGM and PBM are essential for mercury removing from the air by wet and dry deposition [1].

Mercury can undergo methylation by aquatic biota and bioaccumulation in aquatic food chains including fish and shellfish. The consumption of contaminated seafood and rise is the main source of MeHg intoxication, which is known human neurotoxin, especially dangerous for foetus [4]. Organic and inorganic mercury exposure has also been proven to cause elevated risks for cardiovascular diseases and negative impacts to the reproductive and immune systems. Exposure to inorganic mercury by the general population and in occupational settings is primarily through inhaling elemental mercury vapours [5].

Over geological time, Hg existing naturally in the Earth's crust (~80 μg/kg) has been released to the upper layer environment via the weathering of rocks, volcanic and geothermal activity, fires, movement of land water, and oceanic upwelling [5]. These are primary natural sources. Secondary natural sources redistribute mercury among ecosystems due to Hg historical emissions through the biomass burning, the land use changes, and the mechanisms of gaseous mercury exchange between the air and the water, top soil, or snow–ice pack surfaces [6]. There are several estimates on Hg emissions of natural origin in scientific literature, ranging from 3600 Mg/yr to 5300 Mg/yr [6]. In the EU GMOS project, the inventory of these emissions (primary and secondary) were assessed to be 5207 Mg/yr [7]. The major anthropogenic sources of global mercury emissions to the atmosphere are artisanal and small-scale gold mining, combustion of fossil fuels for power and heat generation, industry of non-ferrous metals, cement production and large-scale gold production [8]. According to the AMAP/UNEP inventory, the anthropogenic mercury emissions have been estimated to 1960 Mg/yr [8]. Therefore these emissions can be regarded as the main factor disrupting the natural (biogeochemical) mercury cycling. Due to anthropogenic emissions, current average mercury levels in the atmosphere are about 3–6 times higher than in the preindustrial atmosphere, resulting in a significant increase in human and wildlife exposure [9].

Monitoring of various forms of atmospheric mercury is a key factor in the assessment of the effectiveness of activities aimed at reducing the emission of mercury and assessing the health risk related to that problem. This applies in particular to inhabited areas, with a high Hg emissions levels. Poland, according to EEA data, is the country with the highest Hg emissions in the EU (Figure 1).

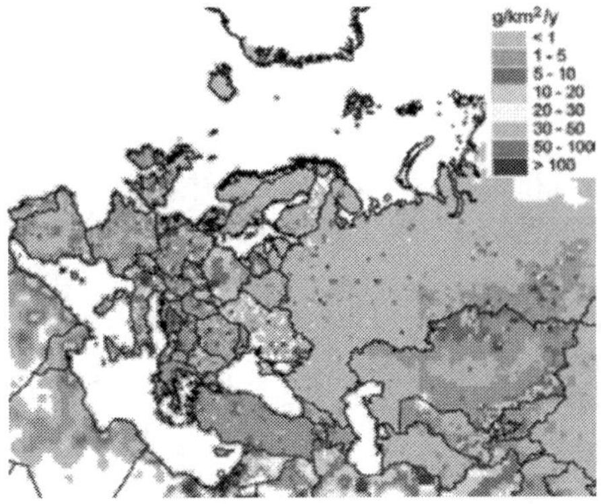

Fig. 1. Spatial distribution of Hg anthropogenic emissions over the EMEP domain in 2015 [10].

The amount of this emissions, according to the data from 2015 was estimated at 11 Mg, which consist18.6% of the whole emission in the EU [11]. The results of the EMEP (European Monitoring and Evaluation Program) modelling show that the most polluted area of Poland is the Silesia Province [12]. These results become a reason for the launching in the period from January 2014 to December 2015 an extended atmospheric mercury research program in the Silesia Province. The project was realized in Institute of Environmental Engineering of the Polish Academy of Sciences (IEE PAS) in cooperation with the Regional Inspectorate for Environmental Protection (RIEP) in Katowice (measurements were carried out at the stations belonging to RIEP, PM samples were partially collected by the RIEP service). The results of TGM and PBM studies were presented in this paper.

2 Sampling sites

The measurements were conducted at five sites located in different parts of the Silesia Province, with a different topographical, emission and land use conditions. Three stations were located in a typical urban areas: (a) in Bielsko-Biała, at Kossak-Szczucka St., city with a population about 173 thousand people, located in the southern part of the Silesia Province, in the foothills of the Beskidy mountains; (b) in Lubliniec at Piaskowa St., city inhabited by a 24 thousand people, located in the western part of the province; (c) in Zabrze, at Skłodowska-Curie St., city with a 177 thousand inhabitants, located in the central part of the province, within the Upper Silesian Agglomeration, which is a densely populated and the most industrialized part of Poland. Zabrze, among all of the 14 cities of the Upper Silesian Agglomeration, has the least developed heating network. The other two measurement sites were located in a non-urban zones in a typical rural areas: (a) in Złoty Potok – a village inhabited by 1200 people, located in the north-western part of Silesia Province, about 20 km south-west of the city Częstochowa, and 45 km north west of the borders of the Upper Silesian Agglomeration; the station is surrounded by farmlands and arable land; and, being located at relatively weakly polluted area, it serves as regional background station for Silesia Province; (b) in Godów, at a Glinki St., a village inhabited by 2000 people, located in the southern part of Silesia Province, just near the Polish-Czech border and the Moravian-Silesian Region; the station is surrounded by meadows, farms and detached houses.

3 Measurement methods

3.1 Automatic measurements of gas-phase mercury

In Zabrze and Złoty Potok the concentration of gaseous mercury was measured continuously by using an automatic Tekran 2537B analyzer (Tekran Instr. Co.). The mercury present in the air was pre-concentrated with forming of Au-Hg amalgam onto a gold trap. After that, a thermal desorption and cold vapour atomic fluorescence spectrometry analysis (CVAFS), at wavelength λ = 253.7 nm was performed. Two Au sorbent tubes were working alternately in the analyzer, which ensured continuity of sampling. In Złoty Potok the TGM concentration was measured by Tekran 2537B, working as a separate module collecting results in every 5 minutes. However, in a RIEP data acquisition centre the measuring results were archived as 1-hour averages. On the other hand in Zabrze, Tekran 2537B is a module of much more advanced measuring system, allowing to the Hg speciation analysis, and it works in a 2 hours cycles (1 hour of sampling and an hour of both PBM and RGM forms desorption). Thus the frequency of results receiving in Zabrze is twice less. All the operational parameters of the analyzers -and their calibration methods were the same. The analyzers were calibrated every day by using an internal Hg^0 permeation tube and periodically by using an external gas standard. The detection limit of Hg^0 for the measurement conditions in both cases was equal 0.06 ng/m^{-3} and the precision of analysis was 2%.

3.2 Manual measurements of gas-phase mercury

Manual measurements of gaseous mercury in Bielsko-Biała, Godów, and Lubliniec were carried out in 24-h intervals with the frequency of minimum 7 measurements per month, equally distributed during the year. Air samples were collected using sorbent tubes (gold traps) and gas aspirators operating with a flow rate of 30 dm^3/h. Besides the current samples, a blank transportation samples were also made. For the analysis, quartz tubes filled with gold-coated diatomite (M165 tubes by Nippon Instruments Co.) were used. The inlets of the tubes were protected with a quartz wool stoppers. MA-2 analyzer (Nippon Instruments Co.) was used for a chemical analysis of collected samples. The analysis was based on a thermal decomposition of samples in zero-air conditions (in temperature >600^{0}C) and pre-concentration of the released Hg on the internal gold trap. After that, mercury in the form of Au-Hg amalgam was desorbed in a high temperature (>600^{0}C) and analyzed by a cold vapour atomic fluorescence spectrometry analysis (CVAFS), at wavelength $\lambda = 253.7$ nm. The detection limit of a mentioned method for 24-h samples, at a flow rate equal 0.5 dm^3/min was 0.15 ng/m^3.

3.3 PM sampling and determination of particulate mercury

Particle mercury bound with the respirable fraction PM2.5 (24-h samples) were analyzed at all stations. In Lubliniec, PM10- and PM1-bound Hg were also determined. The measurements of PBM concentrations were performed with the frequency of minimum 7 measurements per month, equally distributed during the year, except for Godów and Złoty Potok, where the concentrations were determined every day. The samples of PM were collected onto quartz fiber filters using a low volume samplers (LVS) with automatic filters changers and the proper PM sampling heads. Mikro PNS samplers were used in all sites except Lubliniec, where 2-channel (PM1, PM2.5) Hydra Plus LVS (FAI Instruments s.r.l.) was installed and Godów with a high volume DHA-80 sampler (Digitel Elektronik AG) for PM2.5. Immediately before and after sampling, the filters were conditioned and weighed. The procedures for conditioning, weighing, storage, and transport of the analysed samples and a blank sample preparation, were conducted according to the QA/QC procedures of the reference method for gravimetric measurements (PN-EN 12341:2014) and were described in detail in Pyta et al.[13]. The Hg content was determined by using CVAAS method, based on thermal decomposition of PM samples using the MA-2 analyzer [13]. The limits of detection and quantification were 0.094 and 0.283 ng, respectively. The average recovery of SRM was 96%. The repeatability was 3.9% for SRM and its average recovery was 96%.

4 Results

4.1 Gas-phase mercury

Statistical parameters of the measurement series of TGM concentrations were summarized in table 1, with a distinction on a summer season (a non-heating period from April to the end of October) and winter (heating) season (the other 6 months). Phrase "total" denotes a whole measurement period. At the station in Zabrze, a higher TGM concentrations were recorded in summer season than in winter. This was due to the elevation of the air inlet (14 m a.g.l.), which allowed for the recording of concentrations caused by higher local emitters in summer season. In winter, the influence of higher emitters was undetectable due to the thermal inversion phenomenon, which prevents the pollutants to reach the ground level.

Table 1. Total gaseous mercury concentrations [ng/m^3] – the descriptive statistics of 24-h and 1-h measurement series at urban and rural stations in Silesia Province, 2014-2015.

Statistical parameter	**Urban stations**			**Rural stations**	
	Bielsko Biała	**Lubliniec***	**Zabrze****	**Godów***	**Złoty Potok****
Summer season					
Mean ± SD [ng/m^3]	2.0 ± 0.4	1.9 ± 0.3	3.0 ± 5.0	2.0 ± 0.4	1.6 ± 0.4
Range [ng/m^3]	1.3 - 3.9	1.3 - 2.8	1.0 - 199.4	1.3 - 3.8	0.81 – 6.0
Winter season					
Mean ± SD [ng/m^3]	2.0 ± 0.4	2.0 ± 0.5	2.6 ± 1.4	2.0 ± 0.4	1.84 ± 0.5
Range [ng/m^3]	1.4 - 4.4	1.2 - 5.1	1.2 - 32.6	1.3 - 2.8	0.80 - 14.0
Total					
Mean ± SD [ng/m^3]	2.0 ± 0.4	1.9 ± 0.4	2.8 ± 3.6	2.0 ± 0.4	1.7 ± 0.5
Measur. No.	179	178	7739	180	15196

* 24-hour data; ** 1-hour data

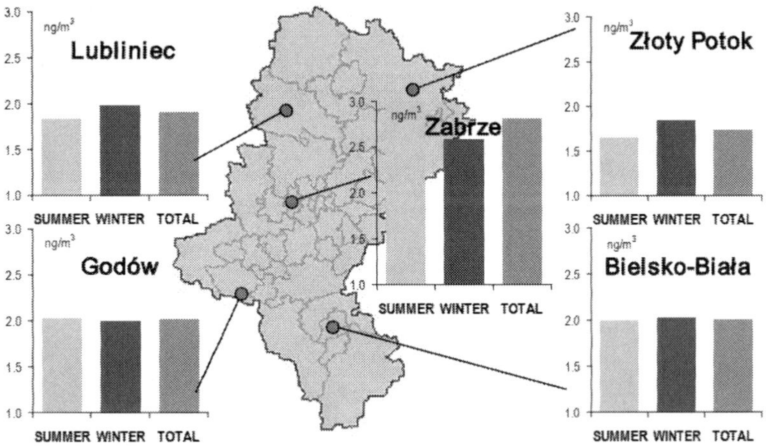

Fig. 2. Total gaseous mercury concentrations [ng/m^3] averaged over summer and winter seasons, and the whole measurement period - Silesia Province, 2014-2015.

4.2 Particle-bound mercury

Table 2 summarizes the statistical parameters of the measurement series of ambient PM2.5-bound Hg concentrations [pg/m^3], and in Table 3 these concentrations were expressed as Hg contents in PM2.5 samples [ng/mg]. This allows to observe the phenomenon of enrichment of the PM samples in Hg in winter season and explains a higher increase in PM-Hg concentrations in this period than it should be resulted in seasonal variability of PM concentrations. The results of additional (simultaneous) measurements of Hg content in PM10 and PM1 samples in Lubliniec were summarized in Table 4.

 https://doi.org/10.1051/e3sconf/20182801030

Table 2. PM2.5-bound mercury concentrations [pg/m³] – the descriptive statistics of 24-h measurement series at urban and rural stations in Silesia Province, 2014-2015.

Statistical parameter	Urban stations			Rural stations	
	Bielsko Biała	Lubliniec	Zabrze	Godów	Złoty Potok
Summer season					
Mean ± SD [pg/m³]	31.1 ± 27.7	21.0 ± 20. 0	27.7 ± 18.6	20.6 ± 14.2	22.9 ± 12.9
Range [pg/m³]	6.4 – 147,0	6.5 - 152.4	7.6 - 121.06	6.0 - 121.2	7.4 - 82.5
Winter season					
Mean ± SD [pg/m³]	75.7 ± 83.1	58.0 ± 55.5	118.7±106.6	63.3 ± 48.8	54.8 ± 44.3
Range [pg/m³]	8.9 - 517.5	11.0 - 306.9	18.8 - 565.7	8.4 - 431.7	7.7 - 281.7
Total					
Mean ± SD [pg/m³]	52.4 ± 64.6	40.8 ± 46.5	69.8 ± 86.5	41.6 ± 41.6	38.5 ± 36.0
Measur.No.	291	245	255	709	652

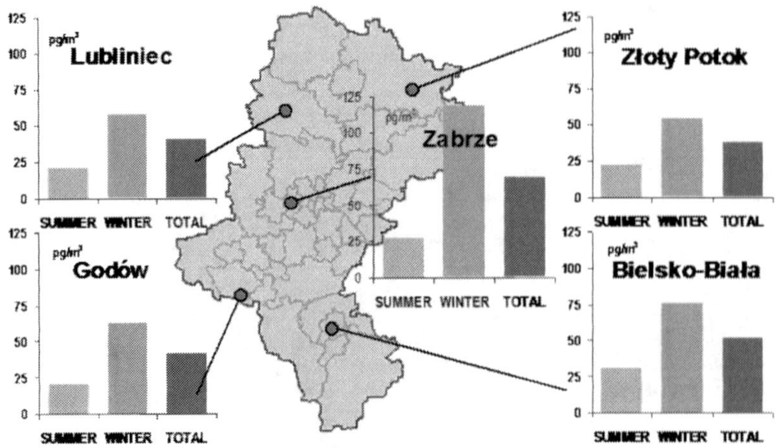

Fig. 3. PM2.5-bound mercury concentrations [pg/m³] averaged over summer and winter seasons, and the whole measurement period - Silesia Province, 2014-2015.

Table 3. Mercury contents in PM2.5 samples [ng/mg] – the descriptive statistics of 24-h measurement series at urban and rural stations in Silesia Province, 2014-2015.

Statistical parameter	Urban stations			Rural stations	
	Bielsko Biała	Lubliniec	Zabrze	Godów	Złoty Potok
Summer season					
Mean ± SD [ng/mg]	1.9 ± 1.5	1.3 ± 0.9	1.7 ± 0.9	1.0 ± 0.6	1.7 ± 1.0
Range [ng/mg]	0.5 - 8.8	0.3 - 5.4	0.6 - 7.4	0.4 - 4.5	0.5 - 7.0
Winter season					
Mean ± SD [ng/mg]	2.1 ± 1.5	1.4 ± 1.1	2.3 ± 2.2	1.4 ± 0.9	2.3 ± 1.4
Range [ng/mg]	0.6 - 9.8	0.5 - 7.1	0.6 - 12.7	0.4 - 5.8	0.5 - 8.9
Total					
Mean ± SD [ng/mg]	2.0 ± 1.5	1.4 ± 1.00	1.9 ± 1.7	1.2 ± 0.8	2.0 ± 1.2

Table 4. PM10 and PM1-bound Hg concentrations [pg/m^3] and Hg contents in these PM fractions [ng/mg] – the descriptive statistics of 24-h measurement series at the station in Lubliniec, 2014-2015.

Statistical parameter	PM10-bound Hg		PM1-bound Hg	
	Concentr. [pg/m^3]	Content [ng/mg]	Concentr. [pg/m^3]	Content [ng/mg]
Summer season				
Mean ± SD	26.99 ± 21.69	1.2 ± 0.8	12.42 ± 8.22	1.7 ± 1.1
Range	9.81-165.34	0.3 - 7.2	4.77-62.93	0.4 - 5.3
Winter season				
Mean ± SD	64.61 ± 61.76	1.4 ± 1.2	24.71 ± 17.32	1.6 ± 0.9
Range	9.5-432.84	0.4 - 8.2	7.6-97.46	0.6 - 5.2
Total				
Mean ± SD	45.99 ± 50.07	1.3 ± 1.00	18.60 ± 14.88	1.7 ± 1.0

5 Conclusions

The highest average TGM concentration in 2-years period between 2014-2015 was recorded in Zabrze, within the Upper Silesian Agglomeration (2.8 ng/m^3). A characteristic feature of the TGM concentrations distribution at this station was a higher TGM level in summer months than in winter season. In winter, the impact of a numerous local emission sources, including high emitters related to commercial power industry, is hindered by the inversion layers. A clearly lower average concentration of TGM (2.0 ng/m^3) was recorded at the second station of urban background in Bielsko-Biała and in the border village of Godów. In both cases, obtained average seasonal concentrations were the same, which may indicate on the other sources of Hg emissions than combustion of fuels for heating purposes, like eg. the advection of polluted air masses. The lowest TGM levels were observed in the suburbs of Lubliniec (1.9 ng/m^3) and at the station of regional background in Złoty Potok (1.7 ng/m^3). These were the only two locations wherein in the yearly distribution of TGM a higher share of the heating season was noted, probably as a result of Hg emissions from the local house furnaces (Lubliniec) or its advection from urban and industrial areas (Złoty Potok). The results obtained for TGM concentrations exceeded the European average level of 1.5 ng/m^3 (AirBase, 2014).

The highest average concentration of PM2.5-Hg was observed in Zabrze (70 pg/m3) and it was 26% higher than in Bielsko-Biała (52 pg/m^3). PM2.5-Hg concentrations at the other 3 sites were very similar and lower by about 40% than in Zabrze (Godów – 42 pg/m^3, Lubliniec - 41 pg/m^3, Złoty Potok - 39 pg/m^3). Comparison of the obtained results with a literature data (for North America and Europe) shows that concentrations of particle-bound mercury in urban and suburban areas remained at the high level, as a result of higher PM2.5 concentrations in Silesia Province. In all of the analyzed locations, particle Hg concentrations were higher in winter season than in the summer, same as the PM2.5 concentrations. Nevertheless, it was noted that during the heating season there was an enrichment of PM by the mercury, significantly higher than it could be expected based on the proportion of the seasonal PM2.5 concentration changes. This is the result of a higher primary Hg emissions from the fuel combustion in this period, as well as lower air temperatures, which can favor the sorption of Hg on aerosol particles.

Measurement results of Hg content in PM10, PM2.5, and PM1 in Lubliniec showed that 77% of PM10-bound Hg mass was associated with the respirable particles (PM2.5) and nearly half mass of PM10-bound Hg was found in PM1 fraction (46%). This distribution of

Hg indicates that the dominant sources of local emissions are the solid fuel combustion processes. The high content of Hg in fine particles of aerosol, that can reach to the lungs (PM2.5), and that, which can penetrate through the pulmonary alveolus to the bloodstream (PM1) are particularly dangerous for health, by inhalation of PM-bound mercury.

The work was carried out thanks to the financial support of the Regional Environmental Protection Fund in Katowice, in cooperation with the RIEP Monitoring Department in Katowice and the RIEP Laboratory in Częstochowa. The TGM concentration data in Złoty Potok came from the RIEP air monitoring station (http://powietrze.katowice.wios.gov.pl).

References

1. S. Lindberg, R. Bullock, R. Ebinghaus, D. Engstrom, X. Feng, W. Fitzgerald, N. Pirrone, E. Prestbo, C. Seigneur, Ambio **36**, (2007)
2. P.S. Weiss-Penzias, E.J. Williams, B.M. Lerner, T.S. Bates, C. Gaston, K. Prather, A. Vlasenko, S.M. Li, J. Geophys. Res. Atmos. **118**, (2013)
3. N. Pirrone, S. Cinnirella, X. Feng, R.B. Finkelman, H.R. Friedli, J. Leaner, R. Mason, A.B. Mukherjee, G.B. Stracher, D.G. Streets, K. Telmer, Atmos. Chem. Phys. **10**, (2010)
4. National Research Council, *Toxicological Effects of Methylmercury* (The National Academies Press, Washington, 2000)
5. World Health Organization, *Elemental mercury and inorganic mercury compounds: human health aspects* (WHO, Concise International Chemical Assessment Document 50, Geneva, 2003)
6. K. Sundseth, J.M. Pacyna, E.G. Pacyna, N. Pirrone, R. J. Thorne, Int. J. Environ. Res. Public Health, **14**, 105 (2017)
7. J.M. Pacyna, O. Travnikov, F. De Simone, I.M. Hedgecock, K. Sundseth, E.G. Pacyna, F. Steenhuisen, N. Pirrone, J. Munthe, K. Kindbom, Atmos. Chem. Phys. **16**, (2016)
8. Arctic Monitoring and Assessment Programme (AMAP), United Nations Environment Programme (UNEP), *Global Mercury Assessment: Sources, Emissions, Releases and Environmental Transport* (UNEP Chemicals Branch: Geneva, 2013)
9. R.P. Mason, J.R. Reinfelder, F.M.M. Morel, Water Air Soil Pollut. **80**, (1995)
10. I. Ilyin, O. Rozovskaya, O. Travnikov, *Assessment of heavy metal transboundary pollution on regional and national scales, transition to the new EMEP grid* (EMEP Status Report 2/2017) http://www.msceast.org
11. European Environment Agency, *European Union emission inventory report 1990–2015 under the UNECE Convention on Long-range Transboundary Air Pollution (LRTAP)* (EEA Report No 9/2017, Publications Office of the European Union, Luxembourg, 2017)
12. I. Ilyin, O. Rozovskaya, O. Travnikov, M. Varygina, W. Aas, *Heavy metals: Analysis of long-term trends, country-specific research and progress in mercury regional and global modelling* (EMEP Status Report 2/2015) http://www.msceast.org
13. H. Pyta, W. Rogula-Kozłowska, B. Mathews, Atmos. Pollut. Res. **8**, (2017)

Identification of PM10 air pollution origins at a rural background site

Magdalena Reizer[1,*], *José A.* G. Orza[2]

[1]Warsaw University of Technology, Faculty of Building Services, Hydro and Environmental Engineering, 00653 Warsaw, Poland
[2]SCOLAb, Física Aplicada, Universidad Miguel Hernández, 03202 Elche, Spain

Abstract. Trajectory cluster analysis and concentration weighted trajectory (CWT) approach have been applied to investigate the origins of PM10 air pollution recorded at a rural background site in North-eastern Poland (Diabla Góra). Air mass back-trajectories used in this study have been computed with the Hybrid Single-Particle Lagrangian Integrated Trajectory (HYSPLIT) model for a 10-year period of 2006–2015. A cluster analysis grouped back-trajectories into 7 clusters. Most of the trajectories correspond to fast and moderately moving westerly and northerly flows (45% and 25% of the cases, respectively). However, significantly higher PM10 concentrations were observed for slow moving easterly (11%) and southerly (20%) air masses. The CWT analysis shows that high PM10 levels are observed at Diabla Góra site when air masses are originated and passed over the heavily industrialized areas in Central-Eastern Europe located to the south and south-east of the site.

1 Introduction

Variability of air pollutant concentrations recorded at a specific site, depending on the species, may vary on time scales from minutes to decades. Over a longer time scale the changes are determined by seasonal or annual cycles in the activity of emission sources, while on a time scale of days or less the concentration variability is closely related to the immediate history of the air before arriving at the sampling point [1]. Back-trajectory analysis comprises a number of methods to identify transport pathways of air masses affecting a study site as well as to determine potential source areas of air pollution. In order to analyse the association between trajectories and concentrations of various species in air arriving at a site, a number of methods to carry out trajectory classifications has been applied. These can generally be split into the following methodological groups: (1) assigning trajectories according to air mass sectors and comparing the air pollution composition in those sectors; (2) clustering the trajectories using a multivariate statistical technique and analysing the concentrations at the receptor site for each trajectory classification; (3) residence time analysis which generates a probability density function identifying the likelihood that an air mass would traverse a given region *en route* to the site

* Corresponding author: magdalena.reizer@pw.edu.pl

of interest over a given time period [2]; this often includes conditional probability densities linking residence times and the concentrations measured at the study site.

There has been a number of studies reporting the use of back-trajectory analyses for determining long range transport of air pollution (see e.g. [3, 4]) or characterizing Saharan dust outbreaks in the Mediterranean Basin (see e.g. [5, 6]) and Canary Islands (e.g. [7]). However, there is a lack of studies of air mass back-trajectories for Central-European countries. In scarce previous studies performed for this region, probable sources of particulate matter (PM) air pollution were identified by trajectory classification into wind sectors at urban site for Zagreb, Croatia [8] as well as back-trajectory analysis for selected days with the highest PM concentrations observed in Polish urban areas [9]. A cluster analysis was also carried out for rural background site in Belsk, Poland [10] as well as for urban sites located in Bucharest, Romania and Szeged, Hungary [11]. Different methods of trajectories interpolation were used for Polish urban sites in Gdańsk and Katowice as well as for rural background site in Diabla Góra [12]. Moreover, cluster analysis, potential source contribution function (PSCF) and concentration weighted trajectory (CWT) methods were used to identify the main atmospheric circulation pathways influencing Black Carbon (BC) [13] and aerosol particle number concentration (PNC) [14] observed in Lithuania. Dvorska et al. [15] applied PSCF method in order to identify potential source areas of persistent organic pollutants (POPs) at rural background site in Kosetice (Czech Republic).

The aim of the present paper is to investigate potential source areas of air pollution contributing to PM10 levels recorded at a Central-European rural background site during the long-term period of 2006–2015. Cluster analysis of back-trajectories has been applied to assess the main transport pathways of air masses, while concentration weighted trajectory approach has been used for identification of potential PM10 source regions.

2 Materials and methods

2.1 Study area

In this study air mass back-trajectories arriving over the monitoring site of Diabla Góra, Poland were investigated. Diabla Góra is a rural background monitoring site operated within the EMEP air quality network (European Monitoring and Evaluation Programme). This site is located in the north-eastern part of Poland (54° 08′ 00″ N, 22° 04′ 01″ E, 157 m a.s.l.), far from dense populated or traffic areas (Fig. 1). The nearest town (>10 000 inhabitants) and road (>50 cars per day) are situated 20 km and 16 km away from this site, respectively.

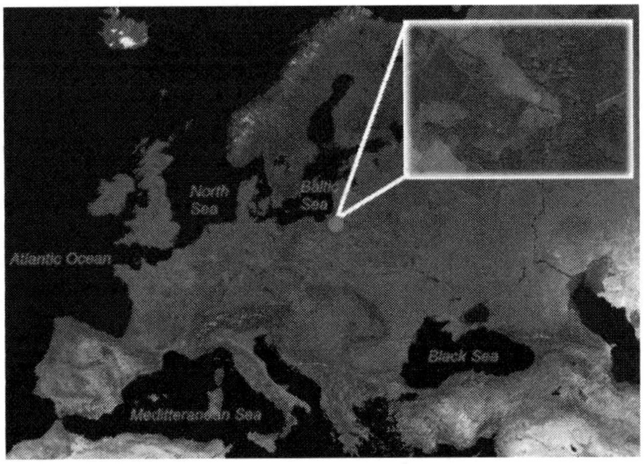

Fig. 1. Location of the study area.

Daily PM10 concentrations were determined by the reference gravimetric method (Standard EN 12341). PM10 data for the period from 1 January 2006 to 31 December 2015 was collected from the EBAS database (http://ebas.nilu.no).

2.2 Back-trajectories calculations

Three dimensional 96-hour back-trajectories of air parcels arriving at Diabla Góra site at the height of 200 m above sea level were computed at 00, 06, 12 and 18 UTC each day during the period 2006–2015. The trajectories were calculated by the PC version of the Hybrid Single-Particle Lagrangian Integrated Trajectory (HYSPLIT) model [16]. The meteorological data used for trajectory calculations was provided by the ECMWF (European Centre for Medium-Range Weather Forecasts).

2.3 Cluster analysis

The k-means procedure, which is the most commonly used form of non-hierarchical clustering for trajectory studies, is an iterative algorithm that uses a specified number of clusters, k, to partition the data by comparing each object to the arithmetic mean of all the members of each of the clusters (cluster centroids). The selection of the optimal number of clusters that best describes the different air flow patterns is performed by computing the percentage change in within-cluster variance, as a function of the number of clusters [17]. The assignment of trajectories to a given cluster is carried out by minimising the internal variability within the group of trajectories and maximising the external variability between different groups based on the trajectory co-ordinates based on the Root Mean Square Deviation (RMSD) of each trajectory from its cluster centroids [2].

In the present study the k-means cluster analysis of back-trajectories following the procedure described in detail by Orza et al. [18] was carried out. Hourly latitude and longitude were used as input variables in the clustering procedures, while the similarity metrics was based on great-circle distances. The selection of the appropriate number of clusters was done by obtaining the best solution for each number of clusters (from a large number of initializations) and then by the analysis of the percentage change in the total RMSD when decreasing one by one the number of clusters.

Daily PM10 data was assigned to a cluster for a given day, if three out of four trajectories for that day fell into the same cluster. Non-parametric pairwise Mann-Whitney tests with the Dunn-Sidàk adjustment for multiple comparisons were used to test the significance of inter-cluster variation in PM10 concentration.

2.4 Concentration weighted trajectory

The concentration weighted trajectory (CWT) is a method of weighting trajectory residence times with associated air pollutant concentrations [19]. This procedure assigns to each grid cell a weighted concentration obtained by averaging sample concentrations associated to trajectories that crossed that grid cell, i.e. each concentration is used as a weighting factor for the residence times of all the trajectories in each grid cell which correspond to that concentration and then it is divided by the cumulative residence time from all trajectories. CWT method employs an arbitrary weight function to minimize the inaccuracy caused by the small number of polluted trajectories.

In summary, weighted concentration fields show concentration gradients across potential sources and helps to identify the relative significance of potential sources [19].

3 Results and discussion

3.1 Cluster analysis

Trajectories arriving at Diabla Góra at 200 m are found to be clustered in 7 groups (Fig. 2). The groups correspond to the major advection patterns at the study site are named according to the main direction they come from.

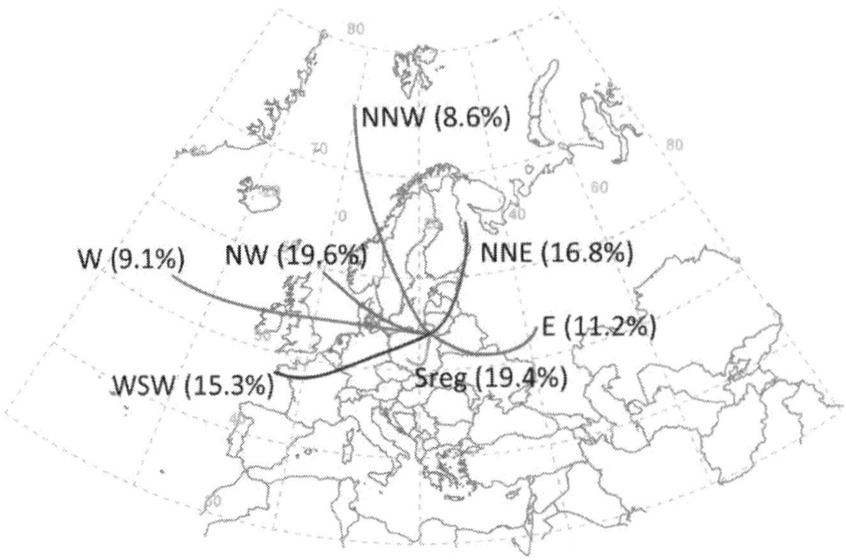

Fig. 2. Cluster centroids of the trajectories reaching Diabla Góra, Poland at 200 m in the period 2006–2015. The percentage of trajectories occurring in each cluster is shown in parentheses.

Most of the trajectories correspond to westerly flows (44% of the cases), bringing air masses from the northern part of the North Sea, through Denmark and North Poland (NW, 19.6%), North France, Central Germany and western part of Poland (WSW, 15.5%), as well as from the Atlantic Ocean, through Great Britain, southern part of Denmark and northern part of Poland (W, 9.1%). The second type of trajectories groups the northerly flows (over 25%), bringing air masses from Scandinavian Peninsula, passing over northern part of Russia, Latvia and Lithuania (NNE, 16.8%), as well as from the southern part of Greenland, through Scandinavia and the Southern Baltic Sea (NNW, 8.6%). The third type of flow includes short (regional) trajectories, bringing air masses from highly polluted Southern Poland (Upper Silesia and Lesser Poland) through central part of the country (Sreg, 19.4%). The last type of trajectories includes the air masses coming from western part of Russia, through the Belarusian-Ukrainian border and the eastern part of Poland (E, 11.2%).

The highest mean PM10 concentrations (26 μg/m^3) observed at Diabla Góra site are associated with slow moving easterly flows (Fig. 3). As these flows are mainly associated to the westward extension of the Siberian high located to the east of the study site, cluster E is also associated with the highest mean air pressure – 1 001 hPa (with maximum of 1 044 hPa and one of the lowest mean air temperatures of 5.0 °C (with minimum of -21.5°C). Quite high PM10 concentrations (25 μg/m^3) are also related with much warmer (9.2°C, with minimum of -11.5°C) slow moving air masses from the South (Sreg). In opposite, the lowest mean PM10 concentrations (10 μg/m^3) are recorded when air masses arrive with fast moving NNW flows (Fig. 3).

Overall, the Mann-Whitney test results have shown that mean PM10 concentrations associated with slow moving easterly and southerly air masses (clusters E and Sreg) were significantly higher compared to those observed for other clusters.

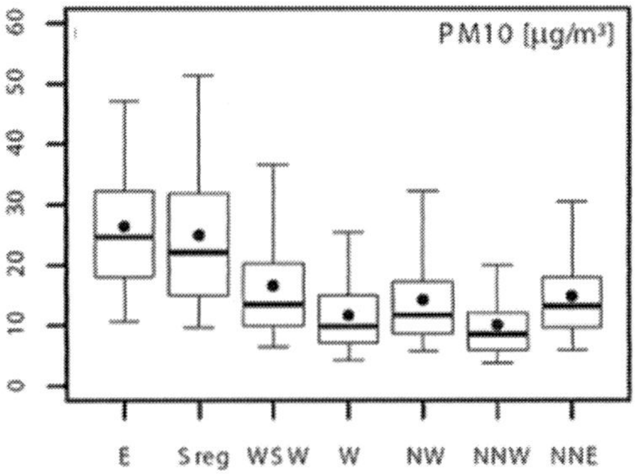

Fig. 3. Boxplots of PM10 concentrations (μg/m^3) for each trajectory cluster arriving at Diabla Góra, Poland. The horizontal lines and dots indicate the median and mean values, respectively, the boxes cover the 25th-75th percentiles. The length of the whiskers is 1.5 times the interquartile range.

3.2 Concentration weighted trajectory

The concentration weighted trajectory map for PM10 in Diabla Góra during study period is shown in Fig. 4.

https://doi.org/10.1051/e3sconf/20182801031

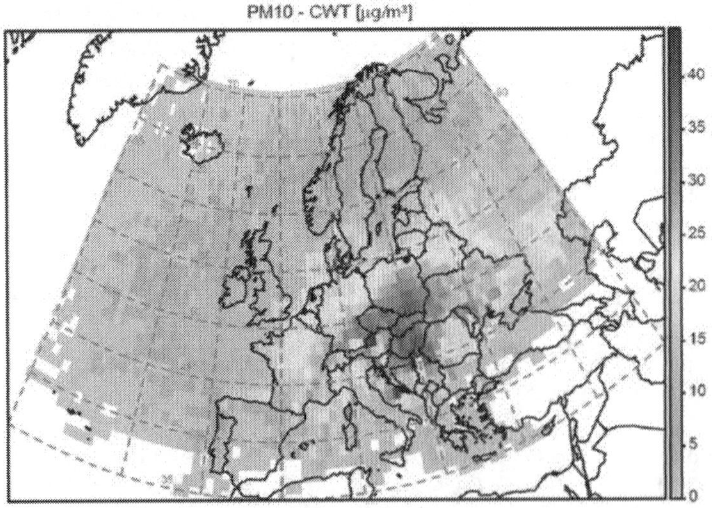

Fig. 4. CWT map for PM10 levels at Diabla Góra, Poland.

Cells with high (above 30 μg/m³) and moderate (above 20 μg/m³) CWT values were found south and south-east of Diabla Góra site, respectively, in good accordance with the previous analysis of PM10 levels by advection pattern. The high CWT values for short lengths of the S and E trajectories indicates stable situations with reduced removal of pollutants and slow air movement over Central-Eastern Europe due to the influence of high-pressure centres. As a result, polluted air masses with a high PM10 content, originating from southern and eastern part of Europe may accumulate during several days under anticyclonic situations and they may remain there for several days before reaching the study site.

4 Summary and conclusions

A database of 96-h back-trajectories arriving to North-eastern part of Poland (Diabla Góra) at 200 m computed for the 10-year period from January 2006 to December 2015 was applied to describe the main flow patterns over the study site, as well as to identify PM10 air pollution origins at the rural background site. A cluster analysis grouped back-trajectories into 7 clusters. Most of the trajectories correspond to fast and moderately moving westerly (45% of the cases) and northerly (25%) flows, followed by the short (regional) trajectories, bringing air masses from the South (20%) and from the East (11%). Mean PM10 concentrations associated with slow moving easterly and southerly air masses are significantly higher compared to those observed for other clusters. The CWT approach has identified the heavily industrialized areas in Central-Eastern Europe as probable PM10 source areas.

This work was partially supported by the Dean of the Faculty of Building Services, Hydro and Environmental Engineering, Warsaw University of Technology within the Grant no. 504/02637/1110/42.000100. The authors also gratefully acknowledge the NOAA Air Resources Laboratory (ARL) for the provision of the HYSPLIT model and the European Centre for Medium-Range Weather Forecasts (ECMWF) for making available the ERA-Interim database.

References

1. J.N. Cape, J. Methven, L.E. Hudson, Atmos. Environ. **34**, 3651-3663 (2000)
2. Z.L. Fleming, P.S. Monks, A.J. Manning, Atmos. Res. **104-105**, 1-39 (2012)
3. K. Dimitriou, P. Kassomenos, Atmos. Environ. **96**, 154-162 (2014)
4. M. Reizer, K. Juda-Rezler, Air Qual. Atmos. Health **9**, 517-531 (2016)
5. X. Querol, J. Pey, M. Pandolfi, A. Alastuey, M. Cusack, N. Pérez, T. Moreno, M. Viana, N. Mihalopoulos, G. Kallos, S. Kleanthous, Atmos. Environ. **43**, 4266-4277 (2009)
6. M. Cabello, J.A.G. Orza, V. Galiano, G. Ruiz, Adv. Sci. Res. **2**, 65-70 (2016)
7. S. Alonso-Pérez, E. Cuevas, X. Querol, Meteorol. Atmos. Phys. **113**, 109-124 (2011)
8. I. Bešlić I., K. Šega, M. Čačković, Z. Bencetić, A. Bajić, Bull. Environ. Contam. Toxicol. **80**, 270-273 (2008)
9. K. Juda-Rezler, M. Reizer, J.-P. Oudinet, Atmos. Environ. **45**, 6557-6566 (2011)
10. A. Pietruczuk, J. Jarosławski, Acta Geophys. **61**, 445-462 (2013)
11. L. Makra, I. Ionel, Z. Csépe, I. Matyasovszky, Sci. Total Environ. **458-460**, 36-46 (2013)
12. J. Godłowska, M.J. Hajto, M. Tomaszewska, Arch. Environ. Prot. **41**, 28-35 (2015)
13. S. Byčenkienė, V. Dudoitis, V. Ulevicius, Adv. Meteorol., 137694 (2014)
14. S. Byčenkienė, K. Plauškaitė, V. Dudoitis, V. Ulevicius, Atmos. Res. **143**, 279-292 (2014)
15. A. Dvorska, G. Lammel, J. Klanova, I. Holoubek, Environ. Pollut. **156**, 403-408 (2008)
16. A.F. Stein, R.R. Draxler, G.D. Rolph, B.J.B. Stunder, M.D. Cohen, F. Ngan, Bull. Am. Meteorol. Soc. **96**, 2059-2077 (2015)
17. S.R. Dorling, T.D. Davies, C.E. Pierce, Atmos. Environ. **26**, 2575-2581 (1992)
18. J.A.G. Orza, M. Cabello, V. Galiano, A.T. Vermeulen, A.F. Stein, *Lagrangian Modeling of the Atmosphere*, 127-141 (2012)
19. A. Stohl, Atmos. Environ. **30**, 579-587 (1996)

E3S Web of Conferences **28**, 01032 (2018) https://doi.org/10.1051/e3sconf/20182801032
Air Protection in Theory and Practice

Factors determining the concentration and chemical composition of particulate matter in the air of selected service facilities

Patrycja Rogula-Kopiec.[1,*], *Józef* Pastuszka[2], *Barbara* Mathews[1], *Kamila* Widziewicz[1]

[1]Institute of Environmental Engineering, Polish Academy of Sciences, M. Skłodowskiej-Curie St. 34, 41-819 Zabrze, Poland
[2]Department of Air Protection, Silesian University of Technology, Center of New Technologies, Konarskiego St. 22B, 44-100 Gliwice, Poland

Abstract. The link between increased morbidity and mortality and increasing concentrations of particulate matter (PM) resulted in great attention being paid to the presence and physicochemical properties of PM in closed rooms, where people spends most of their time. The least recognized group of such indoor environments are small service facilities. The aim of this study was to identify factors which determine the concentration, chemical composition and sources of PM in the air of different service facilities: restaurant kitchen, printing office and beauty salon. The average PM concentration measured in the kitchen was 5-fold (PM_4, particle fraction ≥ 4 μm) and 5.3-fold (TSP, total PM) greater than the average concentration of these PM fractions over the same period. During the same measurement period in the printing office and in the beauty salon, the mean PM concentration was 10- and 4-fold (PM_4) and 8- and 3-fold (TSP) respectively greater than the mean concentration of these PM fractions in outdoor air. In both facilities the main source of PM macro-components, especially organic carbon, were chemicals, which are normally used in such places - solvents, varnishes, paints, etc. The influence of some metals inflow from the outdoor air into indoor environment of those facilities was also recognized.

1 Introduction

The quality of air inside various types of rooms, regardless the time people spend indoors, depends primarily on the activity of indoor occupants, the type of this activity and the effectiveness of ventilation. In kitchen facilities where meals are prepared (cooking, frying) and the coal stove or fireplace is being used, the quality of indoor air is generally worse compared to outdoor one. This mostly applies to particulate matter (PM) air pollution, and especially to its fine particles [1-5].

Fine atmospheric particles are free to migrate inside the buildings, which makes their indoor concentration similar (or higher) compared to the outdoor one. Such tendency may

* Corresponding author: patrycja.rogula-kopiec@ipis.zabrze.pl

or may not occur depending on the presence and effectiveness of internal sources of PM and other pollutants emission (i.e. various gases adsorbed on PM or reacting with each other and/or other PM components), the way in which the room is ventilated (preventing or favoring PM accumulation inside), or the effectiveness of ventilation system [6-10].

Due to the nuisance associated with emission of various air pollutants at workplace and the need to estimate workers exposure to these pollutants the occupational levels of PM during the working hours is currently being monitored at various industrial facilities. Nevertheless, there are many facilities where, despite the fact of heavily contaminated indoor air, there are no legal requirements for the monitoring of pollutants concentrations in their interiors. This anticipated significant air pollution within facilities can be caused either by the building location (for example in highly contaminated region) or the presence of internal pollution sources and in some certain conditions such facilities can be subjected to the excessive accumulation of indoor pollutants. The need to assess the indoor air quality in such facilities is objectively justified when even one from the abovementioned factors occurs.

The Upper Silesian Agglomeration is one of Europe's most polluted region in terms of air quality. The concentrations of PM and PM-bound compounds in some cities of the Agglomeration are several times higher than in large cities of Western, Northern and Southern Europe [11-13]. In comparison to other Polish and European regions, Silesia Agglomeration is also one of the most industrialized and urbanized areas. Within the Upper Silesian Agglomeration, which consists of 14 cities with a total population exceeding 2 million people, more than 60% of employees work in the so-called services. Some of the service facilities which provides a particular service are a significant sources of internal pollutants emission. These are among others printing offices and beauty salons [13, 14]. In such places, due to the type of activity, high ambient concentrations of various organic pollutants can be expected.

Considering the presence of high concentrations of PM in the Silesia urban agglomeration and a relatively high concentrations of PM-bound carbon and some other components of PM [15,16], it was reasonable to investigate the effects of the activities carried out in these facilities on the formation of PM and PM-bound carbonaceous matter, ions and selected metals.

2 Methodology

The study was performed in restaurant kitchen, printing office and beauty salon (Figure 1) located in Bytom (one among14 cities located in the Upper Silesian Agglomeration with an area of 69 km^2 and 170 thousand inhabitants). All service facilities, chosen for this study are located within the same housing estate. There are no power plants, large industrial/manufacturing plants or any major communications arteries close to those facilities.

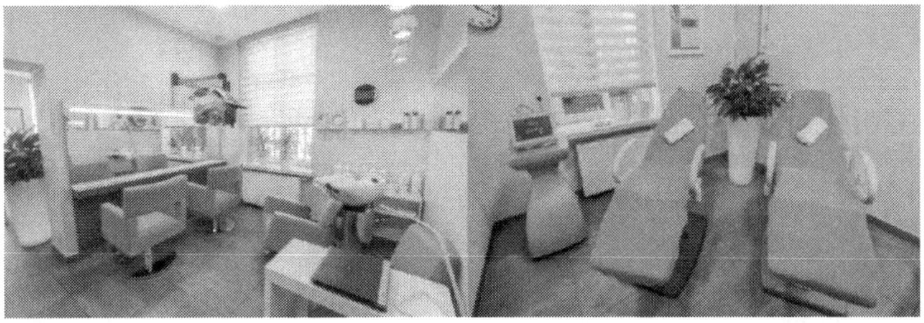

Fig. 1. Selected service facilities

Selected printing office provide printing services like large format printing on a variety types of paper and films, but also other services like thermal treatment (welding) the edges of the selected products (banners, posters). Sampling was conducted in a hall with an area of about 50 m^2, which run printing, at a distance of about 3 m in a straight line from the printing machine. The sampling site was located in a new building, where customer office (about 15 m^2), social room (10 m^2) and a toilet are being placed. The building has a natural ventilation system and two large (PVC) windows on each wall. In the summer, when the measurements were carried out the door between the hall and the office together with the exit door (emergency exits from the hall) and windows were constantly opened. All printouts from the machine are hanging to dry in the open air inside the hall, at the time when the machine is kept on a standby, for example when cyclical printing areas are sprayed with the appropriate solvent.

The beauty salon is located in a residential building, adopted for service activities. Measurements were carried out in one among five rooms belonging to the salon (in each room a different kind of treatments is being run) – the office, where pedicure and manicure are performed. The room has an area of about 30 m^2. The room is equipped with a seat for customers and a beauty table. Measurements were conducted at a distance of approximately 3 m from the table. The building is equipped with a gravity ventilation and PVC windows, most of which are closed, and the room is ventilated only after the working day is end (after 18.00). During work a beautician uses milling machine, nail-files, and a number of cosmetics containing hydrocarbons, alcohols, esters, phenols, and others. Nail coatings and gels are cured under UV lamp.

The restaurant is placed in a ground floor of a relatively new building (built in 2000 yr.) equipped with efficient mechanical ventilation. The restaurant kitchen in a separate room (area of about 40 m^2) which is in contact with the restaurant hall. It is equipped with two electric stoves, two gas stoves and one wood-fired oven, which is designed for baking breads and pizzas. There are 6 people in the kitchen during the working hours. This is an example of a typical kitchen found in a relatively modern restaurants in Polish cities. In

Bytom there are about 50 such objects, while in whole Silesia conurbation more than 1000.

PM samples were collected in the period of 1-30 September 2016 in a 8-hour cycles, usually between 10.00-18.00, therefore during work hours. All measurements were performed by an identical set of two PM_4 (particles with aerodynamic diameter not greater than 4 μm) and TSP (total suspended particles; aerodynamic diameter in the range 1 nm-100 μm) samplers (GilAir PLUS, Ekohigiena, Radom), located inside the facilities as well as in the outside. Altogether 40 samples of PM_4 (20 indoor and 20 outdoor) and 40 samples of TSP (20 indoor and 20 outdoor) were collected in each facility.

Samples were collected on Whatman quartz filters (QMA, ø25 mm) (GE Medical Systems Poland, Life Sciences, Warsaw) using the GilAir Plus samplers (Sensidyne, St. Petersburg, FL). The PM mass was determined by gravimetric analyses using a microbalance (resolution of 1 μg) (Radwag; Radom, Poland) [3,13,15,16]. The contents of OC and EC in PM was determined using a thermal-optical analyzer Lab OC-EC Aerosol Analyzer (Sunset Laboratories Inc.; Portland, OR, USA) and the measurement procedure EUSAAR [5,13,16]; while the ions contents in the PM water extracts was analyzed using ion chromatography (Metrohm AG; Herisau, Switzerland) [3,13,16].

Metal quantitative analysis was preceded by sample mineralization in a mixture of spectrally pure (Merck, Ultrapure) acids. Before mineralization, samples were flooded with 5mL of nitric acid HNO_3 (65%), 1mL of hydrofluoric HF (40%), and 1mL of perchloric acid $HClO_4$ (70%). Mineralization was carried out in a microwave digestion system, Multiwave 3000 from Anton Paar, in sealed polytetrafluoroethylene PTFE vessels under pressure (mineralization furnace parameters: a pressure rise of 0.3 bar/s; 240°C; pressure 60 bar; mineralization time 50 min). PM-bound trace metal concentrations were determined by inductively coupled plasma mass spectrometry (ICP–MS, Perkin Elmer DRC–e 6100) equipped with a dynamic reaction cell. The operating conditions are listed below: ICP RF power: 1125 W; nebulizer gas flow rate: 0.78–0.83 L/min; auxiliary gas flow: 1.15 L/min; plasma gas flow: 15 L/min; sample flow rate: 1 mL/min. Certified multi-elemental standard stock solutions Periodic table mix 1 and Transition metal mix 2 (Fluka) were used as calibration solutions. The analytical method was validated by determination of selected elements content in the Standard Reference Material, SRM 1648a, (urban particulate matter), obtained from the National Institute of Standard and Technology (NIST). The obtained results were corrected by the mean concentrations of metals in blank samples (non-exposed filters). Recovery rates were in the range: 83% (Cd) - 127% (Cr).

3 Results and discussion

In two among three facilities (restaurant kitchen and beauty salon) the concentrations of PM_4 were much higher compared to atmospheric air (I/O>5). These dependences are shown in Table 1. This is a rule and regardless how high concentrations of outdoor PM_4 occur in the following measurement days, the indoor concentrations of those were each time higher.

The effect of fine particles (PM_4) accumulation inside the beauty salon and restaurant kitchen was significant. This does not apply to PM_4 in a printing office, which indoor concentration was on average even slightly lower than in atmospheric air. In all facilities the TSP concentrations were also slightly lower compared to the outdoor ones (I/O<1). A similar effect of PM accumulation inside the buildings was observed earlier in a lecture rooms in Gliwice and Warsaw cities [16,17]. In case of lecture rooms, where there were no internal sources of fine PM or PM precursors, this phenomenon was fully explained by insufficiently effective air exchange in the examined rooms. In considered facilities in addition to insufficiently effective ventilation, one of the reason for the high indoor PM concentrations is the emission of PM and/or its precursors from the internal emission

https://doi.org/10.1051/e3sconf/20182801032

sources. This is evident when analyzing the I/O (Inside/Outside) ratios of PM (Table 1), and selected PM components (Table 2).

Table 1. Mean (±standard deviation) of PM_4 and TSP concentrations inside (I) and outside (O) the tested facilities
and their ratios (I/O).

PM_4		TSP		PM_4	TSP
(O)	(I)	(O)	(I)	I/O	I/O
RESTAURANT KITCHEN n=80, PM_4 - 20(I) and 20(O), TSP-20(I) and 20(O)					
11.28±4.96	96.57±60.37	155.83±301.25	153.1±217.4	**8.6**	**1.0**
PRINTING OFFICE n=80, PM_4 - 20(I) and 20(O), TSP-20(I) and 20(O)					
73.31±204.44	58.46±27.69	169.27±328.36	87.55±46.82	**0.8**	**0.5**
BEAUTY SALON n=80, PM_4 - 20(I) and 20(O), TSP-20(I) and 20(O)					
11.11±4.85	62.66±63.41	169.27±328.36	98.48±108.65	**5.6**	**0.6**

Table 2. The average (I/O) concentration ratios of PM components

Component	Beauty salon		Restaurant kitchen		Printing office	
	PM_4(I/O)	TSP(I/O)	PM_4(I/O)	TSP(I/O)	PM_4(I/O)	TSP(I/O)
OC	6.1	2.4	5.1	3.1	1.2	1.4
EC	3.3	1.0	5.2	1.5	4.1	2.1
Mn	3.2	0.5	7.5	0.9	0.3	0.3
Co	1.4	1	1.1	0.9	0.4	0.6
Ni	0.2	0.3	0.3	0.3	0.2	0.1
Cu	1.9	0.7	1.7	1.1	0.7	0.4
Zn	2	1.1	2.3	1.9	0.3	0.2
Cd	1.2	0.6	1.4	0.7	0.8	1
Pb	1.8	0.4	2.5	0.5	0.7	0.2
Cr	2	0.5	2.6	0.4	0.3	0.3
Na^+	10.8	1	26.1	1.6	0.4	0.3
NH_4^+	4.6	6.8	19.4	4.4	2.2	1.1
K^+	37.1	6.8	135.8	4.4	49.1	2.3

Ca^{2+}	1	2	11.6	28.8	49.1	2.3
Mg^{2+}	25.3	0.7	23.9	3.6	3.3	10.4
F^-	10.1	0.9	20.1	1.5	29.4	2.1
Cl^-	7.4	0.6	2.7	0.6	2.1	1
NO_3^-	1.1	0.7	2	1.1	0.4	0.8
PO_4^{3-}	10.2	1	20.9	2.1	1.2	1.5
SO_4^{2-}	9.1	1	25.7	2.3	2	1.4

Significantly higher levels of PM-bound OC and EC concentrations and some water-soluble ions inside the considered facilities compared to the outside ones 1. had be probably due to indoor emissions; 2. affect the concentration of the measured components in the tested rooms. This is especially evident in case of OC concentrations in the beauty salon and restaurant kitchen; and less obvious is the printing office (Table 2). It could be therefore concluded that a decisive influence of PM-bound carbon emissions (most likely volatile organic compounds) on the formation of OC concentrations has been noted during cosmetic treatments in a beauty salon; during cooking and meal preparation inside restaurant kitchen (this also includes the emission of the primary organic matter) and printing works inside the printing office. A significant differences were found between the indoor and outdoor concentrations regarding both the PM-bound OCs concentrations and its mass share in the PM (Figure 2). Both parameters remain at a very high level in all facilities no matter if OC was associated with fine or coarse particles. Both in the printing office and in the restaurant kitchen the concentration and the mass shares of PM-bound EC are higher than in atmospheric air, which is likely to be influenced by various processes (including combustion) carried out in both facilities. These processes were also associated with generally higher concentrations of PM_4-bound water-soluble ions, i.e. K^+, Ca^{2+}, NH_4^+, Mg^{2+}, F- and SO_4^{2-} in these facilities with respect to their concentrations in atmospheric air (Table 2).The concentrations of PM_4-bound ions were also much higher inside the beauty salon than in the outside. In the beauty salon and in the restaurant kitchen the service activities also influence the differences in the concentrations of Na^+ and NO_3^- between indoor and outdoor environment. The elevated concentrations of some ions inside facilities (like for example SO_4^{2-}, NH_4^+ and NO_3^-) in relation to their outdoor levels can be caused by two phenomena: first of all, due to the occupant activity, and secondly due to the emission of gaseous precursors of those components, like SO_X, NO_X and NH_3 [16,17]. However, independently of the presence of the indoor emission sources, it can be expected that the SO_4^{2-}, NH_4+ and NO_3^- gaseous precursors that migrate into the facilities along with the atmospheric air (or/and those emitted inside), under stable temperature conditions, high humidity and limited movement of the air masses (i.e., the conditions prevailing in all enclosed spaces where people actually live), will soon be transformed/reacted more readily than in atmospheric air, thus forming a primary, next to carbon matter, fine PM component – the secondary inorganic aerosol. It is worth to note that not only in case of PM_4 fraction the water-soluble ions were significantly higher inside the facilities; this also applies to TSP. The highest concentration of TSP-bound Ca^{2+} occurred in the restaurant kitchen, the TSP-bound NH_4^+ and K^+ in the restaurant and beauty salon interiors while in case of TSP-bound Mg^{2+} in the printing office (Table 2). In the case of PM_4, the concentrations of some elements were also definitely higher inside the facilities than in the atmospheric air. This applies especially to Mn, Co, Ni, Cu, Zn, Cd, Pb, Cr (Table 2). These elevated concentrations, aren't however correlated with the higher shares of these metals in the PM_4 mass (Figure 2). In general, the activities carried out in the investigated facilities did not affect the contents of heavy metals and in case of beauty salons also the concentrations of ions derived from the water-soluble TSP compounds.

https://doi.org/10.1051/e3sconf/20182801032

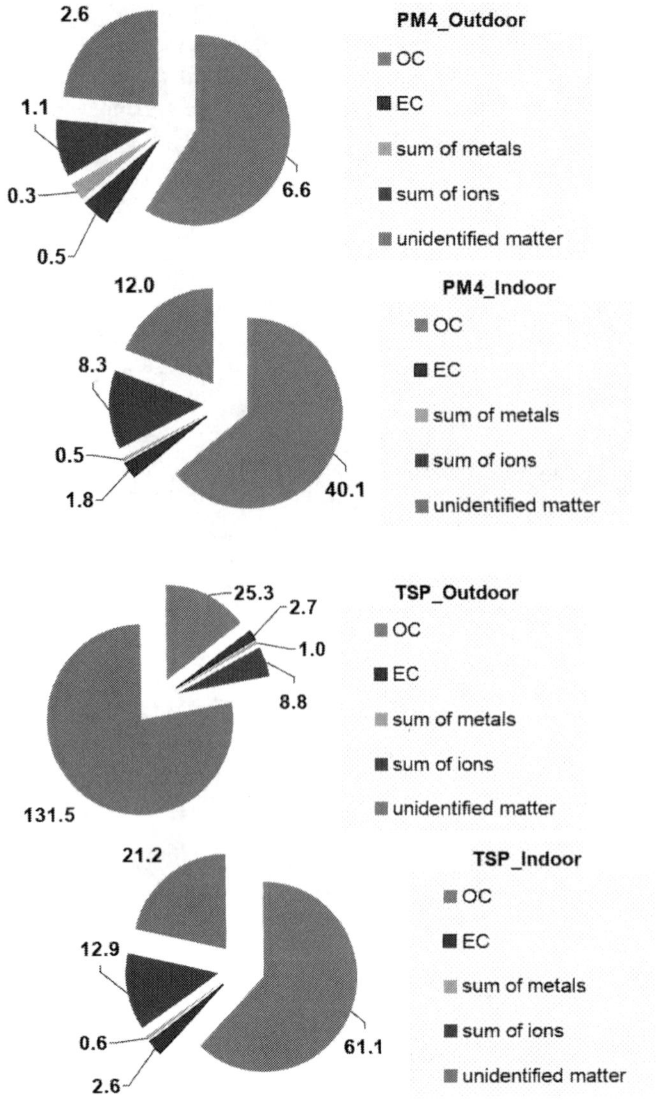

a) beauty salon

https://doi.org/10.1051/e3sconf/20182801032

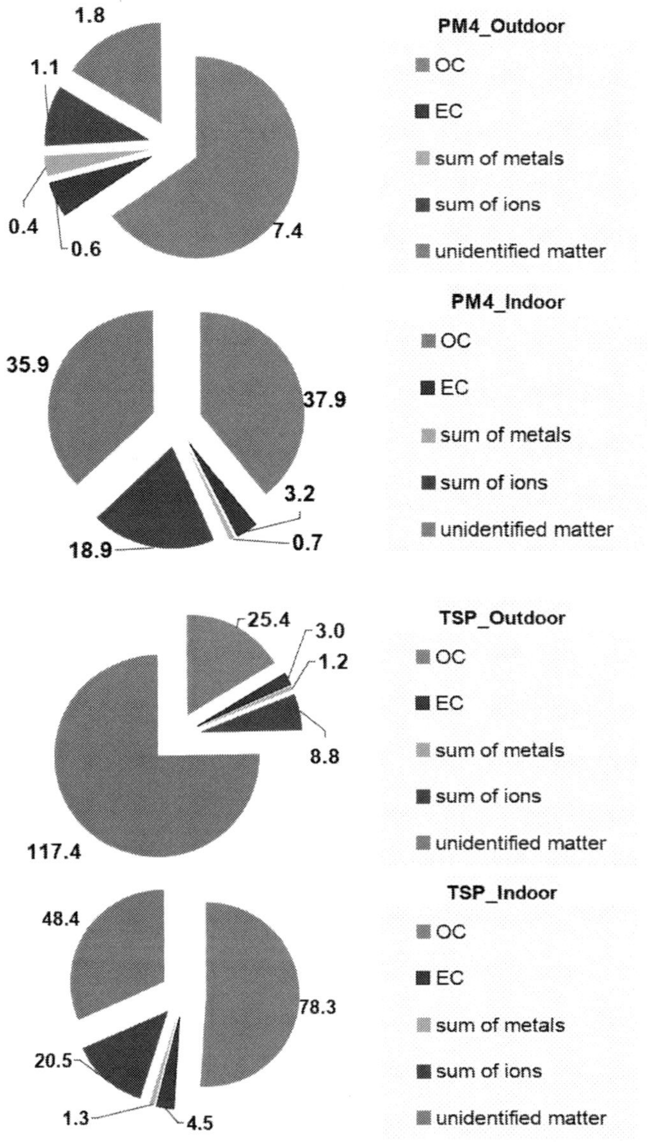

b) restaurant kitchen

https://doi.org/10.1051/e3sconf/20182801032

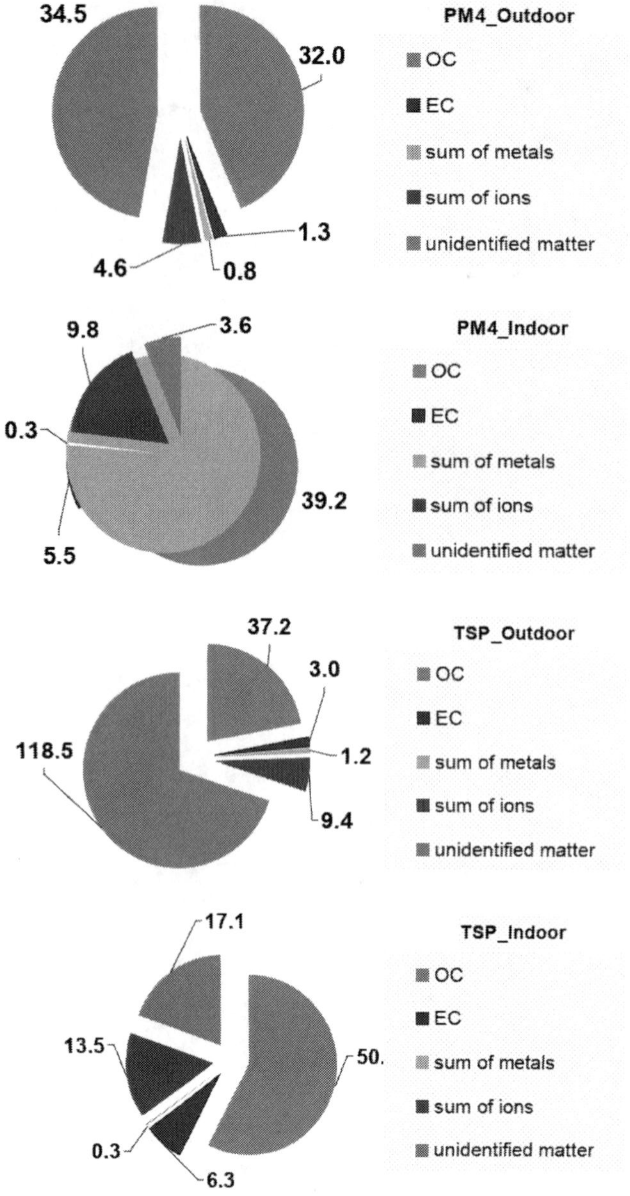

c) printing office

Fig. 2. The average chemical composition of PM4 and TSP in selected service facilities and in atmospheric air during the measurement period (the numbers on the labels indicate the concentrations of selected PM components in μg/m³).

4 Conclusions

The emission of different pollutants during hair and nail treatments, cooking and preparing meals, but also printing and plotting significantly alter the chemical composition of the respirabile (PM_4) as well as the total PM (TSP) inside those facilities in relation to the chemical composition of these PM fractions in atmospheric air.

The observed changes appears by both: an increase in the concentration of PM-bound components (especially carbon compounds and water-soluble ions) as well as their mass share in PM. Nevertheless, the migration of PM and especially its respirable fraction from the outside, is an important and essential source of PM in the indoor air of all tested facilities.

The work has been financed by National Science Centre in Poland under the project no. UMO-2013/09/N/ST10/04224

References

1. Cosmetic Ingredient Review, Int J Tox. **27**(4):1 (2008)
2. P. Pant, R.M. Harrison. Atmos. Environ. **77**:78–97 (2013)
3. W. Rogula-Kozłowska, K. Klejnowski, P. Rogula-Kopiec, L. Ośródka, E. Krajny, B. Błaszczak, B. Mathews, Air Qual Atmos Hlth. **7**:41-58 (2014)
4. J.S. Pastuszka, W. Rogula-Kozłowska, E. Zajusz-Zubek Environ, Monit Assess. **168** (1-4):613-27 (2010)
5. W. Rogula-Kozłowska. Bull Environ Contam Toxicol. **93**:493–502 (2014)
6. G. Gennaro, L. Gennaro, A. Mazzone, F. Porcelli, M. Tutino Atmos Environ. **83**:119-126 (2014)
7. A. Tsigonia, A. Lagoudi, S. Chandrinou, A. Linos, N. Evlogias, E.C. Alexopoulos, Int. J. Environ. Res. Public Health. **7**:314-324 (2010)
8. V.M. Pak, Powers M, Liu J., Workplace Health & Safety **61**(12):522-529 (2013)
9. L.M. Castro, C.A. Pio, R.M. Harrison, D.J.T. Smith, Atmos. Environ. **33**(17):2771-2781 (1999)
10. P. Arezes, J.S. Baptista, M.P. Barroso, P. Carneiro, P. Cordeiro, N.N. Costa, R.B. Melo RB, Sergio A. MiguelA, G. Perestrelo, Occupational Safety and Hygiene II. Taylor & Francis Group (2014)
11. P. Pant, R.M. Harrison, Atmos. Environ. **77**:78–97 (2013)
12. R.W. Atkinson, A. Analitis, E. Samoli, G.W. Fuller, D.C. Green, I.S. Mudway, H.R. Anderson, F.J. Kelly, J Exp Sci Environ Epidemiol. **26**:125–132 (2016)
13. W. Rogula-Kozłowska, Environ Protec Eng. **41**(2):15-29 (2015)
14. P. Rogula-Kopiec, J.S. Pastuszka, W. Rogula-Kozłowska, W. Mucha; Młodzi Naukowcy;**7** (2017)
15. P. Rogula-Kopiec, J.S. Pastuszka, W. Rogula-Kozłowska, W. Mucha, ASEE17, (2017)
16. W. Rogula-Kozłowska, B. Kozielska, G. Majewski, P. Rogula-Kopiec, W. Mucha, K. Kociszewska, J. Environ. Sci. (2017),
17. J.H. Seinfeld, S.N. Pandis, *Atmospheric chemistry and physics: from air pollution to climate change*. John Wiley & Sons (2006)
18. M. Sharma, S. Kishore, S. N. Tripathi, S. N. Behera, J Atmos Chem. **58**:1–17(2007)

Determination of PCDDs in spider webs: preliminary studies

Justyna Rybak[1,*] and *Radosław* Rutkowski[1]

[1]Wrocław University of Science and Technology, Department of Environmental Protection, Wybrzeże Wyspiańskiego 27, 50-370 Wrocław, Poland

Abstract. The application of spider webs for determination of polichlorinated dibenzo-*para*-dioxins (PCDDs) has been studied for the first time. The aim of the studies was to find out if spider webs are suitable for such examinations as it was proved in the previous research they are excellent indicators of air pollutants. Spiders are ubiquitous, thus collection of samples is easy and non-invasive. Studies were conducted within the city of Wrocław and surroundings, one of the biggest and at the same time heaviest polluted city in Poland. Five research sites have been chosen, where spider webs were collected after 60 days of continuous exposure time. Webs belonging to two genera *Tegenaria sylvestris* and *Tegenaria ferruginea* (family Agelenidae) have been chosen as they are large and very dense, thus they are very suitable for such examinations. Webs were found to retain dioxins probably mainly by external exposure. These promising results should be continued and expanded in the future research.

1 Introduction

Air pollutants make people sick, causing breathing problems and increasing the likelihood of cancer [1]. Polichlorinated dibenzo-*para*-dioxins (PCDDs) belong to the group of chlorinated polycyclic aromatic hydrocarbons which are persistent environmental contaminants and as a result they are hazardous to the environment because of their toxic, mutagenic and carcinogenic character. They are by-products of industrial and combustion processes or industrial compounds [2].

Environmental exposure to dioxins contamination has caused health concerns as tetrachloro dibenzo-*p*-dioxin (TCDD) was classified as carcinogenic to humans by the International Agency for Research on Cancer [2, 3]. PCDDs have been identified as residues in hair, wool and human milk [4]. Spider webs have never been studied. They proved to be perfect indicators of air pollutants as web samples are easily collected from different areas. They are widely available and they are usually woven in secluded places preventing them from destroying by weather conditions. It is the universal, organic, environment friendly tool as webs do not need any preparation before exposition to pollutants and do not need to be degraded. Also the exact time of exposition to pollutants can be easily controlled as they can be obtained from rearing of spiders under laboratory conditions. Spider webs were

* Corresponding author: justyna.rybak@pwr.edu.pl

successfully used to evaluate heavy metals and trace element pollution as well as polycyclic aromatic hydrocarbons (PAHs) content in the air [5,6].
The aim of this study was to find out if the spider webs are suitable to determine the environmental exposure to PCDDs.

2 Material and Methods

2.1 Material

The studies were conducted in 2012 at five sites in Wrocław (Fig. 1.). Genera of *Tegenaria ferruginea* (Panzer, 1804) and *Tegenaria silvestris* (L. Koch, 1872) from Agelenidae family have been chosen for studies as their web is very dense and large. *T. ferruginea* occurs in hedgerows, fences, tree crevices and on walls of buildings. Males are mature in summer, whereas females are present all year [7]. *T. silvestris* is usually associated with woodland, but occasionaly it can be found in fences or even houses where it weaves sheet webs. Adults are found most of the year [7]. Spiders belonging to Agelenidae do not eat their own webs when destroyed as such behaviour could influence the further pollution level in the web.
Samples were collected from secluded locations, from buildings and walls. All sites were protected from weather conditions e.g. severe rain. Webs were collected after 60 days from the creation of the new construction (after the removal an old web if such previously existed). Webs of similar sizes, age, and weight were used for analyses [8].

2.2 Sampling preparation

Sample preparation and analysis of PCDDs contained in the samples were made in the Chemistry Department of University of Wrocław and University of Environmental and Life Sciences in Wrocław. Webs were extracted with n-hexane/acetone (50/50, v/v) in a Soxhlet apparatus. The extract was concentrated on the rotary evaporator. Samples were analysed for the 7 PCDD EPA-designated toxic congeners with high resolution gas chromatograph/high resolution mass spectrometer (HRGC/HRMS) following EPA Method 8290 [9]. The recovery of analytes was calculated on the basis of stable isotopically labeled $_{13}$C-PCDDs which was introduced into samples of spider webs after extraction to measure the recovery. Mean recoveries ranged from 53% to 87%, depending on the congener. We calculated the TEQ in spider webs using World Health Organization TEFs.

2.3 Sampling sites

Five sites have been chosen for studies according to the pollution level.

Site 1.
51°04’46.6” N, 17°05’16.5”E Located in the south-east part of the city where tenement houses with domestic solid fuel furnaces predominate. The inhabitants also burn trash occasionally. The area lies distantly from the main communications trails. Spider silk was collected from buildings.
Site 2.
51°06’03.2” N, 17°05’40.9”E Biskupin, the housing estate which is situated in the eastern part of Wrocław, in Śródmieście district near Odra River. Heating system differs according to the wealth of inhabitants. Rarely houses still are still equipped with old furnaces where inhabitants use not only coal as a fuel, but occasionally burn trash. Surroundings are

diversified including areas of allotments, parks, high and low buildings. Spider silk was collected from building walls.

Site 3.

51°06'40.7" N, 17°03'30.7"E Grunwaldzki Square– the city centre in the south east part of Wrocław, one of the main and most important place, main communication node with high traffic load day and night. Trucks and other vehicles cross this section as the ring road of Wrocław is still missing. Webs were collected from walls of the parking neighbouring with main road.

Site 4

51°05'03.8" N, 17°01'18.2"E Fences and hedges along Wiśniowa Avenue, one of the most polluted are in Wrocław, because of heavy traffic which is constantly monitored by local pollution agency. Webs were collected from hedges surrounding the street.

Site 5.

51°31'04.6" N, 17°15'42.3"E Quiet rural area situated near the Landscape Nature Park "Dolina Baryczy" where medical and veterinary solid waste incinerator was built by ECO ABC company. Company is relying on the old equipment and is still working despite inhabitants complaints and severe protests as it burns the waste exceeding recommended norms and standards. Spider silk was collected from hedges and bushes at distance up to 500 m from incinerator.

Fig. 1. The map of study area.

3 Results

We found measurable PCDD in spider webs at all sites. All congeners were present at levels above the LOD (limit of detection) at studied sites (Table 1 and 2). The most toxic congeners, 2,3,7,8-TCDD and 1,2,3,7,8-P_5CDD (TEFs=1), had concentrations of 6.89 pg/g and 10.88 pg/g, at sites 4 and 5 and 7.23 and 24.9 at sites 1 and 5, respectively. At site 5 the most abundant congeners were 2,3,7,8-TCDD and 1,2,3,7,8-P_5CDD, contributing 30.2%, and 67.3% to the TEQ. At site 1 the most abundant congeners were 1,2,3,7,8-P_5CDD and 1,2,3,7,8,9-H_6CDD contributing 39.9%, and 19.2% to the TEQ. At site 2 the most abundant congeners were 1,2,3,7,8-P_5CDD and 1,2,3,7,8,9-H_6CDD, contributing 36.7%, and 20.8% to the TEQ. At site 3 the most abundant congeners were 1,2,3,7,8-P_5CDD and 1,2,3,6,7,8-H_6CDD, contributing 28%, and 24.1% to the TEQ. At site 4 the most abundant congeners were 2,3,7,8-TCDD and 1,2,3,7,8-P_5CDD, contributing 45.2%, and 41.1% to the TEQ. The web samples from site 5 had the total TEQ (36.95 pg/g).

Table 1. Concentration of PCDD in spider webs at sites 1, 2 and 3 (pg/g).

Isomers	TEF	Site 1	TEQ (pg/g)	Site 2	TEQ (pg/g)	Site 3	TEQ (pg/g)
2,3,7,8-TCDD	1	2.21	2.21	1.05	1.05	6.02	6.02
1,2,3,7,8-P5CDD	1	7.23	7.23	3.21	3.21	8.6	8.6
1,2,3,4,7,8-H6CDD	0.1	14.3	1.45	7.54	0.75	27.4	2.74
1,2,3,6,7,8,-H6CDD	0.1	33.7	3.37	15.6	1.56	74.1	7.41
1,2,3,7,8,9-H6CDD	0.1	34.5	3.47	18.2	1.82	49.7	4.97
1,2,3,4,6,7,8-H7CDD	0.01	35.9	0.36	28.3	0.28	89.8	0.89
OCDD	0.0003	223	0.06	210	0.06	181	0.05
Total TEQ	-	-	18.09	-	8.73	-	30.68

Table 2. Concentration of PCDD in spider webs at sites 4 and 5 (pg/g).

Isomers	TEF	Site 4	TEQ (pg/g)	Site 5	TEQ (pg/g)
2,3,7,8-TCDD	1	6.89	6.89	10.88	10.88
1,2,3,7,8-P5CDD	1	6.27	6.27	24.9	24.9
1,2,3,4,7,8-H6CDD	0.1	23.8	0.24	63.4	0,63
1,2,3,6,7,8,-H6CDD	0.1	67	0,67	114	0,11
1,2,3,7,8,9-H6CDD	0.1	46.9	0,47	110	0,11
1,2,3,4,6,7,8-H7CDD	0.01	66.5	0,66	234	0,23
OCDD	0.0003	172	0,05	311	0.09
Total TEQ	-	-	15.24	-	36.95

4 Discussion

This is the first study to examine concentrations of PCDDs in spider webs in different sites across Lower Silesia region in Poland. PCDDs were detectable in spider webs, and we noticed relationships between high values of PCDDs and certain sites. Proximity to medical solid waste incinerator at site 5 was associated with highest concentrations of PCDDs (total TEQ 36.95 pg/g). Proximity to major roadways (site 3) was also associated with higher concentrations of PCDDs (total TEQ 30.68 pg/g). Sites 1 and 3 had lower values of PCDDs but not very low which is probably connected with the proximity of major roadways (site 4; total TEQ 15.24 pg/g) and possibility of trash burning at the houses (site 1; total TEQ 18.09 pg/g). The lowest values of PCDDs were obtained in spiders webs collected at site 2 (total TEQ 8.73 pg/g) where trash burning at houses occurs only occasionally. However, results of these analyses are based on small numbers of samples and need to be explored in a larger number of spider webs at studied sites.

What is interesting, other studies [10] did not observe the association between PCDD in dust (spider webs have never been analysed) and proximity to medical waste incinerators, because such incinerators are usually small, thus emission level is rather low. In our studies site 5 was characterised by elevated level of PCDDs which is probably related with the proximity of incinerator, but on the other hand samples at site 5 were collected in the rural area where trash burning and heating with coal or waste are usually applied by inhabitants, what is more, the backyard trash burning and agriculture burning are also very popular. These emissions could significantly contribute to the obtained results of PCDD concentration.

The emissions of PCDD from petrol and diesel powered vehicles were well documented in the past (sites 3 and 4) as well as the existing relationship between trash burning and elevated dioxins levels (sites 1 and 2) [10, 11].

TCDD is perceived as a human carcinogen according to the International Agency for Research of Cancer. Many studies have demonstrated that exposure to dioxins cause various diseases including endocrine disruption, reproductive problems, neurotoxicity and cancers [12]. The obtained concentrations of PCDDs suggest the possible humans' exposure as they are probably exposed to resuspended, contaminated dust which is was found in spider webs. Because webs showed elevated levels of dioxins there is the evidence that inhabitants have been and are exposed to dust containing dioxins.

5 Conclusion

The spider webs were found to absorb dioxins from ambient air. The recorded high concentrations of PCDDs were highly combustion oriented.The highest values were recorded near the medical solid waste incinerator (site 5) which is not a surprise. The high content of PCDDs was also related with personal furnaces and vehicle emissions. The presented prelimnary results proved that webs can serve as an effective, natural tool for the indication of PCDDs contained in silk. Compared with studies based on other bioindicators such as mosses, lichens and vascular plants research based on spider webs are rare. Especially, webs of Agelenidae show excellent accumulative properties which is a result of their high density, and large size comparing to the webs of other spider families. What is more, representatives of Agelenide are active and present all year which is an additional advantage comparing to other bioindicators which can grow and develop only within the vegetative season. The obtained results suggest the need for further investigation of the possibilities of the broader application of spider webs in PCDDs determination.

The work was realized within the allocation No. 0401/0068/16 awarded for Faculty of Environmental Engineering Wroclaw University of Science and Technology by Ministry of Science and Higher Education.

References

1. E. Błaszczyk, W. Rogula-Kozłowska, K. Klejnowski, I. Fulara, D. Mielżyńska-Švach Air Qual. Atmos. Health **10,** 389 (2016)
2. V. Cirimele, M. Villain, G. Salquèbre, C. Staub, P. Kintz, Forsenic. Sci. Int. **176**, 51 (2008)
3. *International Agency for Research on Cancer, Polychlorinated dibenzo-pdioxins and polychlorinated dibenzofurans*, (IARC Monographs on the Evaluation of Carcinogenic Risks to Humans, IARC, Lyon, 1997)
4. T. Nakao, O. Aozasa, S. Ohta, H. Miyata, Chemosphere **48**, 885 (2002)
5. J. Rybak, Water Air. Soil. Pollut. **226**, 105(2015)
6. J. Rybak, T. Olejniczak, Environ. Sci. Pollut. Res. **21**, 2313 (2014)
7. M.J. Roberts, *Spiders of Britain and Northern Europe* (London: Harper Collins Publishers, 1995).
8. G.C. Hose, J.M. James, M.R. Gray, Environ Pollut **120**, 725 (2002)
9. *U.S.EPA. Method 8290A. Polychlorinated dibenzo-p-dioxins (PCDDs) and polychlorinated dibenzofurans (PCDFs) by high-resolution gas chromatography/high resolution mass spectrometry (HRGC/HRMS) 2007*
10. N.C. Deziel, J.R. Nuckols, J.S. Colt, A.J. De Roos, A. Pronk, C. Gourley, R.K. Severson, W. Cozen, J.R. Cerhan, P. Hartge, M.H. Ward, Sci Total Environ. **433**, 516 (2012)
11. X. Jiang, K. Knutson, L. Zwica L, S.Y. Lee, Q. Chen, B. Hong, X. Zhong, W. Hao, T. Towey, B.W., Gillespie, A. Demond, P. Adriaens, J. Lepkowski, A. Franzblau, D. Garabrant, Organohalogen Compd. **73**, 1571, (2011).
12. P.K. Mandal, J. Comp. Physiol. Biol. **175**, 221 (2005).

Preliminary analysis of odor legal solutions and study on their implementation possibilities in Poland

Izabela Sówka[1*], *Leszek Karski*[2]

[1] Wroclaw University of Science and Technology Faculty of Environmental Engineering, Unit of Engineering and Protection of Atmosphere, Wyb. Wyspianskiego 27, 50-370 Wroclaw, Poland
[2] Inspectorate of Environmental Protection, 52/54 Wawelska Str., 00-922 Warsaw, Poland

Abstract. The problem of odor nuisance requires undertaking legal means that aim towards implementation of regulations in order to improve the odor-related quality of air in selected areas in Poland. So far the works carried out in the country were concluded by drawing up 'The guidelines for the bill on counteracting the odor nuisance'. However, as a result of completed social consultations in Poland, the Ministry of Environment, in 2015, resigned from implementing of so called anti-odor act. Currently, the legislature is taking steps which aim at undertaking specific actions in order to introduce solutions, which would directly regulate the issues of odors and the odor nuisance, to the national system. In the countries of the European Union, the issues related to odors are solved in diversified ways and the system still lacks of a uniform proposition, among others related to odor standards. In connection with the above, actions that are taken on a national level should fundamentally aim at developing national odor standards which would take into account the type / the kind of economic activity being a source of odor emission (e.g. clearly separated for existing objects and planned investments), at establishing a procedure and also legal and operational requirements related to determination and the types of zones with defined values of acceptable concentration and determination of reference methodology in monitoring, and modeling the dispersion of odors e.g. including strictly defined frequency of necessary tests. In addition, the process should be accompanied by a creation of financial mechanisms and streams in range of investments related to the development of technology and methods used to limit emission of odors.

1 Introduction

In the countries of the European Union and generally in the world, the odors, which are the product of direct activity of a man and cause negative effects are commonly classified as contaminants and, throughout the world, are subjected to legal regulations. In such cases the odors are treated as contaminants, which released into the environment may cause a number of negative effects, among others, including: worsening of the quality of the natural environment, damages to property or to the world of plants and life of animals, discomfort;

*Corresponding author: izabela.sowka@pwr.edu.pl

decrease of the level of safety and quality of life, inability to use the property in a normal way, and disrupting the normal manner of conducting an economic activity [1].

Among the societies which reside in areas exposed to emission of odors, the odors may cause a number of negative reactions, beginning from: anxiety, discomfort, irritation, insomnia, lack of appetite, awkwardness (when visiting guests feel the unpleasant odor), up to specific health effects in forms of: illnesses of the respiratory and digestive systems or skin diseases, and also physical or mental deficiencies [1-2]. Mechanisms related to the reaction of an organism to an odor may include, among others: innate aversion to odors, exacerbation of fundamental conditions related to reception/perception of odor (hypersensitivity), diseases caused by stress or even psychogenic diseases [3].

By taking into account the influence of odors on the mental and emotional sphere, the odor nuisance is defined as a discomfort caused by odor emitting substances or circumstances related to a presence of odors, which may negatively influence a single person or a group of people. It is caused by perception of unwanted odors which result in worsening of the mood [4-5].

Besides olfactory factors , the perception of an odor may be also influenced by external factors, including: situations accompanying the higher concentrations of a odor (housing situation, the length of sleep, tiredness, the length of work in 'odor-accompanied' environment, the condition of the environment and spatial development in an area of odor nuisance presence (current level of odor, social and economic structure in the residential area, architecture) and odor-unrelated properties of the source of pollution (e.g. the level of noise, vibrations, concentration of particulate matter). Those factors – next to the concentration, hedonic quality, intensity and the time of exposure to a specific odor – in a significant way influence the individual perception of emitted odor and definition of the odor as a nuisance. Because of , among others, the complicated nature of perception of a smell and its assessment, the regulations related to odor nuisance in the European Union and throughout the world are solved on a national or a regional level. Mandatory regulations define following elements in various ways: criteria and acceptable values, including, in the scope of: concentration of the odor (expressed in ou_E/m^3), reference methods and models and also techniques of control and prevention of odor nuisances [6-8], and the considerations dedicated to odors, covered odor-creating features of gases cover among others: concentration, intensity, character, offensiveness, persistency (CICOP), frequency, intensity, duration, offensiveness and location (so called FIDOL) and hedonic tone [6, 8 -10].

The presented work aims at an analysis of the solutions used in legal systems of selected European Union countries in relation to issues of assessment of odor nuisances and at indication of initial possibility of their application in Poland, including determination of directions of actions aiming at introduction of complex solutions relating to laws regulating odor issues to the national legal system.

2 Legal solutions related to the issues of odor nuisance

Solutions and formal regulations in relation to odors, existing in European systems and throughout the world, have a national or regional range, have no uniform character, are related to various types of odor emission sources and include, among others: Environmental Acts and Regulations, Ambient Air Acts and Regulations, Odor Acts and Regulations, Nuisance Acts and Regulations, Natural Resource Acts and Regulations, Right Farm Acts and Regulations, Agricultural Acts and Regulations, Confined Animal Feeding Operations Acts and Regulations [8-9]. Exception – in a strictly technical aspect (methodological) - is constituted by a specific, commonly applied throughout the world, methodology of odor testing described in the norm: EN 13725:2003 [11], which provides a determined value of odor rate emission as a final result (basing on performed measurement of an odor concentration and the knowledge of the stream of tested odor emitting gases). It is a value, which may be applied to assess and verify the odor level, in relation to, used in specific regulations, criteria concerned with the issues of odor nuisance, including the value of odor emission rate/factor (ou_E/s, Figure 1) and (imission) odor concentration (ou_E/m^3, Figure 2). The review of implemented in European legal systems criteria in relation to assessment of odor influence (assessment of odor nuisance) of chosen sources of odor emission done by [6] indicates that mainly – in operational range – the established criteria include issues related to standards corresponding to the value of odor concentration or to a concentration of single odor emitting substance described by usage of reference tools and methods (measurement methods – including dynamic olfactometry in the standard mentioned above, calculation methods – reference models) in a point of emission (so called maximum emission standard) or in the receptor points (so called maximum impact standard), and in selected scenarios of solutions, in specific distances from tested source (separation distance standard). In addition, in chosen legal systems, there is a specific, relating to the number of odor complaints or annoyance level, standard called maximum annoyance standard, and in the scope of technological solutions in accordance with those recommended in BAT, among others, techniques of odor emission limiting (odor nuisance) – we may consider so called technology standard [6]. Mentioned reference measurement and calculation methods (described by standards) are applied most of all in order to determine the value of emission, and then, the odor concentration in receptor points, in relation to, already described in the legal regulations of specific countries, odor imission limits, including related to jurisdictional mission standard called odor impact criteria (OIC) which embraces the value of odor concentration, the percentile value and the time of averaging, which may be calculated, for example, in dependence of the type and the nature of conducted economic activity (Figure 1-2). According to currently possessed knowledge applying such kind of solution is dominating among 28 countries subjected to analysis, which in their legal systems have regulations related to odors [6].

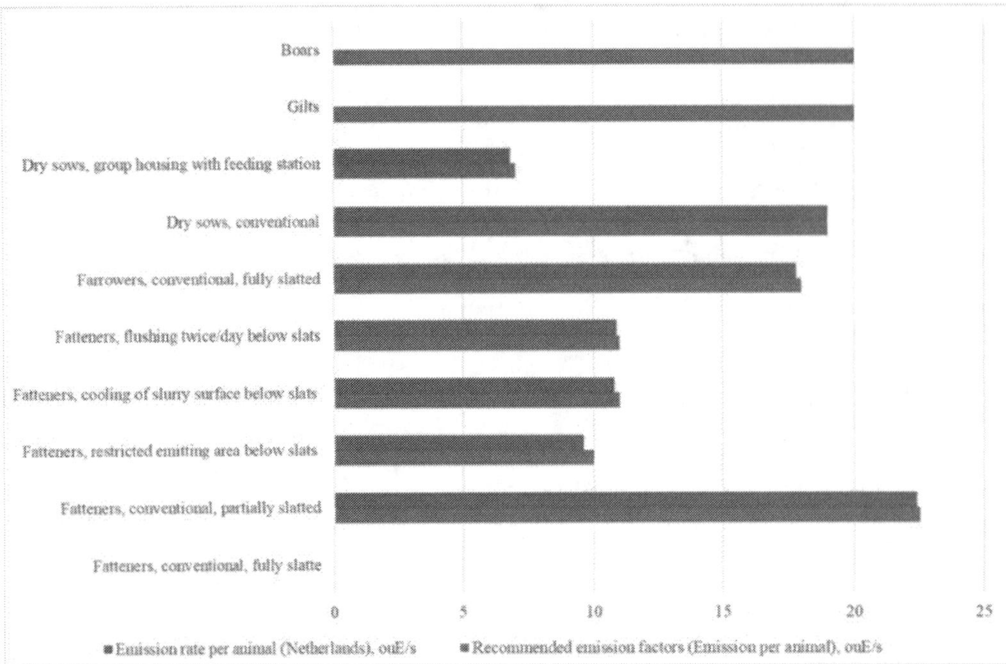

Figure 1. Examples of (odor) emission factors (ou_E/s) for pigs at different stages in the life cycle (based on data published by [12]).

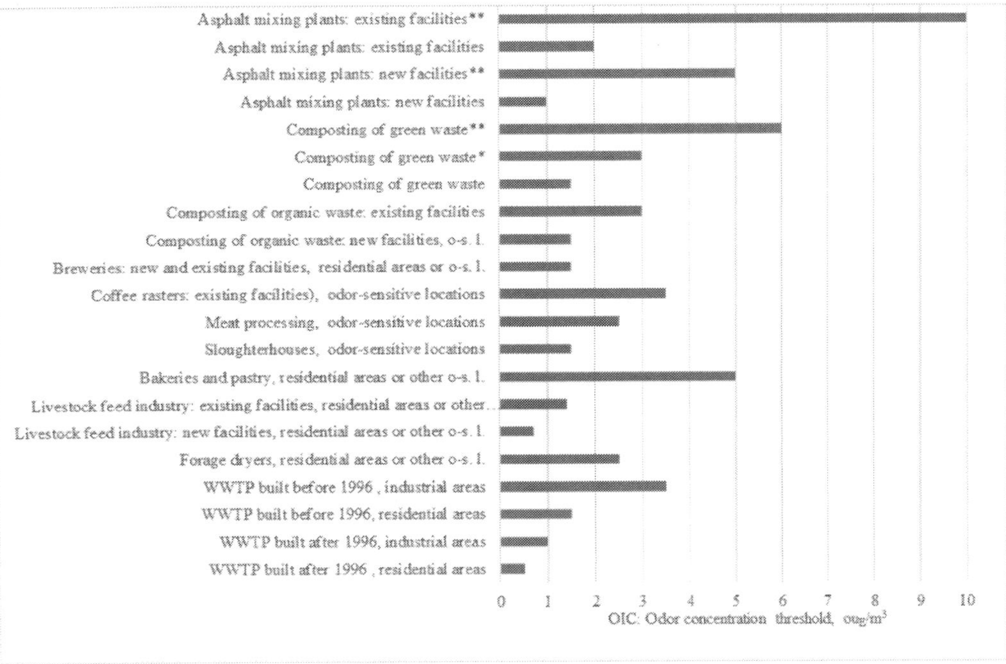

Figure 2. Examples of selected maximum (odor) impact standards (ou_E/m^3) in Netherlands for values of percentile P=98 and average time A_t = 1h with exception: * P=99,5 At = 1h; ** P=99,9 At = 1h; o-s. l: odor sensitive locations (based on data published by [6]).

Another issue in the context of odor-active substances is, therefore, an analysis of implemented solutions in range of measurement methods (for instance, mentioned method of dynamic olfactometry) and mathematical modeling of odor propagation, which include advanced tools – models of odor dispersion in the atmosphere – used to estimate odor concentration in receptor points with a developed meteorological module (preprocessor).

3 Indicating the line for action in relation to a possibility of creating new provisions on a national level

The activities aiming at establishment of laws, undertaken in Poland so far, in relation to odor nuisance, were concluded with drawing up of 'The guidelines for the bill on counteracting the odor nuisance' from 22 December 2010, which were subjected to social consultation. As a result of the consultations the Ministry of Environment, in 2015, resigned from implementing, so called, anti-odor act [13]. Despite the previous ministerial actions, as a result of which there were created, among others, legal regulations corresponding to the issue of odors (table 1) and 'The code of counteracting odor nuisance' that was developed in 2016 [14] there is clearly a need – bodies of Environmental Protection Inspection in 2016 have processed over 1 thousand of complaints, which were related to odor nuisance [15] – to draw up legal regulations on a national level, related to the odor nuisance.

Table 1. Examples of laws corresponding to the issue of odor in Poland (based on [7]).

Provision	**Content of the article**

Civil Code Act from 23 April 1964. (Journal of Laws, 2017, item 459, as amended)	- Art. 144 'The owner of the property, while exercising his rights, shall refrain from activities, which would disturb a usage of neighboring properties in an excessive manner, other than that which results from social and economic designation of the property and local conditions.' [16]
Environmental Protection Act, 27 April 2001 (Journal of Laws, 2016, item 672)	- Art. 362 sec. 1. 'If an entity which benefits from the environment, influences this environment in a negative way, the environmental protection body has a right to, through the means of a decision, impose an obligation of : 1) limiting the influence on the environment, and its endangerment; 2) restoring the environment to a proper condition. sec. 2. In the decision, identified in the sec. 1, the environmental body may set out: 1) the scope of limitation of the influence on the environment, and condition to which the environment shall be restored; 1a) activities aiming at limiting the influence on the environment or at restoring the environment to a proper condition; - Art. 363. 'A head of the commune, a mayor or a president of the city may, through the means of a decision, impose on an individual, whose actions influence the environment negatively, an obligation to undertake activities, in a specified period of time, which aim at 1) limiting the negative influence on the environment and its endangerment; 2) restoring the environment to a proper condition.' [17]
The Act on Provision of Information on the Environment and its Protection, Public Participation in Environmental Protection and Environmental Impact Assessment (Journal of Laws, 2016, item 353).	-Art. 63. sec. 1. 'Obligation to complete an assessment of the influence of an endeavor on the environment, for planned endeavor which may potentially influence the environment in a significant way, may be imposed by, through the means of a resolution, an authority which is competent to issue a decision on environmental constraints, taking into account, in total, flowing criteria: 1) kind and nature of the endeavor, including: (…) d) emission and presence of any other nuisances,..' -Art. 97. sec.1. 'Regional Director of Environmental Protection, through the means of a resolution, may impose an obligation to complete an assessment of the influence of an endeavor on the area of Natura 2000 (Nature 2000), by taking into account following constraints: 1) kind and characteristics of an endeavor, which include (…): d) emission and presence of any other nuisances.' [18]
The Act on fertilizers and fertilizing from 10 July 2007 (Journal of Laws, 2015, item 625, as amended)	Art. 23. sec. 2. 'Fertilizers, including ones marked as 'EC Fertilizer', and substances enhancing the cultivation of plants in a liquid form, shall be transported in closed and sealed packages, containers or tanks.' -Art. 25. sec. 1 Unfermented and fermented liquid manure shall be stored only in sealed containers with a volume which makes it possible to cumulate at least 4-month production of this fertilizer. These containers should be closed, in understanding of provisions based on the art.7, sec. 2, point 2 of the Act from 7 July 1994 [19] related to location and technical conditions which shall be met by agricultural buildings.' [20]
Waste management act from 14 December 2012 (Journal of Laws, 2013, item 21, as amended)	-Art. 16 sets out that the waste management may not cause, among others, nuisances caused by an odor or a noise. -Art. 160 sec. 1. 'Administrator of a waste incineration plant or co-incineration plant is obliged to take necessary precautions, during receipt and thermal transformation of waste, to take necessary precautions which aim at prevention or limitation of a negative influence on the environment, especially in relation to the

	contamination of air, soil, surface and ground waters, odors and noise, and also a direct threat to a human life or health, and to comply with the requirements in scope of thermal transformation of waste. '[21]
Regulation of the Minister of Agriculture and Food Economy on location and technical conditions which should be met by agricultural buildings, from 7 October 1997 (Journal of Laws, 2014, no.0, item 81, as amended)	-Art. 12. 'Agricultural buildings which cause a nuisance in their surroundings, especially including pollination, odors or release of toxic substances, shall be isolated from adjacent areas with a green strip composed of high and average-height growing plants.' - Art. 46. 'The agricultural buildings, inside which there are released substances and odors, which are harmful to health, shall be equipped with efficient ventilation system, during the temporary presence of the staff, which provides that activities related to cleaning, repair and conservation, are performed in a manner which is in accordance with appropriate regulations related to occupational health and safety'. [22]
Regulation of the Minister of Infrastructure on location and technical conditions, which should be met by buildings, from 12 April 2002 (Journal of Laws, 2015, no.0, item 1422)	-Art. 12 sec. 6. 'Livestock or industrial building, by taking into account separate regulations and ones contained within § 13,60 and 271-273, may not be situated in a way, in which the wall with window openings or door openings is in a distance lower than 8 m, from a wall of existing specific buildings located on a construction plot. These buildings include: residential buildings, collective residential buildings or public utility buildings or a building for which there is a final decision about a permit for its construction, subject to sec. 3, point 4.' [23]
Regulation of the Minister of Economy on technical conditions which should be met by liquid fuel depots and stations, long-distance transmission pipelines used to transmit crude oil and oil products, and their location, from 21 November 2005 (Journal of Laws, 2014, no.0, item 1853)	- Art. 6. sec.1 'Containers designated to store crude oil and oil products and also long-distance transmission pipelines and technological pipelines shall be subjected to technical evaluation and leakage tests, provided that separate regulations do not state otherwise.' [24]
Ordinance of the Minister of the Environment on reference values for specific substances in the air, from 26 January 2010 (Journal of Laws, 2010, no.16, item 87)	The ordinance sets out determined reference values for some of the substances in the air, in relation to the health protection. Among 167 substances, there were determined also odor-active substances [25].
Regulation of the Minister of the Environment, from 30 April 2013, on landfill sites (Journal of Laws, 2013, item 523)	-Art. 10 sec.1 'Landfill site shall be surrounded by a green strip composed of trees and shrubs, in order to limit to a minimum, the inconveniences and threats caused by waste located in the landfill site formed as a result of odor and particulate matter emission, carrying the waste by the wind, noise and road traffic, influence of animals, formation of aerosols and outbreaks of fire.' [26]

Analysis of the regulations on an European level clearly indicates, that evaluation of odor nuisance (treated as a phenomenon) should be done by measuring odor concentration in receptor points (ou_E/m^3) – for instance, basing on the emission value determined at the 'source' and analyses with usage of model tests – because considering only one primary compound (and its concentration) as a cause of nuisance in a selected problematic area may be fundamentally inefficient (odors are defined as a mixture of even hundreds of substances

[27-29] especially in case of areas, where a very high number of complaints and protests of local social communities is registered as a result of poor / progressively worsening odor-quality of the air. In Poland, provisions regulating issue of odors in a direct way (e.g. odor standards, measurement and calculation reference methods), would have a relevant utilitarian meaning, especially when considering the high number of registered complaints and court proceedings taking place on a national level. In order to do that it is necessary to undertake actions which include among others: defining odor/odorants as a pollution of the air, establishing national odor standards (e.g. in relation to odor concentration) which take into account a type / kind of economic activity and a nature / kind of installation being a source of odor emission, which would be clearly separated for existing objects and planned investments, developing the procedure, legal and operational requirements related to determination and kinds of zones with defined values of permitted concentration, measurement / assessment of efficiency of deodorization installation and investment project 'acceptance' in this scope – in case of a necessity of its presence, development of complaint system and procedure of complaint analysis, creating market possibilities in order to establish a network of laboratories – accredited entities – in scope of performing e.g. emission measurements (necessity to indicate the measurement output in accordance with the reference methodology in odor monitoring, together with a strictly defined frequency of required tests).

4 Summary

Europe lacks of a uniform for every country of EU, regulations in relation to odors. Legal solutions available throughout the world include as well regulations in the scope of emission standards, acceptable levels, as in the scope of test methods and modeling of odor propagation, and also the assessment of their nuisance level, among others, with usage of sociological survey. In Polish conditions, an issue of a high importance, would be capturing in possible regulations, next to odor standards (emission and imission) – which would also include hedonic quality of odor, or a kind of conducted activity – zones around construction objects which cause an odor nuisance.

Evolution of regulations in relation to odors in Poland, should be supported by creation of mechanisms and financial streams in scope of investments related to development of technology and methods used to limit the emission of odors.

This work was supported by a statutory research No. 0401/0065/16.

References

1. J. Nicell , Atmos. Environ. **43**, 196–206 (2009)
2. A. Jaremków, Ł. Szałata, B. Kołwzan, I. Sówka, J. Zwoździak, K. Pawlas, Polish Journal of Environmental Studies **26**, 127-136 (2017)
3. D. Shusterman, Arch. of Environ. Health **47**, 76–87 (1992)
4. VDI 3883, part I *Effects and assessment of odours. Psychometric assessment of odour annoyance. Questionnaires* (Beuth Verlag, Berlin, 1997).
5. E. Vance, Nature **455**, 726-728, (2008)
6. M. Brancher, K. D. Griffiths, D. Franco, H. de Melo Lisboa, Chemosphere **168**, 1531-1570 (2017)
7. I. Sówka, A. Bokowa, Ł. Pachurka, A. Grzelka., Przegląd Komunalny **4,** 15-18 (2017) (in Polish)

8. L.Woźniak, *Legal regulations on counteracting odor nuisances in selected countries of the European Union together with the commentary about social and economic consequences of their implementation*, Analyses and Topical Papers Unit of the Analyses and Documentation Office, Copyright by Chancellery of the Senate (2014) (in Polish)
9. A. H. Bokowa, Chemical Engineering Transactions **23**, 32-36 (2010)
10. M. Górski, Przegląd Komunalny **4**, 16-17 (2016) (in Polish)
11. EN 13725:2003, *Air Quality. Determination of Odour Concentration by Dynamic Olfactometry*
12. M. Górski, Przegląd Komunalny **2**, 20-21 (2017) (in Polish)
13. EPA: *Odour Impacts and Odour Emission Control Measures for Intensive Agriculture. Final report*, (R&D Reports Series No 14, 2001)
14. *The code of counteracting odor nuisance* (Department of Air and Climate Protection, Polish Ministry of the Environment, Warsaw, September 2016)
15. http://www.gios.gov.pl/pl/aktualnosci/448-odory-a-zakres-dzialan-inspekcji-ochrony-srodowiska (access: 11.11.2017)
16. Civil Code Act from 23 April 1964. (Polish Journal of Laws, 2017, item 459, as amended)
17. Environmental Protection Act from 27 April 2001 (Polish Journal of Laws, 2016, item 672)
18. The Act on Provision of Information on the Environment and its Protection, Public Participation in Environmental Protection and Environmental Impact Assessment (Polish Journal of Laws, 2016, item 353)
19. Construction Act (Polish Journal of Laws, 2013, item 1409, as amended 1)
20. The Act on fertilizers and fertilizing from 10 July 2007 (Polish Journal of Laws, 2015, item 625, as amended)
21. Waste management act from 14 December 2012 (Polish Journal of Laws, 2013, item 21, as amended)
22. Regulation of the Minister of Agriculture and Food Economy on location and technical conditions which should be met by agricultural buildings from 7 October 1997 (Polish Journal of Laws, 2014, no.0, item 81, as amended)
23. Regulation of the Minister of Infrastructure on location and technical conditions, which should be met by buildings from 12 April 2002 (Polish Journal of Laws, 2015, no.0, item 1422)
24. Regulation of the Minister of Economy on technical conditions which should be met by liquid fuel depots and stations, long-distance transmission pipelines used to transmit crude oil and oil products, and their location from 21 November 2005 (Polish Journal of Laws, 2014, no.0, item 1853)
25. Ordinance of the Minister of the Environment on reference values for specific substances in the air from 26 January 2010 (Polish Journal of Laws, 2010, no.16, item 87)
26. Regulation of the Minister of the Environment from 30 April 2013, on landfill sites (Polish Journal of Laws, 2013, item 523)
27. N. Hudson, G.A. Ayoko, Bioresource Technology **99**, 3982-3992 (2014)
28. M. Brattoli, G. De Gennaro, V. De Pinto, L.A. Demarinis, S. Lovascio, M. Penza, Sensors **11**, 5290-5322 (2011)
29. P. Gostelow, S. Parsons, R. Stuetz, Water Research **35**, 579-597 (2001)

Minamata Convention on Mercury. Reporting obligations of the Parties to the Convention and the sources of data existing in Poland

Ewa Strzelecka-Jastrząb[1,*]

[1] Institute for Ecology of Industrial Areas, 6, Kossutha St., 40-844 Katowice, Poland

Abstract. After that, when more than 60 years ago in the Japanese city of Minamata there was caused a mass poisoning of residents by seafood contaminated with mercury, Minamata Convention on Mercury came into force on August 16, 2017. To date, the Convention has been signed by 128 States, the signatories of the Convention and ratified by 83 States - Parties to the Convention. The Convention imposes a number of obligations on the Parties to the Convention, including the reporting obligation. The paper analyses the reporting obligations of the Parties to the Convention, which are in force after the entry into force of the Convention, pursuant to the provisions contained therein. In addition, the existing sources of quantitative data on mercury in Poland are characterized.

1 Introduction

In the 50s of the 20th century, operating in the area of Minamata City, on the island of Kyushu, the chemical factory was emitting into the sea the sewage containing methylmercury. Despite the relatively low concentrations of mercury ions in water, in the meat of large predatory fish and seafood, as further links in the food chain, methyl- and dimethylmercury accumulation have been occurred. Because fish and crabs were the main diet ingredients of the surrounding villages residents, after some time people become poisoned. The first cases were reported in April 1956. Soon the whole "epidemic" of the disease of a previously unknown neurological syndrome was stated, which was called Minamata disease. This disease is a set of symptoms that include damage to the nervous system as a result of mercury poisoning. The disease destroys the brain in particular, causes visual disturbances, motor coordination disorders, mental retardation, and death. In total, by 2001, 2265 victims had been officially recognised as having Minamata disease, 1784 of whom had died [1].

After almost 60 years since these events, at the fifth session of the Intergovernmental Negotiating Committee on mercury in Geneva on 19 January 2013, the Minamata Convention on Mercury was agreed. At the conclusion of the negotiations, the text of the Convention was adopted and opened for signature at the Diplomatic Conference held in Minamata and Kumamoto, Japan, from 9 to 11 October 2013. To date, the Convention has

* Corresponding author: e.strzelecka-jastrzab@ietu.pl

been signed by 128 signatory States, including Poland, on September 24, 2014, and has been ratified 83 States - Parties to the Convention [2].

Pursuant to Article 31 (1), the Convention entered into force on the ninetieth day after the date of the deposit of the fiftieth instrument of ratification, which took place on 16 August 2017. The first Conference of the Parties to the Minamata Convention was held on 24 September 29, 2017 in Geneva.

The Minamata Convention on Mercury is a global treaty of United Nations Environment Programme (UNEP) to protect human health and the environment from the adverse effects of mercury. The Parties to the Convention recognised that mercury is a chemical of global concern owing to its long-range atmospheric transport, its persistence in the environment once anthropogenically introduced, its ability to bioaccumulate in ecosystems and its significant negative effects on human health and the environment.

2 Reporting obligation

The Convention imposes a number of obligations and tasks on the Parties to the Convention, including, under Article 21, reporting obligations that involve inventory of stocks, installations, emissions and releases [3]. Each Party to the Convention shall submit reports on the measures it has taken to implement the provisions of the Convention and the effectiveness of such measures and potential challenges in meeting its obligations under the Convention. These reports are to be submitted to the Conference of the Parties through the Convention Secretariat.

In accordance with Article 21 of the Convention, each Party shall include in its reporting the information as called for in Articles 3, 5, 7, 8 and 9 of the Convention. These articles concern the following:

- Article 3 - mercury supply sources and trade,
- Article 5 - manufacturing processes in which mercury or mercury compounds are used,
- Article 7 - artisanal and small-scale gold mining,
- Article 8 - emissions of mercury or mercury compounds to the atmosphere,
- Article 9 - releases of mercury or mercury compounds to land or water.

The reporting obligations of the countries on mercury supply sources and trade cover the following issues:

- it should be identified individual stocks of mercury or mercury compounds exceeding 50 metric tons, as well as sources of mercury supply generating stocks exceeding 10 metric tons per year, that are located within territory of the country,
- it should be collected information related to the principles concerning of mainly importing, exporting, producing and trading mercury resources, which will confirm the fulfillment of the requirements for mercury supply sources and mercury trade,
- take measures to ensure that, where the Party determines that excess mercury from the decommissioning of chlor-alkali facilities is available, such mercury is disposed of in accordance with the guidelines for environmentally sound management, using operations that do not lead to recovery, recycling, reclamation, direct re-use or alternative uses.

For the purposes of Article 3 of the Convention, mercury is defined as "mercury" include mixtures of mercury with other substances, including alloys of mercury, with a mercury concentration of at least 95 per cent by weight while mercury compounds means mercury (I) chloride (known also as calomel), mercury (II) oxide, mercury (II) sulphate, mercury (II) nitrate, cinnabar and mercury sulphide. Article 3 also specifies that the provisions on sources of mercury supply and trade shall not apply to:

- quantities of mercury or mercury compounds to be used for laboratory-scale research or as a reference standard, or

- naturally occurring trace quantities of mercury or mercury compounds present in such products as non-mercury metals, ores, or mineral products, including coal, or products derived from these materials, and unintentional trace quantities in chemical products, or
- mercury-added products.

Within manufacturing processes in which mercury or mercury compounds are used, according to the provisions of the Convention, should be included chlor-alkali production, acetaldehyde production in which mercury or mercury compounds are used as a catalyst, vinyl chloride monomer production, production of sodium or potassium methylate or ethylate and production of polyurethane using mercury containing catalysts. In relation to these processes:

• it should be identified facilities within its territory that use mercury or mercury compounds for manufacturing processes and submit to the Secretariat, no later than three years after the date of entry into force of the Convention, information on the number and types of such facilities and the estimated annual amount of mercury or mercury compounds used in those facilities,
• it should be collected information on taking measures to address emissions and releases of mercury or mercury compounds from facilities in which mercury or mercury compounds are used in manufacturing processes,
• it should be collected information confirming the withdrawal chlor-alkali production and acetaldehyde production in which mercury or mercury compounds are used as a catalyst according to the presented timetable - respectively the phase-out date 2025 and 2018,
• it should be collected the quantitative data demonstrating the fulfilment of the requirements for mercury use and/or its emissions and releases from the vinyl chloride monomer production, production of sodium or potassium methylate or ethylate and production of polyurethane using mercury containing catalysts.

In the case of vinyl chloride monomer production the requirements mentioned above are to reduce the use of mercury in terms of per unit production by 50 per cent by the year 2020 against 2010 use, in the case of production of sodium or potassium methylate or ethylate the requirements mentioned above are:
- to reduce the use of mercury aiming at the phase out of this use as fast as possible and within 10 years of the entry into force of the Convention,
- to reduce emissions and releases in terms of per unit production by 50 per cent by 2020 compared to 2010,
- to prohibit the use of fresh mercury from primary mining,

as well as in the case of polyurethane production using mercury containing catalysts – to reduce the use of mercury, aiming at the phase out of this use as fast as possible, within 10 years of the entry into force of the Convention.

The reporting obligation may also apply to countries where artisanal and small-scale gold mining is used and where there is used processing in which mercury amalgamation is used to extract gold from ore. The Convention specifies that if at any time the Party determines that artisanal and small-scale gold mining and processing in its territory is more than insignificant then develops and implements a national action plan and submits it to the Secretariat no later than three years after entry into force of the Convention and thereafter, provides a review every three years of the progress made in meeting the obligations and includes such reviews in its reports. It should be assumed that the reporting obligation resulting from the artisanal and small-scale gold mining and processing will not be concern to Poland.

The Convention also imposes a reporting obligation in respect to the point sources of mercury and mercury compounds emissions into the atmosphere. These sources include the following categories: coal-fired power plants, coal-fired industrial boilers, smelting and roasting processes used in the production of lead, zinc, copper and industrial gold, waste

incineration facilities and cement clinker production facilities. In relation to these emission sources, the Convention imposes the following requirements on countries – the Parties to the Convention:
• it should be establish, as soon as practicable and no later than five years after the date of entry into force of the Convention, and maintain thereafter, an inventory of emissions from relevant sources,
• it should be defined which sources will be considered in the country as "relevant" within each source category. Party may establish criteria to identify the sources covered within a source category so long as those criteria for any category include at least 75 per cent of the emissions from that category,
• it should be collected and reported information confirming that action is taken to control emissions of mercury and its compounds from relevant sources,
• it should be collected and reported information confirming that other provisions of the Convention relating to point sources of mercury emissions to the atmosphere are being implemented, and in particular that:
- for new sources, each Party is required the use of best available techniques and best environmental practices to control and, where feasible, reduce emissions, as soon as practicable but no later than five years after the date of entry into force of the Convention. Each Party may use emission limit values that are consistent with the application of best available techniques,
- for existing sources, each Party, taking into account its national circumstances, and the economic and technical feasibility and affordability of the measures, should implement as soon as practicable but no more than ten years after the date of entry into force of the Convention, one or more of the following measures: a quantified goal for controlling and, where feasible, reducing emissions from relevant sources, emission limit values for controlling and, where feasible, reducing emissions from relevant sources, the use of best available techniques and best environmental practices to control emissions from relevant sources,, a multi-pollutant control strategy that would deliver co-benefits for control of mercury emissions or alternative measures to reduce emissions from relevant sources. A national plan setting out the measures to be taken to control emissions and its expected targets, goals and outcomes may also be prepared. Then the plan should be submitted to the Conference of the Parties within four years of the date of entry into force of the Convention.
- reasonable progress must be made in reducing emissions over time by means of measures, wherein Parties may apply the same measures to all relevant existing sources or may adopt different measures in respect of different source categories.

Mercury releases to land and water from point sources located in territory of the country are also subject to reporting obligations. In relation to them:
• it should be demonstrated that there is controlling and, where feasible, reducing releases of mercury and mercury compounds, often expressed as “total mercury”, to land and water from the relevant point sources not addressed in other provisions of the Convention,
• each Party should, no later than three years after the date of entry into force of the Convention and on a regular basis thereafter, identify the relevant point source categories, relating to the release of mercury and mercury compounds to the land and water from relevant point sources not addressed in other provisions of the Convention,
• each Party should establish, as soon as practicable and no later than five years after the date of entry into force of the Convention, and maintain thereafter, an inventory of releases from relevant sources,
• it should be collected and reported information confirming that other provisions of the Convention relating to releases, and in particular that:

- a Party with relevant sources should take measures to control releases of mercury and mercury compounds to land and water. A Party may also prepare a national plan setting out the measures to be taken to control releases and its expected targets, goals and outcomes. Then the plan should be submitted to the Conference of the Parties within four years of the date of entry into force of the Convention.
- the measures taken, in accordance with country needs should include one or more of the following, as appropriate: release limit values to control and, where feasible, reduce releases of mercury and mercury compounds to land and water from relevant sources, the use of best available techniques and best environmental practices to control releases from relevant sources, a multi-pollutant control strategy that would deliver co-benefits for control of mercury releases or alternative measures to reduce releases from relevant sources.

3 Sources of data existing in Poland

At present, the following sources of information and quantitative data necessary for the fulfillment of reporting obligations resulting from the provisions of the Minamata Convention on Mercury are existing and available in Poland:

• European Pollutant Release and Transfer Register (E-PRTR) [4], in the form of a public electronic database. Within the database there are only facilities that meet the criteria of the EC Regulation No. 166/2006 and are in excess of the relevant E-PRTR thresholds, mainly there are only the largest plants in the country. The database includes, among others, mercury and its compounds (as Hg). Pollutant releases on air, water and soil and off-site pollution transfers have been reported. For mercury, the threshold values for pollutant release were set at the following level: releases to air – 10 kg/year, releases to water – 1 kg/year, releases to land - 1 kg/year. Data is reported annually starting from 2007. The latest report from Poland includes data for 2015 and contains 48 facilities reporting mercury emissions to air (total emission of 5.46 tons), 29 facilities reporting mercury releases to water (total release of 440 kg), 0 facilities reporting mercury releases to land and 7 facilities reporting mercury transfer to waste water (total transfer of 898 kg).

• National Database on emissions, managed by National Centre for Emission Management (KOBiZE) [5]. The national database is an IT system containing a secure database that collects data on greenhouse gas emissions and other substances, including mercury. The database contains unique information about emission sources together with location of sources and parameters of operation of individual installations. The database provides accurate information based on actual data provided by the facilities as part of annual reporting. Data may be made available on the basis of a written request under the provisions governing access to environmental data.

• the so-called "fee databases" existing in the Marshal's Offices in the 16 respective voivodships. The databases were created under Article 275 of the Environmental Protection Law and serve to pay the fees for the use of the environment by the facilities that use it. The fee for using the environment is borne by, among others, for the emission of gases or particulates into the air and the release of sewage into the water or to the ground. Facilities, paying a fee also report the total amount of pollutant releases to the air, water or ground. Unfortunately, many facilities, due to the identical rate of fees, report only the total amount of heavy metals emitted into the air, than mercury alone. Data may be made available on the basis of a written request under the provisions governing access to environmental data.

• the national inventory of pollutant emission is prepared by the National Center for Emission Management (KOBiZE) for the purposes of national statistics, European Union requirements and commitments to international organizations within the European Union, Eurostat and the European Environment Agency (EEA) and the United Nations Convention on Long-range Transboundary Air Pollution (LRTAP), United Nations Economic Commission for Europe (UN ECE) and the European EMEP Program. The countrywide inventory includes, among others. heavy metals, including mercury. The latest report includes data for 2014 and 2015 (the total national emission of mercury into the atmosphere were 9.6 tons and 10.6 tons respectively), the description of the methodology used for emissions balancing and analysis of key categories and emission trends from 1990 [6].

4 Summary

After analyzing the existing sources of information and resources of data on mercury, it should be stated that the current status of the availability of data and information needed to meet the reporting obligations of the Minamata Convention on Mercury should be assessed as insufficient. Practically there is no access to data on sources of mercury supply and mercury trade which is most often a trade secret of companies. On the other hand, data on other issues which should be reported are inconsistent and widely dispersed between the different registers and the different authorities. For this reason, further work is needed on the rearrangement of information and the creation of access to reliable data.

References

1. Minamata Disease. The History and Measures, Ministry of the Environment Government of Japan (2002) http://www.env.go.jp/en/chemi/hs/minamata.html
2. www.mercuryconvention.org (last access 10.10.2017)
3. Minamata Convention on Mercury. Text and Annexes, UNEP (October 2013)
4. http://prtr.ec.europa.eu/#/home (last access 10.10.2017)
5. Act on the Greenhouse Gas Emission Management System and other substances (OJ 2013, item 1107 with later amendments), in Polish
6. Poland's Informative Inventory Report 2017, National Centre for Emission Management (KOBiZE) at the Institute of Environmental Protection – National Research Institute, Warsaw (February 2017) http://www.kobize.pl/en/article/krajowa-inwentaryzacja-emisji/id/385/zanieczyszczenia-powietrza

E3S Web of Conferences **28**, 01036 (2018) https://doi.org/10.1051/e3sconf/20182801036
Air Protection in Theory and Practice

Application of fuzzy logic to determine the odour intensity of model gas mixtures using electronic nose

Bartosz Szulczyński[1,*], *Jacek* Gębicki[1], and *Jacek* Namieśnik[2]

[1]Department of Chemical and Process Engineering, Faculty of Chemistry, Gdańsk University of Technology, 11/12 Gabriela Narutowicza Street, 80-233 Gdańsk, Poland
[2]Department of Analytical Chemistry, Faculty of Chemistry, Gdańsk University of Technology, 11/12 Gabriela Narutowicza Street, 80-233 Gdańsk, Poland

Abstract. The paper presents the possibility of application of fuzzy logic to determine the odour intensity of model, ternary gas mixtures (α-pinene, toluene and triethylamine) using electronic nose prototype. The results obtained using fuzzy logic algorithms were compared with the values obtained using multiple linear regression (MLR) model and sensory analysis. As the results of the studies, it was found the electronic nose prototype along with the fuzzy logic pattern recognition system can be successfully used to estimate the odour intensity of tested gas mixtures. The correctness of the results obtained using fuzzy logic was equal to 68%.

1 Introduction

Public opinion on the state of the environment depends on many factors, among which the smell of air we breathe occupies one of the important factors. Today, many areas of human activity, in particular waste water treatment and municipal waste management, are a source of odour nuisances [1,2]. For this reason, measurement and monitoring of odour nuisance is an extremely important issue to be addressed. According to EN 13725:2003, the odour concentration is determined by the dynamic olfactometry method. From a practical point of view, it would be most valuable to directly determine the odour nuisances using the results of analytical air monitoring. The solution to this problem would keep the public about odour air quality. In addition, it will be possible to forecast the extent of significant nuisance of the projected objects based on the results of simulation of odour spread in the environment of emitters.

One of the parameters that characterize the smell is an odour intensity. It is one of the most commonly determined odour parameters. It is defined as the perceived strength of odour sensation that will be triggered by a specific stimulus [3]. Quantitative odour intensity determinations are typically based on sensory analysis in which a sensory panel determines the intensity of the odour using a verbal point scale. An example of such a scale

*Corresponding author: bartosz.szulczynski@pg.edu.pl

is the seven-point scale described in German Standard VDI 3940. Degrees corresponding to the respective odour intensity are shown in Table 1.

Table 1. Odour intensity scale described in German Standard VDI 3940.

Odour strength	**Intensitylevel**
Not perceptible	0
Very weak	1
Weak	2
Distinct	3
Strong	4
Very strong	5
Extremely strong	6

The paper presents the possibility of fuzzy logic application to determine the odour intensity of model ternary gas mixtures using electronic nose prototype. As the ingredients of the mixtures three substances were selected: α-pinene, toluene and triethylamine. These substances have low odour thresholds and have been identified in the air nearby waste water treatment plant or landfill [4-6]. The results obtained using fuzzy logic algorithms were compared with the values obtained using multiple linear regression model and sensory analysis.

2 Electronic nose

In recent years there has been a significant increase in interest in the use of artificial senses in chemical analysis. Particular position is occupied by devices that mimic the sense of human smell - electronic noses. These instruments allow a holistic analysis of the gas mixture composition, without separating it into individual components (as is the case with the use of chromatographic methods) [7]. In terms of construction, they are equipped with a set of several chemical gas sensors. The commercially available volatile organic compounds sensors are the most commonly used sensor types [8,9].

E-noses in their functioning resemble the sense of human smell (Figure 1). Sensors are the analogs of the olfactory receptors. They turning the chemical information into an analytically useful signal. Then the signal is sent to the recognition system, which in the case of human is the brain, and in the case of e-nose is the appropriate mathematical algorithm. The most commonly used data processing methods are: principal component analysis (PCA), principal component regression (PCR), partial least square regression (PLS), fuzzy logic and artificial neural networks (ANN) [10-14].

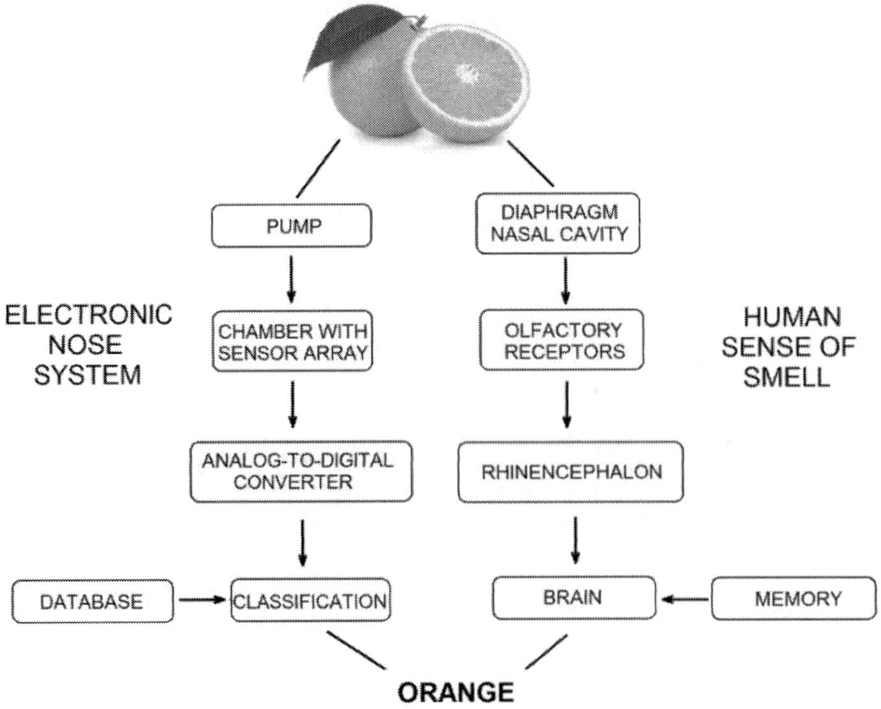

Fig. 1. Diagram of the electronic nose operation along with human sense of smell.

The electronic nose consists of four basic and independent elements:

- sampling system – provide reproducible and stable measurement conditions (gas flow velocity, humidity, temperature) and eliminates all undesirable factors that can affect the sensor response,
- detection system - a set of chemical sensors located in the measuring chamber which exhibit different selectivity and sensitivity to the individual components of the sample, but as a whole they generate a characteristic chemical pattern of the gas mixture (so-called "fingerprint"),
- data processing system – responsible for signal processing,
- pattern recognition system - assigns the received set of signals to one of the pattern classes.

The most commonly used type of sensors in electronic noses are Metal Oxide Sensors (MOS). The sensor signal's dependence on the concentration of the gaseous substance in the sensor environment is described by the formula (1):

$$R_s = B \cdot c^{-\alpha} \quad (1)$$

where: R_s – sensor resistance; B, α – constant values characterizing the sensor; c – substance odour concentration.

Previous studies have shown that the theoretical dependence (2) is linear [8]. For this reason, a multiple linear regression model (3) was proposed combining the odour intensity with the electronic nose sensor signals:

$$I = \mathrm{f}(\log S) \quad (2)$$

$$I = a_0 + a_1 \log S_1 + a_2 \log S_2 + \ldots + a_n \log S_n \quad (3)$$

where: I – odour intensity; S – sensor signal; a – model parameter, n – number of sensors working in electronic nose system.

3 Fuzzy logic

One of the most interesting approaches in the field of e-nose data analysis is fuzzy logic. Classical logic system is based on the two values, mostly represented by 0 and 1, or true and false. The boundary between them is unambiguously defined and unchanging. Fuzzy logic is an extension of the classical approach to approach closer to human brain - it introduces additional values between standard 0 and 1. Blurring the boundaries between them gives the opportunity to come up with values between this interval (eg, almost false, half truth). An example of fuzzy and non-fuzzy sets is shown in Figure 2.

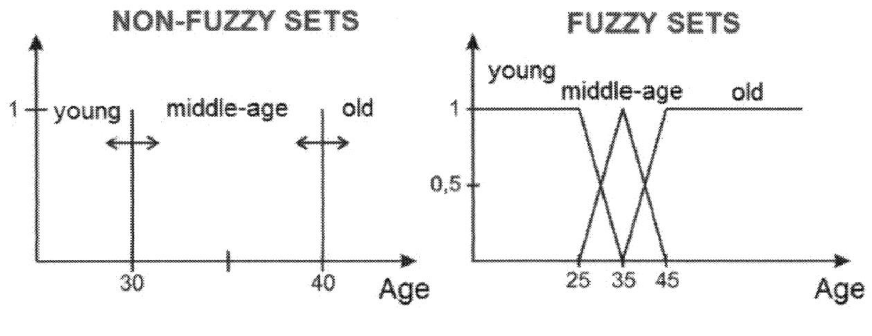

Fig.2. An example of fuzzy and non-fuzzy sets.

Fuzzy logic allows a fuzzy description of real systems and is an alternative to describing systems using classical binary logic. The basis of fuzzy sets theory was introduced by L. Zadech. From this point on, the methods developed on the basis of the theory are of great interest. Their development goes hand in hand with the growing number of applications of fuzzy logic in practice. The proposed scheme of using fuzzy logic to estimate the odour intensity is presented in the Figure 3.

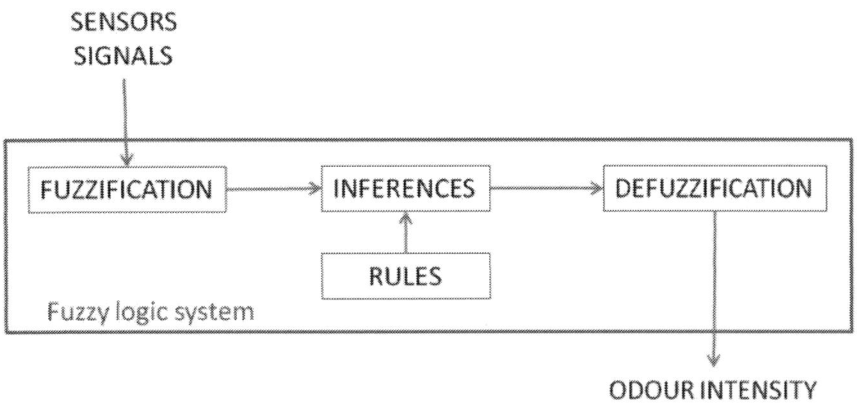

Fig.3. Proposed scheme of using fuzzy logic to estimate the odour intensity using electronic nose.

In the case of non-fuzzy sets, the membership function is rectangular. It is set to 0 (no membership in the set) or one (membership in the set). In case of fuzzy sets, other membership functions, such as trapezoid, triangular, gaussian, sigmoidal are used. In this work, trapezoid functions were used. In the Figure 4, the defined fuzzy set for the e-nose sensor signal value is shown as an example. The chosen values were selected using optimization methods.

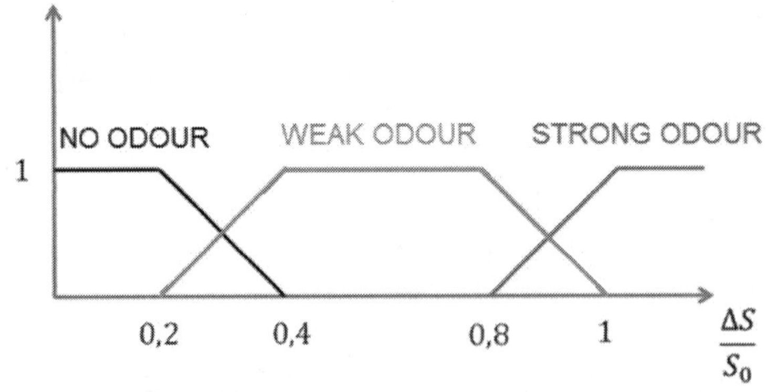

Fig.4. An example of fuzzy and non-fuzzy sets.

For the fuzzy system's output variable, seven fuzzy sets defined by the names in Table 1 were created. The output value is calculated by the fuzzy system, using a set of rules, based on the input values. Thus, the determined odour intensity corresponds to a specific set of sensor signal values. In the application process, a set of rules based on conjunctive operations is applied. An example of the rule is shown in formula 4:

IF (S_1∈NO ODOUR) **AND** ... **AND** (S_8∈NO ODOUR) **THEN** OI∈(Not perceptible) (4)

Fuzzy inference process in the developed model proceeds in three stages (Figure 3). At the entrance of the model are introduced eight input variables (each sensor signals). In the fuzzification block, the degree of belonging of individual values to fuzzy sets is calculated.In the next block, the resulting function of the model output is calculated. In this process, the rule base is used. At the defuzzification stage, the resulting affinity function is the basis for calculating the value of the output variable.The center of gravity mechanism was used for this purpose.

4 Experimental

In order to determine the usefulness of the fuzzy logic to determine the odour intensity using the electronic nose, 22 aqueous solutions were made. The samples 1,6 and 11 were prepared in such a way that their odour intensity was equal to 1 according to the VDI 3940 standard. Five aqueous solutions characterized by a 2-step dilution were prepared for each of the investigated substances. Prepared solutions were used for determination of odour intensity of each sample, which was performed by a team of assesso. Obtained results were utilized to determine the dependency of odour intensity versus logarithm of concentration of particular odorous substance in water (according to the Weber-Fechner law). These dependencies were used to determine concentration of particular substances in deionised water corresponding to odour intensity equal to 1. These concentrations were respectively: 0.6 ppm v/v toluene, 0.1 ppm v/v α-pinene and 1.5 ppm v/v triethylamine. Other samples were prepared in appropriate volume proportions defined by the position of the sample in the ternary system (Figure 5).

 https://doi.org/10.1051/e3sconf/20182801036

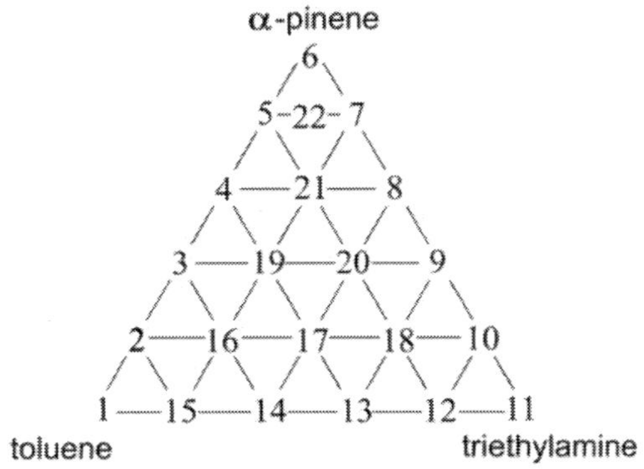

Fig.5. A ternary plot illustrating the composition of the tested mixtures.

The odour intensity of these samples was determined in three ways:

1. using a sensory panel (sensory analysis) according to the procedure described in [15],
2. using electronic nose and fuzzy logic,
3. using electronic nose and multiple linear regression model (3). Model parameters and their statistical significance are presented in Table 2. The model was used for further calculations after removing non-significant members. The final form is represented by the equation (5) - the determination coefficient of the model was equal to 0.72:

$$I = -8.45 - 6.35 \log S_1 + 14.43 \log S_4 \tag{5}$$

Table 2. Calculated model parameters and their statistical significance.

Coefficient	Calculated value	Statistical significance
a_0	-12.53	yes
a_1	-3.22	yes
a_2	1.98	no
a_3	-0.44	no
a_4	15.22	yes
a_5	-2.55	no
a_6	0.44	no
a_7	-0.39	no
a_8	2.55	no

The analyzes were carried out using an electronic prototype equipped with a measuring chamber containing 8 MOS sensors manufactured by Figaro Engineering inc: TGS2104, TGS2106, TGS2180, TGS2600, TGS2602, TGS2201A, TGS2201B and TGS2611. The measurement system is shown in Figure 6.

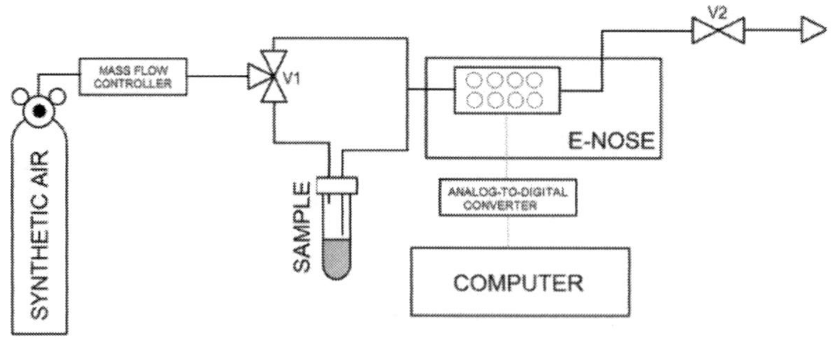

Fig.6. Electronic nose measurement system.

Synthetic air flow through a system at a constant flow rate of 330 cm^3min^{-1}. It was controlled by a mass flow controller. The headspace analysis of prepared samples was conducted. By changing the position of the valve V1, the air flowed through the sample, which then passed to the measurement chamber. The electronic nose worked in the stop-flow mode [16]: the sample flow time was 40 seconds and the stop time of the mixture in the sensors chamber – 20 seconds (after closing V2 valve). After this time the purified air was returned to the measurement chamber for regeneration of the sensors. Signals from the sensors were recorded using an ADC (Simex SIAi-8) and saved on the computer. Data analysis and other calculations were performed using RStudio Desktop (v. 1.0.143) software.

5 Results and discussion

In Figure 7 are presented statistical differences between the determined odour intensity values (in pairs: sensory analysis – fuzzy logic and sensory analysis – multiple linear regression).

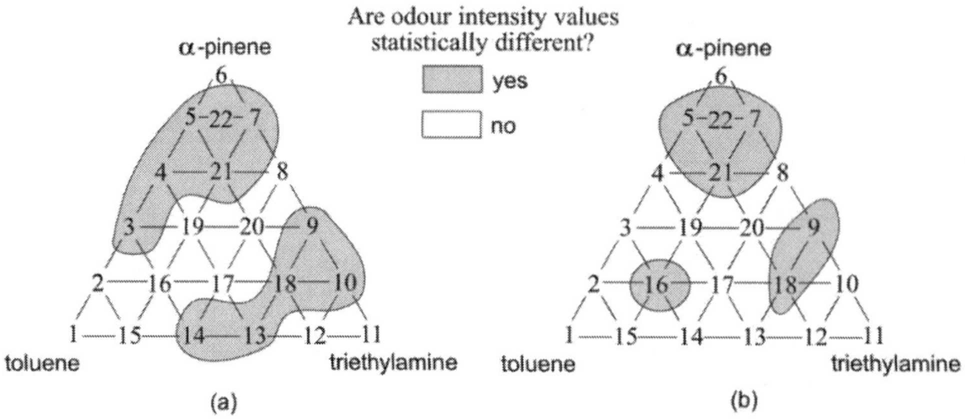

Fig.7. Ternary plots of gas mixtures odour intensity statistical differences: (a) determined by the multiple linear regression model, (b) fuzzy logic, compared to the values obtained by sensory analysis.

As the results of the studies, it was found that the electronic nose prototype along with the fuzzy logic pattern recognition system can be successfully used to estimate the odour intensity of α-pinene, toluene and triethylamine as compounds potentially responsible for the presence of odour nuisances in the vicinity of the landfill or wastewater treatment plant.

Determining the odour intensity using both: fuzzy logic and multiple linear regression model does not fully match the expected results (obtained by sensory analysis). Comparing the number of correct results showed that the correctness of results was better for the fuzzy logic model (68%) compared to the multiple linear model (50%).However, it is important to note that the creation of rules for fuzzy logic is much more complicated than the determination of linear model parameters and it requires some expert knowledge.

Observed differences in odour intensity determination may be due to the presence of odour interactions between the gas mixture components [12]. Conducted studies have shown that the greatest differences were observed for samples with high α-pinene content. In this case, the discrepancy may be caused by masking of unpleasant odours by the nicely scented α-pinene.

The investigations were financially supported by the Grant No. UMO-2015/19/B/ST4/02722 from the National Science Centre (Poland).

References

1. I. Sówka, Ł. Pachurka, Y. Bezyk, A. Grzelka, U. Miller, Ochrona Środowiska i Zasobów Naturalnych **28**, 16-21 (2017)
2. I. Sówka, U. Miller, P. Sobczyński, Eviron. Prot. Eng. **42**, 143-152 (2016)
3. J. Kośmider, B. Mazur-Chrzanowska, *Odory* (Wydawnictwo Naukowe PWN, Warszawa, 2012)
4. J.J. Fang, N. Yang, D.Y. Cen, L.M. Shao, P.J. He, Waste. Manage. **32**, 1401-1410 (2012)
5. P. Lewkowska, B. Cieślik, T. Dymerski, P. Konieczka, J. Namieśnik, Environ. Res. **151**, 573-589 (2016)
6. E. Davoli, M.L. Gangai, L. Morselli, D. Tonelli, Chemosphere **51**, 357-368 (2003)
7. F. Rock, N. Barsan, U. Weimar, Chem. Rev. **108**, 705-725 (2008)
8. B. Szulczyński, J. Gębicki, Environments **4**, 21 (2017)
9. K. Arshak, E. Moore, G.M. Lyons, J. Harris, S. Clifford, Sensor Rev. **24**, 181-198 (2004)
10. J. Gębicki, B. Szulczyński, M. Kamiński, Meas. Sci. Technol. **26**, 125103 (2015)
11. W. Wojnowski, T. Majchrzak, T. Dymerski, J. Gębicki, J. Namieśnik, Meat Sci. **131**, 119-131 (2017)
12. B. Szulczyński, J. Namieśnik, J. Gębicki, Sensors **17**, 2380 (2017)
13. M. Badura, A. Szczurek, P.M. Szecówka, Sens. Actuators B Chem. **188**, 815-823 (2013)
14. S.M. Scot, D. James, Z. Ali, Microchim. Acta **156**, 183-207 (2006)
15. J. Gębicki, T. Dymerski, S. Rutkowski, Environ. Prot. Eng. **40**, 103-116 (2014)
16. M. Maciejewska, A. Szczurek, L. Ochromowicz, Sens. Actuators B Chem. **141**, 417-423 (2009)

Energy and emission aspects of co-combustion solid recovered fuel with coal in a stoker boiler

Ryszard Wasielewski[1,*], *Krzysztof* Głód[1] , and *Jolanta* Telenga-Kopyczyńska[1]

[1]Institute for Chemical Processing of Coal, 41-803 Zabrze, Zamkowa 1 St, Poland

Abstract. The results of industrial research on co-combustion of solid recovered fuel (SRF) with hard coal in a stoker boiler type WR-25 has been presented. The share of SRF in the fuel mixture was 10%. During the co-combustion of SRF, no technological disturbances or significant reduction in energy efficiency of the boiler were noted. Obtained SO_2, NO_x and CO emissions were comparable with coal combustion but dust emissions increased. During combustion of the coal mixture with a 10% share of SRF in the test boiler WR-25, the emission standards established for the combustion of the dedicated fuel were met. However, comparison of obtained emission results with the emission standards established for co-incineration of waste, revealed the exceedance of permissible levels of HCl, dust, heavy metals, dioxins and furans. Additionally, the residence time of flue gases in over 850°C conditions for the test boiler WR-25 was too short (1.3 seconds) in refer to the legislative requirements (2 seconds) for the thermal conversion of waste.

1 Introduction

Energy recovery of wastes by using them for production of stable quality parameters fuels and their energy effective and environmentally safe application is a very important issue in the contemporary waste management system. For several years both development of technology of fuel production from wastes and its use has been observed [1-5]. Fuels from waste are widely used in energy-intensive branches of industry, such as production of clinker or paper, but also for production of electrical energy and heat. Numerous research studies are also conducted to broaden the knowledge on phenomena accompanying the process of energy recovery from these fuels [5-6]. Growth of the market for these fuels is also connected with the development of international SRF (solid recovered fuels) classification [1-2].

Searching for new sales markets, the increasing domestic potential of producing fuels from waste looks with hope toward the power engineering [7,8]. One of the barriers is however, current technical level of the boiler installations, including in particular flue gas purification systems [9,10]. Many boiler houses in Poland are equipped with coal boilers with mechanical grate. However, not too much experience has been gained so far in relation to SRF co-combustion in such installations.

* Corresponding author: rwasielewski@ichpw.pl

The Institute for Chemical Processing of Coal conducted industrial tests of SRF co-combustion with hard coal in WR-25 stoker fired boiler type installation.

The purpose of the research was assessment of the waste and coal co-combustion impact on performance and operational parameters of the boiler, and also determining the possibility to meet the legislative requirements related to the environment protection by the power engineering installation.

2 Materials and methods

The research was of comparative nature and its scope included conducting two parallel series of energy-emission measurements for: combustion of hard coal and combustion of a mix of coal and 10% of SRF.

Scope of measurements included all emission determinations for co-combustion of waste, as well as verification of process conditions necessary for thermal conversion of waste.

2.2 Characteristics of the installation

The WR-25 boiler (of the 23.26 MW_t rated capacity) is a high temperature water-tube boiler with forced draught and blow. Combustion chamber has the following horizontal dimensions: 5.30 m x 3.90 m. The base fuel for the boiler is fine hard coal. By-products of combustion are: slag and fly ash. The WR-25 boiler is equipped with a two-stage dust removing system from flue gases, consisting of series circuit cyclone dust collectors. Flue gases are discharged to the atmosphere via an emitter of 120 m height and 5.2 outlet diameter.

2.3 Research methodology

Samples of hard coal, SRF and their blends were subject to technical and elemental analysis. Furthermore, bulk density and content of heavy metals were determined. Respective determinations were performed in accordance with the research methodology specified in Polish Standards and research procedures of the Institute for Chemical Processing of Coal.

Samples of slag and fly ash were taken every 60 minutes. After averaging, the samples were tested for combustibles content and for heavy metals content. The research was performed in accordance with the research methodology specified in Polish Standards and research procedures of the Institute for Chemical Processing of Coal.

Determinations of dust concentrations, speed and density of flue gas were performed in accordance with the requirements of the PN-Z-04030-7:1994 standard. Determinations were made by means of the EMIOTEST 2598 gravimetric dust meter, aspiration probe with external filtration, and "S" type impact-pressure tube.

Gas concentration in flue gases was determined by means of GASMET DX 4000 gas analyser with hot measurement path, using the FTIR method, among others concentrations of the following gases were determined: NO, NO_2, SO_2, CO_2, CO, HCl and HF. Composition of flue gas was measured continuously.

Determination of VOCs (Volatile Organic Compounds) was made by means of a portable organic substances analyser of the AWE–PW series.

Sampling procedure and determination of heavy metals concentrations and emissions in flue gases (Cd, Tl, Sb, As, Cr, Co, Cu, Mn, Ni, Pb, V) were made in accordance with the PN-EN 14385:2005 standard.

https://doi.org/10.1051/e3sconf/20182801037

Samples for determinations of mercury (Hg) concentration and emission were taken in accordance with the PN-EN 13211+AC:2006 standard.

Sampling of polychlorinated dibenzodioxins and polychlorinated dibenzofurans PCDD/DE was made in accordance with the requirements of the PN-EN 1948-1: 2006 standard. The sample was chemically analysed in the accredited laboratory of the Institute of Public Health Ostrava.

To determine how long the flue gases remain in the area of temperature exceeding 850°C (in accordance with the requirements specified in the Journal of Laws of 2016, item 108) identification measurements were conducted of flue gases temperature distribution in the WR-25 boiler combustion chamber. Measurement points were located in the representative zones of the combustion chamber on the left and right wall, with consideration to technical capabilities of making measurement nozzles. Layout of measurement points in the boiler is schematically presented in figure 1.

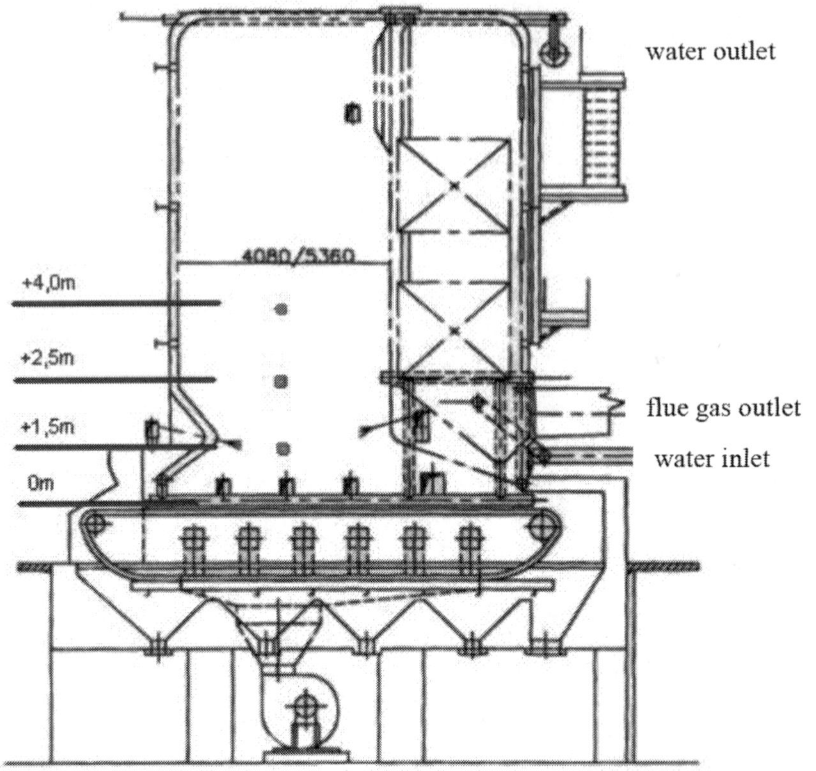

Fig. 1. Diagram of temperature measurement points layout in the WR-25 boiler combustion chamber.

Measurements were conducted by means of temperature sensors – NiCr–NiAl thermocouples of K type and Agillent recorder. Temperatures were recorded at the same depth in the combustion chamber to which the temperature sensor was inserted. Quantity of flue gas flowing through the combustion chamber was determined by recalculating the flux of gases determined during the measurement of flue gas dusting into conditions in the chamber.

3 Results and discussion

3.1 Properties of combusted materials

SRF was a loose material, shredded to aggregate of granulation below 40 mm. It was a mix of fractions of plastics, rubber, fabric, paper, cardboard and wood coming from automotive waste processing, and coal slurry with a small quantity of halloysite.

Results of physico-chemical properties determination of all combusted materials are presented in tables 1-2.

Table 1. Physico-chemical properties of combusted materials.

Parameter	Symbol	Unit	Hard coal	SRF	Blend of coal and 10% SRF
Moisture content	W_t^r	[%]	8.0	3.1	13.6
Ash content	A^d	[%]	21.5	24.4	22.0
Volatile matter	V^{daf}	[%]	36.45	89.40	40.90
Coal content	C^d	[%]	67.50	56.50	68.80
Hydrogen content	H^d	[%]	4.10	7.15	4.31
Sulphur content	S^d_t	[%]	0.50	0.25	0.45
Nitrogen content	N^d	[%]	1.20	0.78	1.28
Chlorine content	Cl^d	[%]	0.20	0.86	0.39
Fluorine content	F^d	[%]	0.02	0.08	0.01
Caloryfic value	Q^a_s	[kJ/kg]	26754	29216	28406
Gross caloryfic value	Q^r_i	[kJ/kg]	23861	27357	23797
Bulk density	$(BD)^r$	[kg/m^3]	877	186	812
Bulk density	$(BD)^d$	[kg/m^3]	807	180	702

Where, indices *d, daf, r, a* from tab. 1 mean: dry, dry ash free, as received, analytical, respectively.

SRF was characteristic of good energy parameters, similar to the base hard coal. However, SFR contained double the quantity of volatile particles. Despite low density, after SRF (with 10% weight share) was mixed with hard coal – density of the blend was close to the one of coal. It shall be noted that SRF contained 4 times more chlorine than the examined hard coal.

Table 2. Metals content in combusted materials.

Symbol	Unit	Hard coal	SRF	Blend of coal and 10% SRF
Pb^d	[mg/kg$_{d.m.}$]	15.30	102.00	27.70
Cd^d	[mg/kg$_{d.m.}$]	0.166	2.180	<0.204
Cu^d	[mg/kg$_{d.m.}$]	29.4	2207.0	39.8
Cr^d	[mg/kg$_{d.m.}$]	22.1	345.0	24.2
Ni^d	[mg/kg$_{d.m.}$]	14.4	160.0	19.6
Hg^d	[mg/kg$_{d.m.}$]	0.039	0.245	0.069
V^d	[mg/kg$_{d.m.}$]	47.4	23.0	37.1
Sb^d	[mg/kg$_{d.m.}$]	1.44	245.00	26.50
Co^d	[mg/kg$_{d.m.}$]	7.5	95.3	15.3
As^d	[mg/kg$_{d.m.}$]	1.800	1.860	<0.714
Mn^d	[mg/kg$_{d.m.}$]	188.0	443.0	195.0
Sum	[mg/kg$_{d.m.}$]	327.54	3624.59	386.19

Where, superscripts *d* mean dry state and subscripts *dm*, dry matter state, respectively.

 https://doi.org/10.1051/e3sconf/20182801037

SRF contained a lot more heavy metals in comparison to hard coal, in particular: copper, chromium, nickel and cobalt, which is related to the origin of waste materials it was made of. A small share of alternative fuel in the blend (approx. 10%) results in the fact that the content of heavy metals in the blend and their content in coal are on comparable level.

3.2 Establishing duration of the flue gases residence in the temperature exceeding 850°C

Measurement results of the flue gases temperature distribution in combustion chamber of the WR-25 boiler are presented in figure 2.

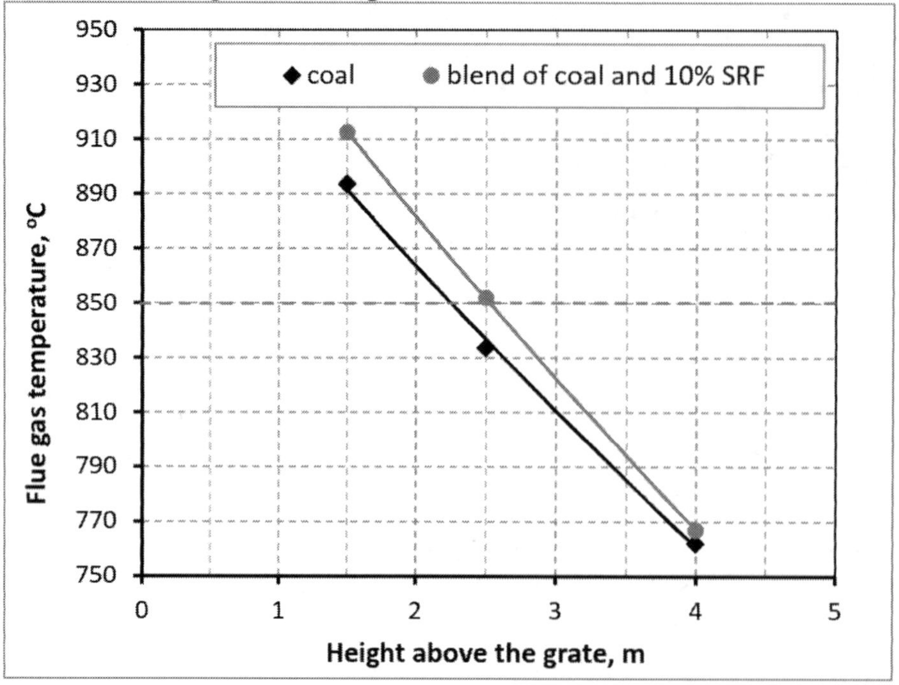

Fig.2. Distribution of average flue gases temperature during combustion of hard coal and blend of coal and 10% SRF.

During co-combustion of fuel mix containing 10% SRF, the average temperature of flue gases was raised in all measurement points of the combustion chamber. Both SRF and blend of coal and 10% SRF contained more volatile particles in comparison to coal which resulted in the fact that substances released in the degassing process were combusted in higher zones of the combustion chamber.

To determine the mean duration of the flue gases residence in the combustion chamber, the following assumptions were made:

- piston flow model of the flue gas flux flow in combustion chamber,
- temperature of flue gas was taken as a positive value recorded in points between the grate and 4.0 m level,
- as the average flow distance of 2.35 m was taken for combustion of coal and 2.5 m for combustion of the coal and waste mix,
- the actual flux of flue gas volume in the conditions of the determined average flue gas temperature in the combustion chamber was calculated on the basis of the flue gas volume flux behind the boiler determined during tests.

Calculation results are presented in table 3.

Table 3. Statement of measurement results of flue gas flow in the WR-25 boiler during tests.

Parameter	Unit	Hard coal	Blend of coal and 10% SRF
Velocity of flue gas in the combustion chamber	[m/s]	2.08	1.92
Flue gas flux (real conditions)	[m^3/h]	154904	142953
Flue gas density	[kg/m^3]	0.29	0.30
Height of the zone above 850 °C	[m]	2.35	2.50
Residence time of flue gas in zone above 850 °C	[s]	1.1	1.3

Calculation results show that the tested WR-25 boiler failed to maintain the legislative requirements for thermal conversion of waste during tests. The WR-25 boiler structure is adapted for a specific type of fuel (hard coal) and it does not allow for environmentally safe co-combustion of waste.

3.3 By-products of combustion/co-combustion

Tables 4 and 5 present test results of by-products of hard coal and SRF combustion/co-combustion.

Table 4. Combustible particles content in slag and fly ash obtained during the tests.

Parameter	Unit	Hard coal		Blend of coal and 10% SRF	
		ash	fly ash	ash	fly ash
Flammable parts	[%]	5.78	29.73	15.63	33.38

Both the slag and fly ash from co-combustion of SRF have higher content of combustible particles in comparison to by-products of hard coal combustion. This contributes to the increase of losses in the boiler energy balance. It shall be noted, however, that no legislative obligation to maintain a adequate level of process by-products after-combustion has been specified.

Table 5. Metals content in ash and fly ash.

Symbol	Unit	Hard coal		Blend of coal and 10% SRF	
		ash	fly ash	fly ash	ash
Pb^d	[mg/$kg_{d.m.}$]	68.0	154.0	32.9	128.0
Cd^d	[mg/$kg_{d.m.}$]	0.41	3.48	<0.27	5.04
Cu^d	[mg/$kg_{d.m.}$]	56.5	321.0	730.0	356.0
Cr^d	[mg/$kg_{d.m.}$]	87.9	148.0	224.0	143.0
Ni^d	[mg/$kg_{d.m.}$]	656	205	90	215
Hg^d	[mg/$kg_{d.m.}$]	0.006	1.000	0.009	3.346
V^d	[mg/$kg_{d.m.}$]	162.00	3.71	144.00	353.00
Sb^d	[mg/$kg_{d.m.}$]	3.79	36.5	10.1	169.0
Co^d	[mg/$kg_{d.m.}$]	22.6	65.7	56.2	118.0
As^d	[mg/$kg_{d.m.}$]	<1.7	41.5	1.7	25.3
Mn^d	[mg/$kg_{d.m.}$]	757	697	866	560
Sum	[mg/$kg_{d.m.}$]	1815.904	1676.89	2155.179	2075.686

Where, superscripts *d* mean dry state and subscripts *dm*, dry matter state, respectively.

Both slag and fly ash from combustion of coal blend with 10% SRF contain in total higher quantities of heavy metals than products of coal combustion which results directly from their higher content in the combusted material.

3.4 Statement of energy research results

Specific parameters used for establishing the boiler energy balance is presented in table 6.

Table 6. The WR-25 boiler energy balance.

Parameter		Unit	Hard coal	Blend of coal and 10% SRF
Water flow through the boiler		[Mg/h]	318.98	324.35
Thermal input of the boiler		[MW_t]	17.352	18.835
Mean fuel flux		[Mg/h]	2.90	3.08
Flux of chemical energy of fuel		[MW]	19.221	20.388
Losses	physical outlet of flue gas	[MW] [%]	1.47 7.63	1.43 7.02
	of not combusted CO in flue gas	[MW] [%]	0.01 0.04	0.01 0.03
	enthalpy and content of combustible particles in fly ashes and slag	[MW] [%]	0.62 3.24	1.06 5.20
	to the environment (radiation and convection)	[MW] [%]	0.23 1.19	0.24 1.17
Boiler energy efficiency (gross) Uncertainty		[%]	87.26 ±1.22	86.63 ±1.51

No process difficulties were noted in the operation of WR-25 boiler installation during any of the two tests. The WR-25 boiler worked with similar efficiency, and with minimally lower (by 0.6%) performance arising mainly from unburned slag and fly ash.

3.5 Emission of pollution to the air

Table 7 presents emissions from combustion of coal and co-combustion of coal with 10% SRF in relation to emission standards which shall be maintained during combustion of hard coal in WR-25 boiler and during co-combustion of waste.

Table 7. Comparison of pollution emissions* during combustion of hard coal and a blend of coal with 10% SRF with the emission standards valid for the WR-25 boiler installation during combustion of hard coal and co-combustion of waste.

Emitted substance	**Unit**	**Measured emission**		**Permitted emission**	
		combustion of hard coal	**combustion of blend of coal and 10% SRF**	**combustion of hard coal**	**co-combustion of waste and coal**
SO_2	[mg/m^3_u]	953.4	921.4	1500	1386.5
NO_x	[mg/m^3_u]	407.4	356.0	400	391.1
Dust	[mg/m^3_u]	141.6	243.7	400	47.2
CO	[mg/m^3_u]	39.0	39.0	-	41.6

HCl	[mg/m^3_u]	82.5	208.6	-	77.1
HF	[mg/m^3_u]	1.40	1.00	-	1.45
TOC	[mg/m^3_u]	9.3	1.9	-	9.7
Cd + Tl	[mg/m^3_u]	0.0005	0.0012	-	0.05
Hg	[mg/m^3_u]	0.00014	0.00080	-	0.05
Sb+As+Pb+Cr+ Co+Cu+Mn+Ni+V	[mg/m^3_u]	0.236	0.507	-	0.5
PCDD/DF	[ng/m^3_u]	0.017	0.232	-	0.1

* Substance concentration in flue gas in stipulated conditions recalculated into 6% oxygen content in flue gas.

During combustion of hard coal, the NO_X emission exceeded minimum permitted levels however, in the longer-term measurement period, with little adjustment of the boiler process parameters setup (supply of fuel and air for combustion) – this value may be maintained on the required level.

During combustion of hard coal blend with 10% SFR in the WR-25 boiler, the emission standards valid for this installation were complied with in relation to combustion of fuel used so far. Emissions of SO_2 and NO_X were on the same level, and even slightly lower, compared to hard coal combustion. In case of SRF co-combustion, emission of dust was considerably increased in comparison to hard coal combustion. It shall be stressed however, that results of these measurements are within the range of permitted emission standards for the tested installation.

No considerable changes in emission volume of the monitored pollutions were observed during any of the two measurement tests, which proves stability of the combustion process.

It shall be noted that comparing the emission test results to combustion of solid fuels is solely of theoretical nature, because in the current legal situation SRF is treated as waste other than hazardous and it is identified by 19 12 10 code and its use for power engineering depends on meeting the requirements for thermal conversion of waste. It is connected with much broader scope of emission monitoring and much stricter level of the required emission standards.

Comparing the obtained emission levels to the values permitted for co-combustion of waste, it shall be stated that during co-combustion of SRF with coal, the following concentrations were exceeded:

- hydrogen chloride – almost three-fold,
- dust whose level emission level significantly deviates from the permitted value,
- heavy metals: Sb+As+Pb+Cr+Co+Cu+Mn+Ni+V, very little,
- polychlorinated dioxins and furans (more than two-fold), which arises from relatively high chlorine content in SRF in comparison to coal.

The presented data arise from the fact that the WR-25 boiler installation in which the tests were made is not equipped with the flue gas purification system adapted to the emission requirements for combustion/co-combustion of waste.

4 Conclusion

The industrial research showed that it is possible to meet the emission requirements for WR-25 boiler in relation to the base fuel (hard coal) in spite of its partial replacement with SRF.

However, undertaking combustion/ co-combustion of SRF it shall be kept in mind that it is a waste material and such activity is subject to legislative requirements applicable for thermal processing of waste.

Considering the aforementioned circumstances it was stated that combustion of a mix of hard coal with 10% SFR share in the WR-25 boiler causes that emission standards are exceed in relation to: HCl, dust, heavy metals, and dioxins and furans.
Also the time of the flue gases residence in the area where the temperature exceeds 850°C is too short (1.3 sec.) to satisfy the legislative requirements specified for thermal processing of waste (2 sec.)

SRF co-combustion with hard coal in the WR-25 boiler requires modernisation of the combustion chamber design (e.g. change in heat exchangers location) and in the flue gas purification system (e.g. adding a more effective dust removal system, and system of heavy metals and dioxins removal).

References

1. A. Sobolewski, R. Wasielewski, K. Dreszer, S. Stelmach, Przem Chem, **85**, 1080-1084 (2006)
2. R. Wasielewski, A. Sobolewski, Przem Chem, **90**, 1000-1005 (2015)
3. Refuse derived fuel, current practice and perspectives. Final Report (European Commision – Directoriate General Environment, 2003)
4. S. Thiel, K. J. Kozmensky, Waste Manage Res., **4**, 392-403 (2012)
5. L. Lombardi, E. Carnevale, A. Corti, Waste Manage, **37**, 26-44 (2015)
6. T. Hilber, J. Maier, G. Scheffknecht, M. Agraniotis, P. Grammelis, E. Kakaras, T. Glorius, U. Becker, W. Derichs, H.P. De Jong, L. Torri, J. Air Waste Manag Assoc, **57**, 1178-1189 (2007)
7. L. Del Zotto, A. Tallini, G. Di Simone, G. Molinari, L. Cedola, Energy Procedia, **81**, 319-338 (2015)
8. P. Krawczyk, J. Szczygieł, Rynek Energii, **109**, 91-96 (2013)
9. G. Wielgosiński, O. Namiecińska, Nowa Energia, **2-3**, 33-40 (2015)
10. R. Wasielewski, B. Tora, Polit Energ, **11**, 129-138 (2008)

Environmental impact of emissions from incineration plants in comparison to typical heating systems

Grzegorz Wielgosiński [1] *, *Olga* Namiecińska [1], and *Justyna* Czerwińska [1]

[1] Lodz University of Technology, Faculty of Process and Environmental Engineering, ul. Wólczańska 175, 90-924 Łódź, Poland

Abstract In recent years, five modern municipal waste incineration plants have been built in Poland. Next ones are being constructed and at the same time building of several others is being considered. Despite positive experience with the operation of the existing installations, each project of building a new incinerator raises a lot of emotions and social protests. The main argument against construction of an incineration plant is the emission of pollutants. The work compares emissions from municipal waste incineration plants with those from typical heating plants: in the first part, for comparison large heating plants equipped with pulverized coal-fired boilers (OP-140), stoker-fired boilers (three OR-32 boilers) or gas blocks with heat output of about 100 MW have been selected, while the second part compares WR-10 and WR-25 stoker-fired boilers most popular in our heating industry with thermal treatment systems for municipal waste or refuse-derived-fuel (RDF) with similar heat output. Both absolute emission and impact - immission of pollutants in vicinity of the plant were analyzed.

1 Introduction

Among all waste generated in Poland the biggest problem is the municipal waste. Due to the complex and unstable composition its processing is a huge problem. According to the Central Statistical Office, the total amount of produced waste is about 10-12 million Mg a year [1]. Out of many municipal waste management methods the most effective is thermal treatment. In many EU countries incineration accounts for up to 60% of the methods used, but in Poland this level does not exceed 10% [2]. In recent years, five modern municipal waste incineration plants have been built (Białystok, Bydgoszcz, Konin, Kraków, Poznań), the next one in Szczecin will start at the end of this year. Adding to that the existing incineration plant in Warsaw, which has been operating since 2001, the capacity of Polish incineration plants is approximately 1 million Mg of waste per year [3].

The currently functioning model of waste management is based on the regional waste treatment facilities (so called Waste-to-Energy Plants - WtEs), which are mechanical-biological waste treatment installations (MBTs). In recent years, more than 150 such installations have been built with a capacity of nearly 11 million Mg, generating 2.5 to 3.5 million Mg of combustible fractions, called alternative fuels, refuse-derived-fuel - RDF or pre-RDF. It is important that since January 1, 2016, an alternative fuel fraction, due to its high calorific value, cannot be deposited [4]. This fraction can and should be incinerated in

*Corresponding author: grzegorz.wielgosinski@p.lodz.pl

waste incineration plants and co-incineration plants, in cement plants or used in heat or power plants. In the last case, these installations, due the rigorous emission regulations, must be equipped in sophisticated flue gas treatment system and this is a serious financial problem [5].

Any new proposal to build an installation for thermal treatment of municipal waste and RDF fractions is subject to numerous protests. The most important argument against construction is the emission of pollutants. However, in accordance with the requirements of the Environmental Protection Act, the operation of the installation should not cause exceedance of emission standards that are very stringent in the case of waste incineration and environmental quality standards. Therefore the legislation guarantees a low environmental impact [6].

It was decided to analyze the absolute amount of pollutant emissions as well as the amount of pollutants emitted in vicinity of the waste incineration plants as compared to similar emissions from heating plants that are common in our country. In the analyzed heating plants (heat and power plants) there are both stoker-fired boilers as well as pulverized coal-fired boilers and relatively rarely used gas blocks. The first part of the work focused on the comparison of commonly used boilers fired with hard coal (pulverized coal-fired boilers OP-140, and stoker-fired boilers OR-32) and gas blocks with an installation for thermal treatment of municipal waste and RDF fraction of similar heat output. In the second part, calculations for WR-10 and WR-25 stoker-fired boilers are compared with the systems for thermal treatment of RDF or pre-RDF fractions.

2 Research methods

2.1 Assumptions for calculations

Two cases were investigated: the first is a comparison of the emissions of OP-140 pulverized coal-fired boilers (the smallest currently produced) commonly used in power plants, OR-32 stoker-fired boilers and gas blocks of similar total heat output, with the emissions from thermal waste treatment facility - (Waste-to-Energy - WtE) - a classical incineration plant and thermal waste treatment installation fired with alternative RDF fuel (WtE-RDF). For comparison, power plants operating in high-efficiency cogeneration - like in incineration plants, were selected. In the second case, a comparison was made between emissions from WR-10 and WR-25 stoker-fired boilers common in the Polish heating industry (respectively 30% and 10% of all heating boilers in Poland [7]) and the thermal Refuse Derived Fuel - RDF treatment installations with similar heat output. This is important because in the light of not yet implemented in the Polish law but the existing directive on medium combustion plants (MCPs) the present WR type boilers do not have any chance of meeting the requirements of this directive [8] and will have to be replaced by new installations [9].

In the first part of the calculations, a nominal heat output of the boiler equal to 105 MW_t was assumed. This value results from the capacity of one of the smallest pulverized coal-fired boilers produced and used in Poland, i.e. OP-140. This variant was compared with a stoker-fired boiler also fired with hard coal - in this case the assumed heat output can be provided by three OR-32 boilers. It was assumed that the analyzed boiler should be a steam boiler so that by analogy to the waste incinerator it would work in cogeneration. As a third variant to compare, a gas boiler also of output 105 MW_t was selected. Total capacity of the boiler in the case of waste incineration plant also remained at the same level. At the assumed heating value of mixed municipal waste at an average level of about 7.5 MJ/kg, this means the incineration plant capacity of 440,000 Mg/year, i.e. twice as large as the largest Polish incineration plant for mixed municipal waste in Cracow with a capacity of 220,000 mg/year. In addition, the RDF incineration plant has been analyzed, assuming that it will incinerate approximately 235,000 Mg of RDF per year with a net calorific value of 14 MJ/kg and will have a nominal heat output of 105 MW_t.

The basic characteristics of the installation for the five variants analyzed are presented in Table 1.

Table 1. Basic characteristics of the installation for five variants analyzed in first comparison.

Boiler	Unit	OP-140	3 x OR-32	Gas block	WtE	WtE-RDF
Heat output	MW_t	105.0	105.0	105.0	105.1	105.0
Efficiency	%	90	85	92	86	86
Flue gas temperature	°C	130	140	160	120	120
Fuel net calorific value	MJ/kg *) MJ/m^3	21.0	23.0	34.5*)	7.5	14.0

The second part analyzes pollutant emissions from some of the WR-10 and WR-25 stoker-fired boilers most frequently used in the Polish heating industry. Heat output of the WR-25 boiler is about 30 MW_t and, as in the previous case, the efficiency of the RDF thermal treatment system has been adjusted to the same level. With an assumed net calorific value of RDF equal to about 14.0 MJ/kg, this means a capacity of 67,500 Mg/year. In the case of the WR-10 boiler, the nominal heat output is about 12 MW_t which, at the calorific value of RDF, results in capacity of about 27,000 Mg/year. The calorific value of coal in both cases was assumed to be 21 MJ/kg. The characteristics of individual boilers are shown in Table 2.

Table 2. Basic installation characteristics of WR-10 and WR-25 boilers compared to the corresponding WtE-RDF plants in second comparison.

Boiler	Unit	WR-25	WtE-RDF	WR-10	WtE-RDF
Heat output	MW_t	30.0	30.1	12.0	10.0
Efficiency	%	84	86	82	86
Flue gas temperature	°C	130	4120	130	120
Fuel net calorific value	MJ/kg	21.0	14.0	21.0	14.0

2.2 Maximum and average annual fuel consumption

Then, based on the data in Table 1 and 2 (for both variants), the maximum instantaneous fuel consumption and the average annual fuel consumption was calculated. Next, using the Recknagel relationships [10] the maximum flue gas flow rate was calculated for the operating conditions of the system and converted into assumed conditions (temperature: 273 K, pressure: 1013 hPa, dry gas, assumed oxygen content in the exhaust gas: 3% for the gas block, 6% for coal-fired boilers and 11% for waste incineration plants in both variants).

Results of these calculations are given in Tables 3 and 4. The values of excess air ratio (oxygen concentration in the flue gas) and the flue gas humidity were based on the data obtained from potential installation suppliers.

Table 3. The most important operating parameters of the system for five variants analyzed.

Boiler	Unit	OP-140	3 x OR-32	Gas block	WtE	WtE-RDF
Maximum fuel consumption	Mg/h *) m³/h	20.0	19.3	11 909*)	58.7	31.4
Annual fuel consumption	Mg/year *) m³/year	150 000	145 013	89 319 47 1*)	440 003	235 500
Excess air ratio	-	1.3	1.6	1.1	2.3	2.3
Oxygen concentration in flue gas	%	4.51	7.50	1.74	10.10	10.95
Flue gas humidity	%	7.5	5.5	6	16	16
Flue gas flow rate	m³/h	229 409	296 945	198 448	526 976	436 391
	m³ᵤ/h	158 042	166 937	125 825	335 303	252 744

Table 4. The most important operating parameters for WR-10 and WR-25 boilers as compared to the corresponding WtE-RDF.

Boiler	Unit	WR-25	WtE-RDF	WR-10	WtE-RDF
Maximum fuel consumption	Mg/h	6.1	9.0	2.5	3.6
Annual fuel consumption	Mg/year	45 918	67 500	18 815	27 000
Excess air ratio	-	1.3	2.3	1.3	2.3
Oxygen concentration in flue gas	%	4.51	10.95	4.51	10.95
Flue gas humidity	%	7.5	17	7.5	17
Flue gas flow rate	m³/h	70 227	125 080	28 776	50 032
	m³ᵤ/h	48 380	48 295	29 736	28 977

2.3 Calculation of maximum emission

For the two analyzed cases, a maximum variant was assumed for the calculations, which means that it was assumed that for the maximum instantaneous fuel consumption there was a maximum flue gas flow rate and maximum emission limit values, i.e. not exceeding emission standards (emission standards given in Table 5 according to the Ordinance of the Minister of the Environment on emission standards for certain types of installations, sources of fuel combustion and incineration or co-incineration equipment [11] which adopted the regulations of IED [12]. A total of 7,500 hours of operation per year identical for all analyzed variants was assumed.

Table 5. Applicable emission standards for the analyzed variants.

Boiler	Unit	OP-140	3 x OR-32	Gas block	WtE	WtE-RDF	WR-10	WR-25
SO_2	mg/m³ᵤ	250	1300	35	50	50	1300	1300
NO_x	mg/m³ᵤ	200	400	200	200	200	400	400
TSP (Total suspended particles)	mg/m³ᵤ	25	100	5	10	10	100	100

For these assumed maximum values, in the next part of the calculations, the maximum emission of the three main pollutants from the combustion process, i.e. sulfur dioxide (SO_2), nitrogen oxides (NO_x) converted to nitrogen dioxide, and both instantaneous and annual particulate matter emissions were estimated. Results of these calculations are compiled for the first and second analyzed case, for all variants, in Tables 6 and 7, respectively.

Table 6. Maximum and annual emissions for five variants of the first case.

Boiler	Unit	OP-140	3 x OR-32	Gas block	WtE	WtE-RDF
SO_2	kg/h	39.51	217.02	5.28	16.77	12.64
NO_x	kg/h	31.61	66.77	30.20	67.06	50.55
TSP	kg/h	3.95	16.69	0.75	3.35	2.53
SO_2	kg/year	296 328	1 627 633	39 635	125 739	94 779
NO_x	kg/year	237 063	500 810	226 485	502 955	379 116
TSP	kg/year	29 633	125 203	5 662	25 148	18 956

Table 7. Maximum and annual emissions for WR-10 and WR-25 boilers compared to the corresponding WtE-RDF.

Boiler	Unit	WR-25	WtE-RDF	WR-10	WtE-RDF
SO_2	kg/h	62.89	2.41	38.66	1.45
NO_x	kg/h	19.35	9.66	11.89	5.80
TSP	kg/h	4.84	0.48	2.97	0.29
SO_2	kg/year	471 706	18 111	289 927	10 866
NO_x	kg/year	145 140	72 443	89 208	43 466
TSP	kg/year	36 285	3 622	22 302	2 173

2.4 Calculations of pollutant immissions

According to the Environmental Protection Act in force in Poland, the emission of pollutants from the installation should not only meet the emission standards (if any), but also the emissions should not exceed environmental quality standards (i.e. in this case the permissible concentrations of pollutants in the atmospheric air - reference values). Therefore, in addition to the comparison of the emissions from the thermal waste treatment system with the emissions from heating installations pollutant immisions from the analyzed installations were also compared.

In the first part pollutant immissions from the OP-140 pulverized coal-fired boilers, three OR-32 stoker-fired boilers and gas block of a similar total heat output (about 105 MW_t) were compared with immissions from the thermal waste treatment facility (WtE) - a classical incineration plant and thermal waste treatment facility fired with alternative RDF fuel (WtE-RDF).

The second part analyzes immissions of pollutants from one of the WR-10 and WR-25 stoker-fired boilers in comparison with the thermal RDF treatment facility of similar heat output.

Calculations of immission pollutants were made using the Polish methodology given in the regulation of the Minister of Environment on reference values for some substances in the air, based on Pasquill atmospheric diffusion model [13,14]. Emissions of sulfur dioxide, nitrogen oxides and particulate matter given in Tables 6 and 7 as well as flue gas parameters given in Tables 3 and 4 were taken for calculations.

In order to maintain identical conditions of pollutant diffusion in the atmosphere, the emitter height was assumed to be 60 m and its diameter was chosen to be about 20 m/s. This gave a diameter of 1.9 m for the gas block, 2.0 m for the OP-140 boiler, 2.3 m for three OR-32 boilers, 2.8 m for the RDF incinerator and 3.1 m for the mixed municipal waste incineration plant. In the second case an emitter of 50 m was adopted. As in the previous example, the diameter of the emitter was different: for the WR-10 boiler it was 1.2 m, and for the corresponding RDF boiler with heat output of about 12 MW_t - 1.6 m, while for the WR-25 boiler - 1.8 m and 2.4 m for the corresponding RDF boiler with heat output of about 30 MW_t. That gave the same outlet velocity for both cases compared - about 7.7 m/s for the WR-25 boiler and about 7 m/s for the WR-10. Results of the modeling of spread of pollutants in the form of calculated concentrations in one-hour and average annual immissions are presented in Tables 8 to 11.

Table 8. Maximum calculated one-hour concentrations in the immission within the range of an emitter for five variants analyzed.

Pollutant	Unit	OP-140	3 x OR-32	Gas block	WtE	WtE-RDF	Permissible value
SO_2	μg/m³	62.041	269.368	8.288	13.835	12.236	350
NO_x	μg/m³	49.636	82.876	47.404	55.324	48.933	200
PM10	μg/m³	3.101	10.358	0.589	1.382	1.225	280

Table 9. Maximum calculated average annual concentrations in the immission within the range of an emitter for five variants analyzed.

Pollutant	Unit	OP-140	3 x OR-32	Gas block	WtE	WtE-RDF	Permissible value
SO_2	μg/m³	1.924	8.492	0.257	0.437	0.392	20
NO_x	μg/m³	1.539	2.613	1.470	1.747	1.576	40
PM10	μg/m³	0.096	0.327	0.018	0.044	0.039	40
Dust precipitation	g/m² year	2.574	9.067	0.489	1.250	1.107	200

Table 10. Maximum calculated one-hour concentrations in the immissions within the range of an emitter for WR-10 and WR-25 boilers compared to the corresponding TWTF-RDF.

Pollutant	Unit	WR-25	WtE-RDF	WR-10	WtE-RDF	Permissible value
SO_2	μg/m³	354.119	20.738	319.180	12.655	350
NO_x	μg/m³	108.955	83.123	98.165	50.618	200
PM10	μg/m³	13.626	2.065	12.260	1.265	280

Table 11. Maximum calculated average annual concentrations in the immissions within the range of an emitter for WR-10 and WR-25 boilers compared to the corresponding TWTF-RDF.

Pollutant	Unit	WR-25	WtE-RDF	WR-10	WtE-RDF	Permissible value
SO_2	μg/m³	11.216	0.575	8.942	0.349	20
NO_x	μg/m³	3.482	3.304	2.750	1.398	40
PM10	μg/m³	0.435	0.057	0.343	0.035	40
Dust precipitation	g/m² year	13.276	1.595	9.650	0.972	200

At the same time, when analyzing the data in Tables 5 to 8, one can see that in most cases the pollutant emissions do not exceed air quality standards (permissible one-hour concentrations and permissible average annual concentrations) as required by article 144 of the Law on Environmental Protection [15].

As can be easily observed, most pollutants (both SO_2 and NO_x, as well as particulate matter) emitted to the atmosphere are generated by stoker-fired boilers. This is due to the relatively liberal emission standards for the existing sources of heat output below 50 MW_t. The smallest amount of pollutants is emitted to the air by gas blocks. A comparison of emissions of sulfur dioxide and particulate matter from waste incineration plants (both mixed and RDF) with the emissions from coal-fired heat plants (equipped with both pulverized coal- and stoker-fired boilers) is far better in favor of the waste incineration plants. Only in the case of emissions of nitrogen oxides, the emissions from the incineration plant are comparable to the emissions from the stoker-fired boilers. The other boiler rooms (with pulverized coal-fired boilers and gas block) are better. This is due to the relatively low calorific value of both mixed municipal waste (7.5 MJ/kg) and RDF (14 MJ/kg), and therefore the need to burn much more fuel which results in significant flue gas emissions with identical emission standards applicable to all installations, i.e. 200 mg/m^3_u.

3 Results

Pollutant emissions for both analyzed cases are compared in Tables 6 and 7. Similarly, the results of calculations of the spread of pollutants from the analyzed sources in the form of calculated pollutant concentrations in imission are summarized in Tables 8 to 11.

4 Conclusions

The calculations show that most of the pollutants are generated by coal-fired stoker-boilers and least by gas blocks, while the emission from pulverized coal-fired boilers is slightly lower than the emission from stoker-fired boilers, because other emission standards are decisive in this case. It is worth noting that emissions from the waste thermal treatment plant (both mixed and RDF) are slightly higher than those from the gas block. The comparison of emissions of pollutants from particular coal-fired boilers commonly used in Polish heating industry is far worse than that of incineration plants for municipal waste (mixed and RDF).

It is therefore safe to say that despite many protests and public reluctance to build a Municipal Waste Thermal Treatment Plant (Waste-to-Eneregy - WtE) or WtE-RDF facilities, they are a significantly smaller source of environmental pollution than the well-known and widely accepted coal-fired heating systems.

In conclusion, this work was aimed at comparing the impact of emissions from boilers commonly used in the heating sector with the municipal waste incineration plant (both mixed and RDF). From the presented results it can be clearly seen that thermal treatment of municipal waste and WtE-RDF has no significant environmental impact. This method is generally used as an indispensable part of the municipal waste management system, proven and functioning in the most industrialized countries not only of the European Union [2].

References

1. Statistical Yearbook Environmental Protection - Polish State Statistical Office, Warsaw 1990-2016
2. G. Wielgosiński, Thermal Treatment of Municipal Solid Waste - selected problems. (Nowa Energia Publishing Office, Racibórz 2016)
3. O. Namiecińska, G. Wielgosiński, A. Targaszewska, Nowa Energia, **1**, 18-22 (2017)
4. Ordinance of the Minister of Economy of 16 July 2015 on the acceptance of waste for landfill (Dz. U. 2015, poz. 1277)

5. G. Wielgosiński., O. Namiecińska, Nowa Energia, **2**, 11-20 (2016)
6. G. Wielgosiński., Nowa Energia, **4**, 8-16 (2015)
7. W. Bujalski, J. Lewandowski, Perspectives of implementation of MCP Directive standards in heating industry in Poland. Proceedings of Warsaw Climate and Energy Talks, (2014)
8. Directive (EU) 2015/2193 of the European Parliament and of the Council of 25 November 2015 on the limitation of emissions of certain pollutants into the air from medium combustion plants - O.J. 2015, L 313, 1-19
9. B. Regulski, W. Bujalski, The heat supply sector in Poland, Proceedings of Warsaw Climate and Energy Talks, (2014)
10. H. Recknagel, E. Sprenger, W. Hönmann, E.R. Schramek, Taschenbuch für Heizung und Klimatechnik. - R. Oldenburg Verlag, München, (1994)
11. Ordinance of the Minister of the Environment of 4 November 2014 on emission standards for certain types of installations, fuel sources and combustion or co-incineration facilities (Dz. U. 2014, poz. 1546)
12. Directive 2010/75/EU of The European Parliament and of The Council of 24 November 2010 on industrial emissions (integrated pollution prevention and control) - O.J. 2010, L 334, 17-119
13. Ordinance of the Minister of the Environment of 26 January 2010 on reference values for certain substances in the air (Dz. U. 2010, Nr 16, poz. 87)
14. S. Douw, M. Rohit, Air Pollution Modeling and its Application. - (Springer 2014)
15. Act of 27 April 2001 on the Environmental Protection Law (Dz. U. 2017 poz. 519, 785, 898, 1089)

Smog episodes in the Lodz agglomeration in the years 2014-17

Grzegorz Wielgosiński[1,*], *Justyna* Czerwińska[1], *Olga* Namiecińska[1] and *Robert* Cichowicz[2]

[1] Lodz University of Technology, Faculty of Process and Environmental Engineering, Wólczańska St. 175, 90-924 Lodz, Poland
[2] Lodz University of Technology, Faculty of Architecture, Civil and Environmental Engineering, Al. Politechniki St. 6, 90-924 Lodz, Poland

Abstract. In recent years, in the winter season we are alarmed about the poor air quality in Poland and significantly exceeded permissible concentrations of certain pollutants, especially PM10 and PM2.5, which are a result of so-called low emissions. The authors analyze smog episodes in the Lodz agglomeration by comparing the recorded values of selected pollutant concentrations at monitoring stations of the Regional Inspectorate for Environmental Protection in Lodz with the meteorological conditions prevailing at this time. The analysis covers data from the years 2014-2017.

1 Introduction

Currently, the most important or most hazardous pollutants emitted into the atmosphere are sulfur dioxide, nitrogen oxides, carbon monoxide and particulate matter, i.e. PM2.5 and PM10. Poland is one of the most polluted countries in the European Union in terms of the amount of these compounds in the atmospheric air [1-4]. At the same time, according to the latest report of the World Health Organization (WHO) [5,6], out of the top ten polluted cities seven are in Poland. These are Rybnik, Wodzisław Śląski, Opoczno, Sucha Beskidzka, Godów and Kraków, Skawina and Nowy Sącz. Unfortunately, apart from the above mentioned cities, also in other regions of Poland there are numerous places with exceeded permissible pollutant concentrations in the atmospheric air [7-10]. In summer it is usually tropospheric ozone, while in winter this are mainly particulate matter PM10 [27] and carcinogenic benzo(a)pyrene. Elevated levels of these substances are primarily caused by low emissions, mainly road transport and municipal sector (individual heating systems). Significant part of urban air pollution is also caused by the lack or building up of the pre-existing 'urban ventilation corridors' [11]. Compact and high-rise buildings cause the accumulation of harmful substances in central areas of cities. In addition, the concentration of pollutants is also enhanced by climatic conditions such as low rainfall and windless weather. In this case smog is formed [12-16] - a phenomenon which was particularly burdensome last winter (2016/2017). At the same time, it should be borne in mind that the main sources of particulate matter include road transport, municipal combustion processes, agriculture, CHP and power plants, and numerous manufacturing processes in industry

* Corresponding author: grzegorz.wielgosinski@p.lodz.pl

https://doi.org/10.1051/e3sconf/20182801039

[17-21]. All of these elements make up the so-called 'low emission', i.e. emission of pollutants from the emitters of a height not exceeding 40 m [22-25].

The presence of particulate matter in the atmosphere is a phenomenon very bad for human health. Large amounts of suspended particulate matter can cause atherosclerosis and respiratory system diseases (children and the elderly are particularly vulnerable) [26-30].

Literature distinguishes two kinds of smog - one is a photochemical smog, also known as 'Los Angeles smog' because of its main area of occurrence. The time of its formation covers the period from June to September with light winds and high temperature (over 30°C). Its main cause is the presence of nitrogen oxides and hydrocarbons from automotive exhaust occurring in the atmosphere. In the presence of solar radiation, these compounds react producing strong oxidants, mainly ozone. As a result of these reactions a brownish mist can be observed floating over the city, which greatly limits the view [31,32].

Another type of smog is a sour smog, also called 'London smog'. In contrast to photochemical smog, it is formed at low air temperatures near 0°C. Factors causing this phenomenon are the lack of air mass movement at high humidity and temperature inversion, as well as high concentration of sulfur, nitrogen and carbon oxides in the air [33].

In recent years it has been noticed that the smog formed in Poland in winter is neither photochemical nor acid smog. This smog, which can be called a 'Polish smog', consists mainly of PM10, PM2.5, PM1 and polycyclic aromatic hydrocarbons, including benzo(a)pyrene. The resulting smog is a consequence of the processes of burning solid fuels in low efficiency furnaces which are not equipped with exhaust gas treatment systems. As a consequence during winter this phenomenon is intensified, which is related to weather in our climate [34]. It is most often generated at temperatures below 0°C in windless anticyclone weather under the conditions of temperature inversion.

2 Description of the problem

The Lodz agglomeration (with a population of about 1.1 million people) is located in central Poland in the Lodz Region (18 218.95 km^2, population 2 488 417 and population density 136.6 inhabitants/km^2). Its central city is Lodz (with an area of 29,325 km^2, population 693 797 and population density 2366 inhabitants/km^2) [35]. In this area majority of the industry of the whole Lodz province is concentrated. The towns belonging to the Lodz Agglomeration are (Fig 1): Lodz, Pabianice, Zgierz, Aleksandrów Łódzki, Ozorków, Konstantynów Łódzki, Głowno, Koluszki, Brzeziny, Tuszyn, Stryków and Zgierz [36].

The analysis of changes in air pollution in the Lodz agglomeration was based on data from 2 measurement stations (Fig. 2) [37] performing automatic measurements, and located in Lodz: Station 1, i.e. Lodz-Widzew (launch date: 1 November 2003) and Station 2, i.e. Lodz-Gdańska 16 (launch date: 1 January, 2012).

The first station, i.e. Station 1, is located in the eastern part of the city at 1/3 Czernika str. It covers the area of Lodz-Widzew district. This is a container automatic measurement station located at 235 m a.s.l. to measure the following parameters: SO_2, NO_2, CO, PM2.5, PM10, NO, NO_x and O_3. In its neighborhood in the north, south and west, there is a scattered multi-family housing. In the east, the station neighbors a park and recreational areas. The main sources of emission in the vicinity of the station include a large municipal heat and power plant (EC-4) and car transport.

The second analyzed station, i.e. Station 2, is located in the central part of the city at 16 Gdańska Street. It covers the area of Lodz-Śródmieście district. It is also a container automatic measurement station located at 206 m a.s.l. and measures such parameters as SO_2, NO_2, CO, benzene, PM2.5, PM10, NO, NO_x and O_3. In its neighborhood in the north, south and east, there is a compact multi-family housing. In the west, the station neighbors commercial, service, school and office buildings. The main sources of emissions in the

 https://doi.org/10.1051/e3sconf/20182801039

vicinity of the station are heating/fuel combustion systems in commercial premises, offices and apartments and car transport.

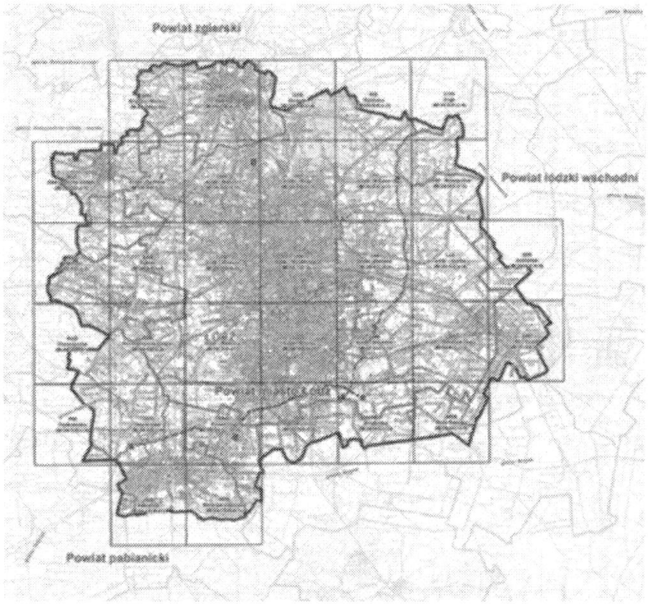

Fig. 1. Lodz agglomeration

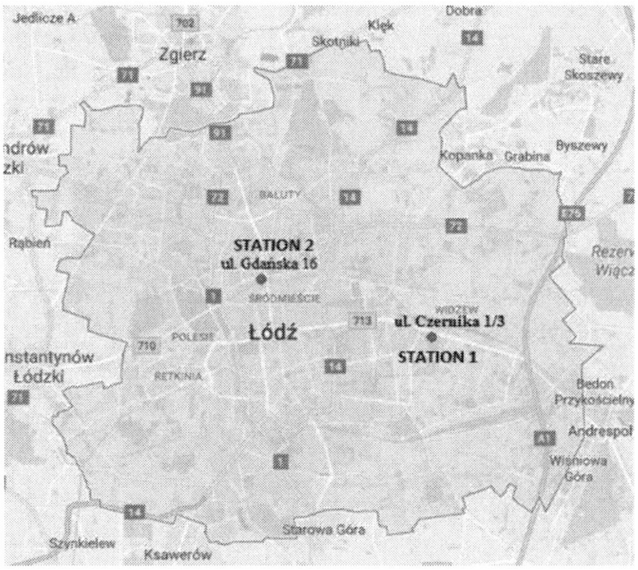

Fig. 2. Location of measurement stations in the city of Lodz

3 Results

Changes in the concentration of the two most important pollutants being the main constituents of the so-called 'Polish smog', i.e. suspended particulate matter PM10 and PM2.5, were analyzed at two measurement stations in the Lodz agglomeration in the years

2014-2017. The effects of atmospheric conditions on changes in PM10 and PM2.5 concentrations during the recorded smog episodes were analyzed.

Figure 3 shows the concentration distribution of PM2.5 and PM10 depending on the atmospheric pressure for the period of smog which was registered on 5 to 7 January 2016. When analyzing the data, it can be seen that the highest PM10 concentration was recorded at atmospheric pressure of 973 hPa and the lowest at 974 hPa. For PM2.5 no influence of atmospheric pressure was observed. In both cases the highest values were read on Station 2, i.e. in the city center.

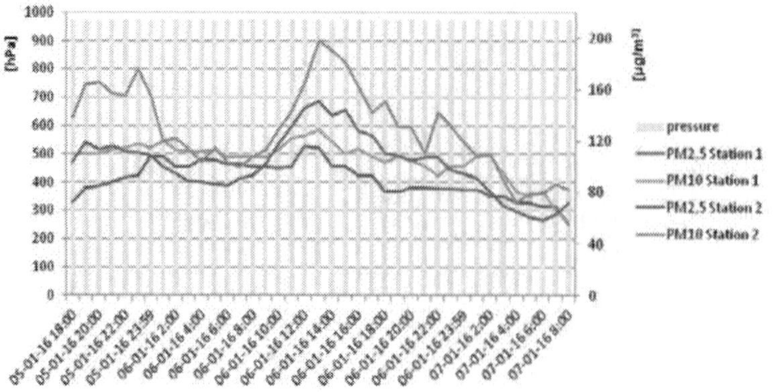

Fig. 3. Dependence of PM10 and PM2.5 concentrations on pressure

Figure 4 shows the distribution of PM2.5 and PM10 concentrations depending on the temperature for the period of smog which was recorded on 5 to 7 January 2016. From this characteristic it can be seen that smog appeared in the temperature range from -10.4°C to -5°C. This is a novelty compared to the classic type of smog and it can be assumed that this is an obvious example of the so-called 'Polish smog' - particulate matter recorded at temperatures below zero.

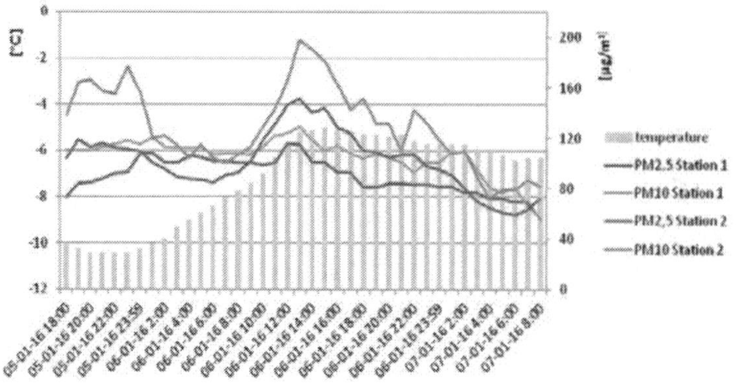

Fig. 4. Dependence of PM10 and PM2.5 concentrations on temperature

Figure 5 shows the distribution of PM2.5 and PM10 concentrations depending on wind speed for the period of smog which was measured on 5 to 7 January 2016. The figure shows that in the analyzed time the wind speed varied from 1.4 to 2.5 m/s. As expected, there is a positive effect of wind speed on the state of air pollution, i.e. the higher the wind speed, the more accurately the city is ventilated and the concentrations of pollutants are smaller.

https://doi.org/10.1051/e3sconf/20182801039

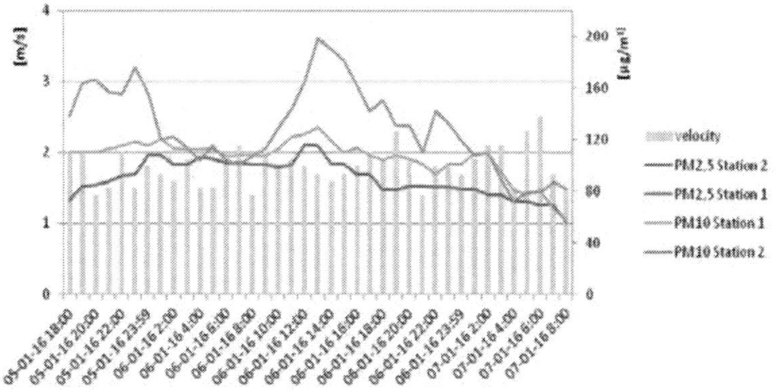

Fig. 5. Dependence of PM10 and PM2.5 concentrations on wind speed

Figure 6 shows the distribution of PM2.5 and PM10 concentrations depending on the absolute humidity for the period of smog which was measured on 5 to 7 January 2016. When analyzing the data, it can be seen that higher concentration values were measured on Station 2, i.e. in the city center. In addition, the maximum humidity varied between 80-90% and as expected, the higher humidity the more particle matter trapped in the air and the higher the particulate matter concentration.

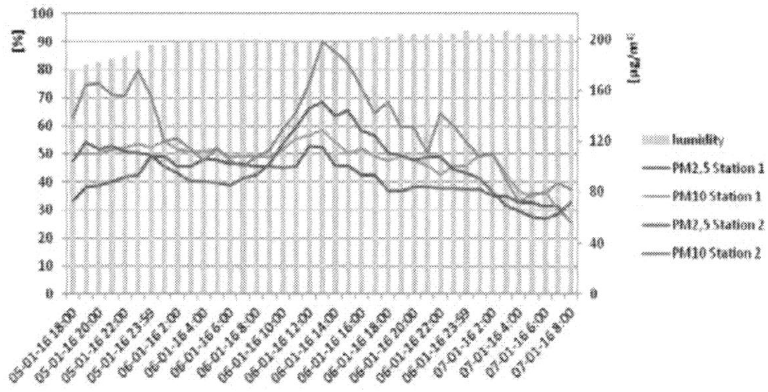

Fig. 6. Dependence of PM10 and PM2.5 concentration on absolute humidity

All drawings (Figure 3-6) show a clear (expected) correlation between PM10 and PM2.5 concentrations and the dependence of the recorded concentrations on stations 1 and 2.

Summing up data on air pollution in the Lodz agglomeration, Table 1 presents information on the number of days during which smog episodes occurred, as recorded on the first or second station analyzed.

Table 1. Number of smog days in 2014-17 recorded on two analyzed stations.

YEAR	STATION 1	STATION 2
2014	28	59
2015	27	59
2016	19	55
2017	28	44

Station 2, located in the city center, recorded significantly more smog days, with exceeded permissible levels of suspended particulate matter in the air than Station 1 located outside the strict city center, in the area of new housing estates. This confirms the thesis that the main source of air pollution in the Lodz agglomeration are low emissions from individual heating systems which dominate the city center. In the area of new housing estates heated by a central system (CHP plant) air quality is much better, even in periods that are particularly conducive to smog formation.

When analyzing the monitoring data, we attempted to find a correlation between meteorological parameters (pressure, temperature, wind speed and air humidity) and immission of pollutants - in this case PM10. This is illustrated in Figures 7 to 10. Data are presented for Station 1 in 2016.

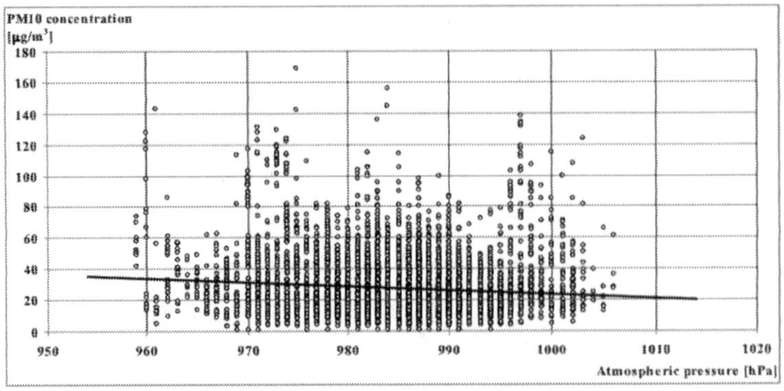

Fig. 7. Dependence of PM10 concentration on atmospheric pressure

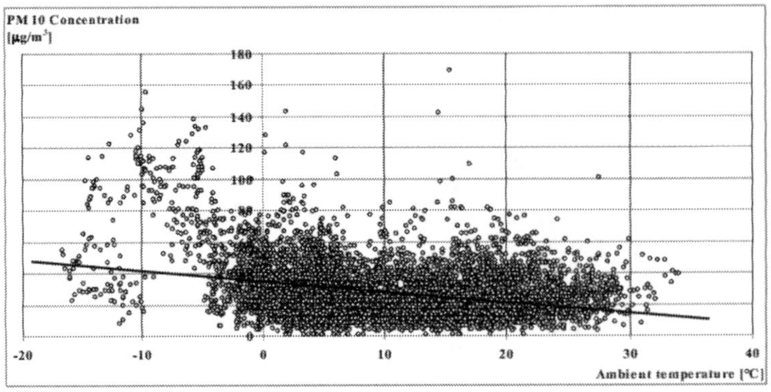

Fig. 8. Dependence of PM10 concentration on air temperature

https://doi.org/10.1051/e3sconf/20182801039

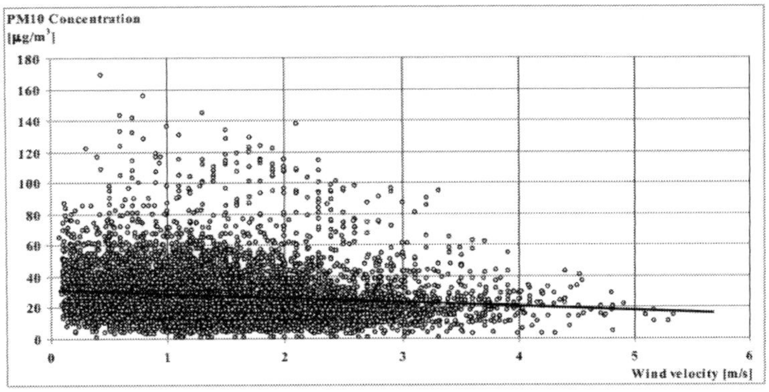

Fig. 9. Dependence of PM10 concentration on wind speed

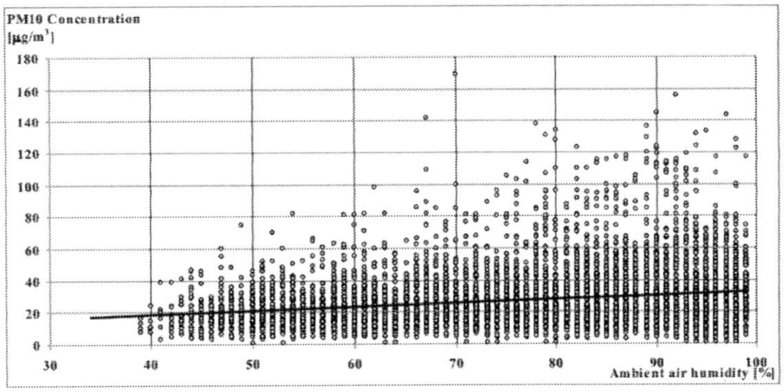

Fig. 10. Dependence of PM10 concentration on air humidity

As it is easy to see, there is a correlation between some meteorological factors and the concentration of PM10. The concentration of PM10 decreases with increasing atmospheric pressure (Fig. 7), temperature (Fig. 8) and wind speed (Fig. 9), while rises with air humidity increase (Fig. 10).

It is no use to make a detailed analysis of correlation and its relevance between PM10 concentration and air temperature, atmospheric pressure, wind speed and air humidity, for the calculated correlation coefficients are very small for this amount of monitoring data and the variability of the parameters. It is important, however, that such dependencies exist, and although the influence of wind speed is obvious, but the dependence on air temperature and atmospheric pressure clearly indicates the difference in recorded smog phenomena from the described in the literature so-called London type smog.

4 Conclusions

Numerous literature reports [3,4] indicate that air quality in Poland is poor. Especially it is observed in the south of the country and in the cities with a residential center built in the early 20th century, heated by individual heating systems. Smog events in Poland appear each year in the winter season. Particularly PM10 and PM2.5 concentrations are exceedingly high, with relatively low concentrations of sulfur dioxide, nitrogen oxides and carbon monoxide.

The paper analyzes smog episodes in the Lodz agglomeration in the years 2014-17. The analysis is based on results from two stations of automatic air quality monitoring located in Lodz. It was found that there the recorded concentrations of PM10 and PM2.5 depended on certain meteorological parameters. At both stations it was observed that PM10 and PM 2.5 concentrations decreased with the increase of air temperature, atmospheric pressure and wind speed, and rose with increasing air humidity. This confirms the thesis on the difference between smog recorded recently in Poland and typical cases described in the literature, i.e. acidic (London) smog and photochemical (Los Angeles type) smog. The 'Polish smog' is characterized by the presence of very high concentrations of particulate matter at high-pressure weather, negative temperatures and at light winds.

This type of weather usually occurs in Poland in winter, when strong Russian anticyclone stabilizes the clear, sunny weather with low winds and low temperatures, leading to frosty air from eastern directions. This is conducive to the formation of temperature inversion, which is usually maintained all the day and night, for several days.

The analysis of atmospheric pressure and presence of smog days and PM10 and PM2.5 concentrations recorded on both monitoring stations also allows us to conclude that the primary source of smog is the combustion of solid fuels in individual heating systems. This way of heating apartments is predominant in the center of Lodz, so it is not surprising that much higher concentrations of particulate matter and a bigger number of days with exceeded air quality standards are recorded at Station 2 (Gdańska Street) compared to Station 1 (Czernika Street) which is located outside the center in the area of new housing estates heated in the central system, i.e. by the CHP plant.

So we might risk a statement that one of the methods for fighting poor air quality in Poland, with frequent smog episodes in winter, should be the elimination of individual heating systems fueled with solid fuels and connecting residential areas in the city center to the heating system provided by CHP plants.

References

1. K. Ćwiek, G. Majewski, Przegląd Naukowy- Inżynieria i Kształtowanie Środowiska, **67**, 54-66 (2015).
2. M. Sobolewski, Infos, **13**, 217 (2016).
3. European Envoronmental Agency, Air quality in Europe - 2016 report
4. European Envoronmental Agency, Air quality in Europe - 2017 report
5. http://www.nytimes.com/interactive/2013/10/15/business/international/europe-air-quality.html
6. http://wiadomosci.gazeta.pl/wiadomosci/7,114883,20072482,33-z-50-miast-ue-z-najgorszym-powietrzem-jest-w-polsce-na-pierwszym.html
7. H. Shi, Y. Wang, J. Chen, D. Huisingh, J. Clean. Prod., **112**, 1261-1271 (2016).
8. H.S. Kim, J.B. Huh, P.K. Hopke, T.M. Holsen, S.M. Yi, Atmos. Environ., **41**, 6762-6770 (2007).
9. J. Cembrzyńska, E. Krakowiak, P.Z. Brewczyński, Medycyna Środowiskowa, **4**, 31-38 (2012).
10. A. Fortelli, N. Scafetta, A. Mazzarella, Atmos. Environ., **143**, 218-228 (2016).
11. S. Saberian, A. Heyes, N. Rivers, Resour. Energy. Econ., **49**, 165-185 (2017).
12. M. Zhou, G. He, M. Fan, Z. Wang, Y. Liu, J. Ma, Z. Ma, J. Liu, Y. Liu, L. Wang, Y. Liu, Environ. Res., **136**, 396-404 (2015).
13. J. Wang, K.W. Hipel, Y. Dang, Expert Syst. Appl., **87**, 240-251 (2017).
14. H.E. Wichmann, Int. J. Hyg. Envir. Heal., **206**, 505-520 (2004).

15. Y.D. Wang, X.K. Fu, W. Jiang, T. Wang, M.H. Tsou, X.Y. Ye, Comput. Environ. Urban., **66**, 110-116 (2017).
16. R. Cichowicz, G. Wielgosiński, W. Fetter, Environ. Monit. Assess., **189**:605 (2017).
17. W. Meanhault, R. Vermeylen, M. Claeys, J. Vercauteren, E. Roekens, Sci. Total Environ., **562**, 550-560 (2016).
18. M. M. T. de Rooji, D.J.J. Heederik, F. Borlée, G. Hoek, I.M. Wouters, Environ. Res., **153**, 161-170 (2017).
19. D. Mira-Salama, C. Grüning, N. R. Jensen, P. Cavalli, J.P. Putaud, B.R. Larsen, F. Raes, H. Coe, Atmos. Res., **88,** 294-304 (2008).
20. R. Cichowicz, G. Wielgosiński, A. Targaszewska, Ecol. Chem. Eng. S. **23**, 1, 49-60 (2016).
21. K. Juda-Rezler, P. Manczarski, Nauka, **4**, 97-106 (2010).
22. R. Cichowicz, G. Wielgosiński, Ecol. Chem. Eng. S., **22**, 2, 189-200 (2015).
23. R. Cichowicz, G. Wielgosiński, Ecol. Chem. Eng. S., **22**, 4, 513-525 (2015).
24. P. Michalik, Ochrona Środowiska i Zasobów Naturalnych, **40** (2009).
25. Z. Yang, J. Wang, Environ. Res., **158**, 105-117 (2017).
26. L. Megido, B. Suárea-Peña, L. Negral, L. Castrillón, S. Suáarez, Y. Fernándes-Nava, E. Marañón, Chemosphere, **162**, 73-79 (2016).
27. J. Chen, H. Chen, Z. Wu, D. Hu, J.Z. Pan, Inform. Syst. J., **64**, 281-297 (2017).
28. T.Y. Wong, J. Food Drug Anal., **25**, 235-244 (2017).
29. S. Mishra, Indian Heart Journal, **69**, 425-429 (2017).
30. A. Watkins, M. Danielewitz, M. Kusha, S. Massé, B. Urch, K. Quadros, D. Spears, T. Farid, K. Nanthakumar, Can. J, Cardiol., **29**, 734-741 (2013).
31. C. Muiwijk, P.J.C. Schrijvers, S. Wuerz, S. Kenjereš, Atmos. Environ., **147**, 470-484 (2016).
32. X. Wang, J. Chen, T. Cheng, R. Zhang, X. Wang, J. Environ. Sci., **26**, 1894-1908 (2014).
33. B. Toczko, Biuletyn kwartalny Rządowego Centrum Bezpieczeństwa, **18**, 18-21 (2017).
34. K. Kożuchowski, Klimat Polski - nowe spojrzenie, PWN, Warszawa 2011
35. http://geoportal.pgi.gov.pl/atlasy_gi/lodz
36. http://geoportal.pgi.gov.pl/atlasy_gi/lodz/mapy
37. http://www.wios.lodz.pl/Stacje_pomiarowe,211

Particle size distribution of the radon progeny and ambient aerosols in the Underground Tourist Route "Liczyrzepa" Mine in Kowary Adit

Katarzyna Wołoszczuk[1,*], *Krystian* Skubacz[2]

[1]CENTRAL LABORATORY FOR RADIOLOGICAL PROTECTION, Department of Individual Monitoring and Calibration, Konwaliowa 7, PL03194 Warsaw, Poland, woloszczuk@clor.waw.pl
[2]CENTRAL MINING INSTITUTE, Silesian Centre for Environmental Radioactivity (BCR), Plac Gwarków 1, 40-166, Katowice, Poland, kskubacz@gig.eu

Abstract. Central Laboratory for Radiological Protection, in cooperation with Central Mining Institute performed measurements of radon concentration in air, potential alpha energy concentration (PAEC), particle size distribution of the radon progeny and ambient aerosols in the Underground Tourist- Educational Route "Liczyrzepa" Mine in Kowary Adit. A research study was developed to investigate the appropriate dose conversion factors for short-lived radon progeny. The particle size distribution of radon progeny was determined using Radon Progeny Particle Size Spectrometer (RPPSS). The device allows to receive the distribution of PAEC in the particle size range from 0.6 nm to 2494 nm, based on their activity measured on 8 stages composed of impaction plates or diffusion screens. The measurements of the ambient airborne particle size distribution were performed in the range from a few nanometres to about 20 micrometres using Aerodynamic Particle Sizer (APS) spectrometer and the Scanning Mobility Particle Sizer Spectrometer (SMPS).

1 Introduction

1.1 Radon and its progeny

There are three natural radioactive series in the environment, in each there is an isotope of radium and its decay products radon (^{222}Rn), thoron (^{220}Rn) or actinon (^{219}Rn). Among them in the air dominates the first one, radioactive noble gas called radon, which is colourless, odourless and tasteless. The half-lifes of radon, thoron and action differ by four orders of magnitude. It is respectively: 3.8 days, 56 s, and 4 s. This is very important from the viewpoint of their distribution in the air and the development of hazards that are significant especially in places with relatively poor ventilation such as caves, underground mines and buildings. Taking into account the exposure due to by these isotopes, the most significant is usually radon, and less often thoron.

* Corresponding author: woloszczuk@clor.waw.pl

Radon decay products, also called "radon progenies" are no longer gases but are solid particles. Radon decay products are divided into two groups: the "short-lived" radon daughters ^{218}Po (3.05 min), ^{214}Pb (26.8 min), ^{214}Bi (19.7 min), and ^{214}Po (164 µs) with half-lives below 30 min; and the "long-lived" radon decay products ^{210}Pb (22.3 a), ^{210}Bi (5.01 d), and ^{210}Po (138.4 d) [1]. The prevalent contribution to the dose give short-lived progenies rather than radon itself.
After the formation from the radon isotope by decay, the freshly generated radionuclides react very fast (<< 1 s) with trace gases and molecules of water vapor and become small particles, called clusters or unattached fraction with diameters from 0.5 to 5 nm, composed primarily of the ^{218}Po [1]. The process was illustrated in Fig. 1. In aerosol-rich air dominates attached fraction- radioactive aerosols with diameter up to 1000 nm with predominant size between 50 and 500 nm.
The dose due to radon progeny depends on the particle deposition and site of their deposition in the respiratory tract which is closely related to the particle size distribution, Deposition of aerosols in the respiratory tract depends mainly on their size, breathing way and ventilation rate, and is due to the forces of inertia, sedimentation and diffusion. At a higher ventilation rate via mouth, the aerosol deposition in the respiratory system is mainly connected to the inertia forces. This mechanism occurs particularly for particles with diameters greater than 4 μm. The role of sedimentation increases with a longer retention time of air in the respiratory system and for particles with diameters in the range of 0.1-4 μm. The diffusion process prevails for particles with smaller size [2]. Very dangerous are so-called ultrafine particles with sizes below 100 nm and high diffusion coefficient because their dose conversion factors can be several dozen higher as compared to 100-500 nm particles. The high dose conversion factors also correspond to coarse particles close to size of 10 μm but their concentration in the air is usually small.

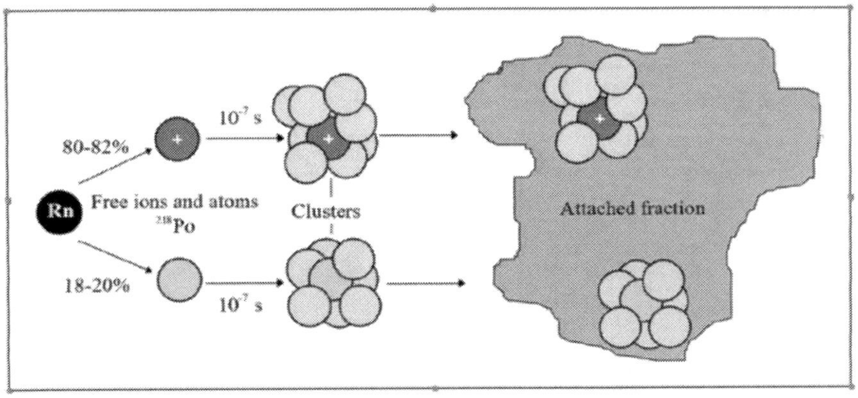

Fig. 1. Basic processes of radon decay product behaviour in air defining the concept of "unattached" and "attached" fractions [1]

It is estimated that equivalent dose to the lung from radon gas is only about one hundredth of that from its short-lived progeny which constitute the main health hazard to the human respiratory tract in the radon environment [3]. The annual effective dose, a measure of this hazard, is calculated as a product of the potential alpha energy concentration in the air (PAEC), exposition time (t), and a dose conversion factor (DCF).

$$E = (PAEC \cdot t) \cdot DCF \qquad (1)$$

where:

 https://doi.org/10.1051/e3sconf/20182801040

E - the annual effective dose [Sv/year],
PAEC - potential alpha energy concentration [J/m^3]
t - exposition time [h/year]
DCF - dose conversion factor [Sv/(J·h/m^3)]

There are two approaches to the estimation of DCF- epidemiological and dosimetric. Epidemiological model introduces only two conversion factors, 1425 mSv/(J·h/m^3) for the occupational hazard and 1100 mSv/(J·h/m^3) for the general population. The conversion factors are determined from the epidemiological studies of uranium miners. The second uses the biokinetic and dosimetric models developed by the International Commission on Radiological Protection (ICRP). This model is used to determine the deposition of inhaled radionuclides in the various regions of the respiratory system as a function of inhaled particle size [4].

1.2 Study site

Measurements were conducted in the Underground Tourist- Educational Route 'Liczyrzepa' Mine in Kowary Adit (Fig.2). These are the remnants after exploitation of iron ore and uranium. From 1947 to 1957, the mine was involved in uranium extraction and it operated as a subsidiary of Industrial Plants R-1, otherwise known as National Detached Company. More than 20 tunnels were dug during those times. The mine is today converted into a tourist attraction. The long tour through one of the underground workings is a sort of multimedia museum with exhibitions of the mining history [5].

Fig. 2. Underground Tourist- Educational Route 'Liczyrzepa' Mine in Kowary Adit [5]

2 Measurement methods

2.1 RPPSS

Central Laboratory for Radiological Protection is equipped with the Radon Progeny Particle Size Spectrometer (RPPSS) that is able to determine effective dose related to radon progeny according to the dosimetric model recommended by International Commission on Radiological Protection (ICRP) [4], taking into account the size distribution of particles compound of these progeny and ambient aerosols. This device was constructed and manufactured in the Australian Radiation Protection and Nuclear Safety Agency (ARPANSA). It allows to measure the progeny activities in 8 stages that correspond to particle size in the range from 0.6 nm to 2494 nm. Then the activity size distribution and a size-weighted dose conversion factors for breathing rates related to both occupational and environmental exposures are calculated by fitting of the experimental data to an analytical function.

RPPSS is designed to work in two modes- grab and continuous sampling mode. In continuous mode the RPPSS as a portable device analyzes potential alpha energy concentration on each stage of the sampling head, and finally the activity-size distributions and weighted dose conversion factors are derived from these results. Grabbing mode is used with the Canberra 8224 Analog Multiplexer (AMX) and Accuspec MCA to provide simultaneous 8-channel alpha spectroscopy. This allows to provide the activity size distributions individually for each radionuclide.

RPPSS comprises 8 stages operated in parallel: one open face stage, 4-stage diffusion battery system and 3-stage inertial impactor system. The eight sampling heads are counting simultaneous while the pump is sampling. The derived PAEC values for each stage are used as input for two independent deconvolution algorithms - Twomey [6] and Expectation Maximisation [7] to derive the particle size distributions. The software also includes radon progeny dose conversion factors for the ICRP Respiratory Track Model as implemented in the computer code RADEP (RAdon Dose Evaluation Program).

2.2 APS and SMPS spectrometers

Two particle spectrometers were applied to measure particle concentrations. The Aerodynamic Particle Sizer (APS spectrometer), made by TSI, USA (Fig. 3), was used to measure the concentration of particles that have aerodynamic diameters in the range from about 0.5 μm to 20 μm. The entire range of measurement is divided into 52 channels, and evaluation of particle size relies on measuring particle flight time using two laser beams and comparing the indications with the calibration curve.

To test the finer particle size distribution, the Scanning Mobility Particle Sizer (SMPS spectrometer), also made by TSI, USA, was applied, comprising an EC-3080 Electrostatic Classifier (EC), DMA-3081 or DMA-3085 Differential Mobility Analyzer (Long DMA or Nano DMA) and CPC-3785 Condensation Particle Counter (CPC). At the air inlet to the electrostatic classifier, an impactor with a diameter of 0.0508 cm was mounted and the sample air passed through the spectrometer at a flow rate of 0.6 dm^3/min (Nano DMA) or 0.3 dm^3/min (Long DMA). The adoption of such parameters allowed the detection of particles with sizes ranging from 4-106 nm using the Nano DMA or 15-698 nm with the Long DMA. The aerosol concentration, measured by the SMPS spectrometer, is a function of the mobility diameters, while, in the case of the APS spectrometer, of the aerodynamic diameters. The aerodynamic diameter that corresponds to an ambient particle is defined as a diameter of a sphere particle having the unit density of 1 g/cm^3 and the same settling velocity as the particle. The following nonlinear relationship between these two diameter types was applied [8] to compare distributions, measured by both spectrometers, and integrate the results:

$$d_p = d_a \left(\frac{C_C(d_a)}{C_C(d_p)} \right)^{1/2} \left(\frac{\chi \rho_o}{\rho_p} \right)^{1/2} \qquad C_C(d) = 1 + \frac{\lambda}{d} \left(2.34 + 1.05 e^{-0.39 \frac{d}{\lambda}} \right) \qquad d = d_a \text{ or } d = d_p \tag{2}$$

where d_a is the aerodynamic diameter, d_p and ρ_p the diameter and density of the particles, respectively, χ the aerodynamic shape factor, C_c is so-called slip correction factor, ρ_o unit density of 1 g/cm^3, and λ the mean free path of the gas molecules. The dimensionless aerodynamic shape factor χ is a correction factor to take into account the influence of the shape on particle motion according to Stokes's law.

3 Measurements results

During measurements of radon concentration in air, potential alpha energy concentration (PAEC), particle size distribution of the radon progeny and ambient aerosols were determined. These results are presented in Table 1-3 and in Fig. 3-6. The parameters allowed to determine the dose conversion factor (DCF), and equilibrium factor (F).
The measured concentrations of PAEC are high, especially compared to 0.05 μJ/m^3, which is not normally exceeded in open air [9,10]. For occupational exposure, taking into account the annual working time of 1800 hours and the dose conversion factor indicated in the legislation, the doses clearly exceed 6 mSv, i.e. the limit for exposure category A. However it is not greater than the boundary exposure limit.
The concentrations of environmental aerosols are significantly lower than in an active coal mine, where their concentration can reach $140 \cdot 10^9$ particles/m^3 [11]. The ultrafine particles with diameter up to 100 nm predominate in the distribution (table 2) predominate, and the contribution of coarse particles larger than 2.5 μm does not exceed 0.1% in total concentration. However, they cannot be omitted taking into account the total particle surface area where their share is about 20%.

Table 1. Measurements results

Measurement day	Radon concentration in air [Bq/m^3]	Potential alpha energy concentration [μJ/m^3]	Humidity [%]	Temperature [°C]	Air pressure [hPa]
1st day	1054	3.56	95	8.3	942
2nd day	1081	4.82	82	9.5	944

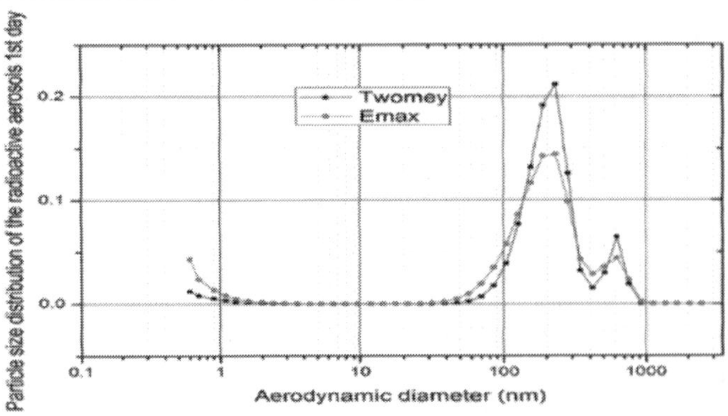

Fig. 3 Particle size distribution of the radon progeny (1st day)

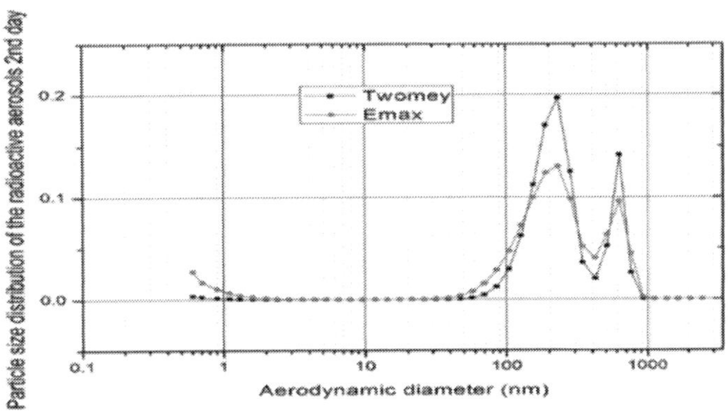

Fig. 4 Particle size distribution of the radon progeny (2nd day)

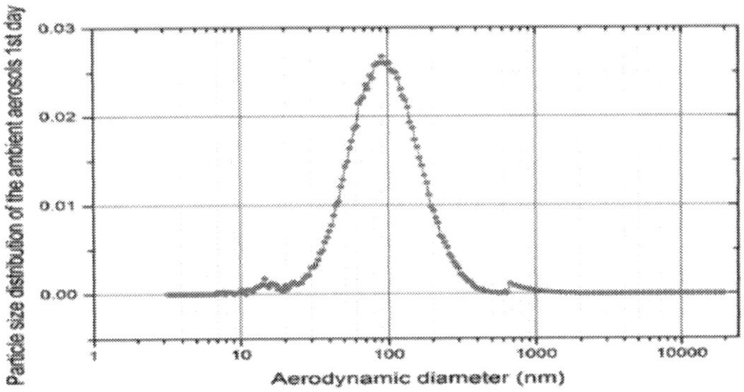

Fig. 5 Particle size distribution of the ambient aerosols (1st day)

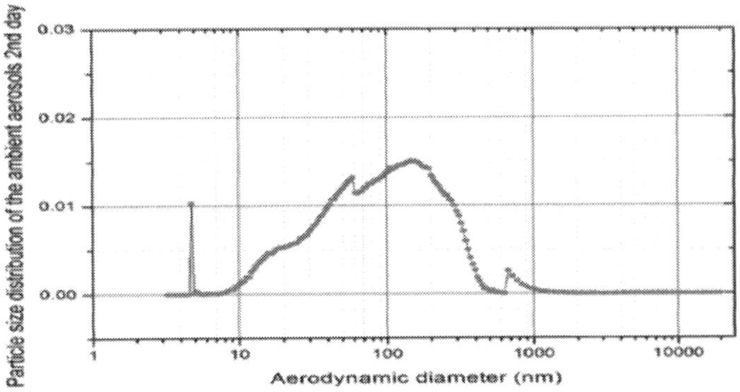

Fig. 6 Particle size distribution of the ambient aerosols (2nd day)

Table 2. Average total concentration of particles and their number contribution to the specific class of aerosol particles.

Measurement day	Total concentration [particles/m³]	Class of aerosol particles [μm]		
		LB–0.1	0.1–2.5	2.5–10.0
		[%]	[%]	[%]
1st day	$2.7 \cdot 10^9$	73	27	<0.1
1st day	$4.0 \cdot 10^9$	53	47	<0.1

Table 3. Average total concentration of particles and their surface area contribution to the specific class of aerosol particles.

Measurement day	Total concentration [particles/m³]	Class of aerosol particles [μm]		
		LB–0.1	0.1–2.5	2.5–10.0
		[%]	[%]	[%]
1st day	$2.7 \cdot 10^9$	18	62	20
1st day	$4.0 \cdot 10^9$	4	78	18

Based on measurements results dose conversion factors (DCF) for the occupational hazard (DCF OCC), for the general population (DCV Env) and equilibrium factor (F) were determined (Table 4). DCF values were also calculated based on the measurement results related to size distribution of ambient aerosols taking into account the method described in the publication [10]. The calculated coefficients differ significantly, although the trend is similar, and on the second measurement day they were higher in each case than these obtained in the first measurement day. There should be emphasized, however, that the DCF coefficients for RPPSS results were calculated for the range up to 2.5 μm, while the measuring range of SMPS + APS equipment also included particles larger than 2.5 μm. For such particles, the surface area that increases the possibility to catch the radon progeny is significant (Table 3), and the DCF values for such particles are significantly higher than for the size of 100-500 nm.

Table 4. Measurements results

Day of measurement	Method	DCF Occ [Sv/(J·h/m³)]	DCF Env [Sv/(J·h/m³)]	F [%]
1st day	RPPSS	1.50	0.82	60
2nd day		1.14	0.78	80
1st day	SMPS+APS	4.11	3.34	-
2nd day		3.92	2.94	-

5 Conclusions

Obtained particle size distribution of radon progeny shows three main peaks with the values about 1nm, 180 nm and 570 nm. The first peak corresponds to the unattached fraction which in this case was about 10 %. Second peak corresponds to the dominating ambient fraction in the distribution (Fig. 3-4), third peak correspond to the biggest particles.

Based on particle size distribution of radon progeny DCF were determined. In this case obtained values DCF are similar to values defined in Polish Atomic Law which are 1.4 Sv/(J·h/m^3) and 1.1 Sv/(J·h/m^3) respectively for the occupational hazard (DCF OCC), for the general population (DCV Env). However, the DCF values, calculated based on the measurement results related to size distribution of ambient aerosols were much higher, despite the both size distributions in the common range were nearly the same. That difference was caused probably by analyzing different size range related for RPPSS results and results related to the SMPS and APS spectrometers that also included particles larger than 2.5 μm.

References

1. J. Porstendörfer, *Properties and behavior of radon and thoron and their decay products in the air*. J. Aerosol Sci. **25** (1994)
2. K. Willeke, *Generation of aerosols and facilities for exposure experiments, inhalation of aerosols: particle deposition and retention*. Ann Arbor Science Publishers Inc. (1980)
3. *Measurement and calculation of radon releases from NORM residues*. — Vienna: International Atomic Energy Agency, Technical Reports Series No.**474** (2013)
4. *Human Respiratory Tract Model for Radiological Protection*. ICRP Publication 66. Ann. ICRP **24**, (1994)
5. http://www.jeleniastruga.pl
6. S. Twomey, *Comparison of constrained linear inversion and iterative algorithm applied to the indirect estimation of the particle size distribution*. J Comput Phys **18** (1975)
7. E.F. Maher, N. Laird, *Algorithm reconstruction of particle size distributions from diffusion battery data*. J Aerosol Sci **16** (1985)
8. C.W. Hinds, *Aerosol technology, properties, behaviour, and measurement of airborne particles*: 2nd ed. New York: John Wiley & Sons, Inc; (1999)
9. S.A. Durrani, R. Ilić, *Radon measurements by etched track detectors: applications in radiation protection, earth sciences and the environment*. World Scientific Publishing Co. Pte. Ltd, (1997)
10. J. Porstendörfer, C. Zock, J. Wendt, A. Reineking, *Radon parameters in outdoor air*. International symposium on technologically enhanced natural radiation, Rio de Janeiro, (1999)
11. K. Skubacz, Ł. Wojtecki, P. Urban, *The influence of particle size distribution on dose conversion factors for radon progeny in the underground excavations of hard coal mine*. Journal of Environmental Radioactivity, DOI 10.1016/j.jenvrad.2016.05.020 (2016)

Changes in Polish law related to the implementation of COUNCIL DIRECTIVE 2013/59/EURATOM of 5 December 2013

Katarzyna Wołoszczuk[1,*], *Krystian* Skubacz[2], and *Zuzanna* Podgórska[1]

[1]CENTRAL LABORATORY FOR RADIOLOGICAL PROTECTION, Department of Individual Monitoring and Calibration, Konwaliowa 7, PL03194 Warsaw, Poland, woloszczuk@clor.waw.pl

[2]CENTRAL MINING INSTITUTE, Silesian Centre for Environmental Radioactivity (BCR), Plac Gwarków 1, 40-166, Katowice, Poland, kskubacz@gig.eu

Abstract. Radon is an invisible, naturally occurring radioactive noble gas. According to the WHO report, it is the most important cause of lung cancer after smoking [1]. Recent epidemiological studies show that a statistically significant increase in the risk of lung cancer already occurs as a result of prolonged exposure to radon inside rooms where the concentration is at 100 Bq/m^3 and increases by 16% per 100 Bq/m^3 increase (considering a prolonged exposure period). For this reason, the Council Directive 2013/59/Euratom (BSS) [2] establishing the basic safety standards for the protection against the hazards related to ionizing radiation, that was passed in 2013, pays particular attention to issues related to radon exposure . BSS reduce radon concentration limits in workplaces to 300 Bq/m^3. According to the regulations in force, the BSS must be implemented in the Polish Atomic Law no later than 6th February 2018.

1 Radon

Radon is an invisible, odourless, and tasteless naturally occurring radioactive noble gas. It could be found in all components of the natural environment - rocks, soil, atmospheric air and water. Radon has no stable isotopes. Thirty seven different radioactive isotopes of radon have been described. The most important, however, from radiation protection point of view, is ^{222}Rn. It has a half-life of 3.8 day and is generated by the radioactive decay of radium ^{226}Ra – the decay product of uranium ^{238}U in the earth's crust. Radon has a relatively high solubility in water so it can also be found in ground and drinking water. It is a part of natural radioactive chain and as an alpha-emitter leads to creation new particles called short-lived radon progeny. Radon is a noble gas so it is non-reactive but it merges into radioactive aerosols by combining with ambient particles in the air.

The concentration of radon in the air is not constant and depends on many different factors as geological structure (uranium content, permeability), location. Seasonal changes have been observed as well.

* Corresponding author: woloszczuk@clor.waw.pl

Radon enters the building through cracks in the foundation and accumulates inside enclosed rooms. Therefore, the indoor radon concentrations are several times higher than outside. High levels of radon in the water supply can also increase indoor radon concentrations, as well as some kinds of building materials (a.e. granites). In Poland the mean radon concentration is about 10 Bq/m^3 in the open air whereas the average indoor concentration is about 40 Bq/m^3 [3]. Radon levels are usually higher in basements, than on the upper floors of buildings.

According to the WHO (World Health Organization) report radon is considered to be the main cause of lung cancer among non-smokers. As estimated, between 3–14% of all lung cancer is caused by radon exposure. It depends on the average level of radon concentration and the smoking prevalence in a country [1]. The report released by the President of the National Atomic Energy Agency in 2016 indicates, that the statistical resident of Poland received an annual effective dose of 3.55 mSv. About 69% of the total dose is associated with natural sources and almost 34% from exposure to radon and its decay products [4].

2 Atomic law in Poland before the implementation of the Council Directive 2013/59/EURATOM.

The Atomic LawbAct of 29 November 2000, which entered into force on 1st January 2002, is a main parliament act governing all nuclear activities- nuclear safety and radiological protection of workers and general public in Poland. The most recent consolidated text of the Atomic Law Act was published in the Dziennik Ustaw on 17.03.2017.

Main provisions of the Atomic Law Act regard activities with the use of ionizing radiation or activities which involve exposure to ionizing radiation, powers of the President of National Atomic Energy Agency to control and supervise these activities. The Act also identifies other tasks of the PAA President, assessment of national radiation situation and procedures, that should be applied in radiation emergency.

The basic national legal regulation establishing the doses limits is the Decree of the Council of Ministers of 18 January 2005 [5]. According to this regulation the annual dose limit for general members of the public, expressed as effective dose, is 1 mSv, and should include external and internal exposure. However, the exposure due to occurrence of radon and its progeny in the residential buildings is excluded as established by the Atomic law act.

3 Basic Safety Standards

On 5 December 2013, Council Directive 2013/59/Euratom establishing the basic safety standards for the protection against the hazards related to ionizing radiation was passed. The new Directive repeals the existing legal acts contained in Directives 89/618/Euratom, 90/641/ Euratom, 96/29/Euratom, 97/43/Euratom and 2003/122/Euratom as of 6 February 2018 and at the same time indicates that date as the deadline for the implementation by the Member States of the European Union of the parliamentary acts, executive and administrative provisions necessary to comply with it.

Contrary to previous legal regulations, a number of provisions in the new directive concern the so-called "radon hazard" associated with the occurrence in the air of radon ^{222}Rn and its short-lived progeny such as ^{218}Po, ^{214}Pb, ^{214}Bi and ^{214}Po. Although for the sake of simplicity it is spoken in the context of "radon hazards", the directive explicitly states that this means exposure to "short-lived radon progeny" because their contribution to the dose is predominant and reaches 99%.

The risk associated with radon is especially for those places where ventilation is poor, for example such as caves or underground mines. The largest exposed population, however, are people living or working in rooms, especially in the winter when elevated concentrations of radon occur due to poor ventilation. The presence of radon in buildings results primarily from its penetration from the ground.

In order to ensure the safety and optimization of radiological protection, Member States should establish national reference levels for radon concentrations in rooms. In the case of residential buildings, the reference levels for the average annual radon concentration in air may not be higher than 300 Bq/m^3. A similar value is recommended in the Directive for radon concentrations in indoor workplaces. In this case, however, an increase in the national reference level is allowed if this is reasonably justified by existing national conditions, and the relevant document will be submitted to the European Commission.

By assuming a reference level, it is important to be aware that a statistically significant increase in the risk of lung cancer already occurs as a result of prolonged exposure to radon inside rooms where the concentration is at 100 Bq/m^3. The Directive also indicates that this risk is significantly increased if the radon hazard is associated with smoking.

The implementation of the new directive will require, in addition to the amendments to the legal regulations, the cooperation of many entities including ministry responsible for health and construction, the chief sanitary inspectorate, investors, building owners, employers and measurement laboratories.

4 Propositions of changes in The Atomic Law Act.

4.1 Definitions

The 8th of March 2017 the bill of The Atomic Law Act was published. In this version new definitions were apply, among others that "radon" means the radionuclide ^{222}Rn and its progeny, and "exposure to radon" means exposure to radon progeny. Apparently the simple definition changes a lot in approach to radon measurements by indicating potencial alpha energy concentration (PAEC). In compliance with the new regulations to estimate the annual average radon concentration, it is necessary to perform at least one month countinous measurement. In this case the most appropriate and easy is to use alpha track detectors CR-39 chips and diffusion chambers (Fig. 1). After exposure, the CR-39 chips are etched in 6.25N NaOH at 70 °C for four hours. Typical and clear alpha tracks can be observed with a microscope. The relationship between alpha track density and radon accumulation concentration is proportional (r>0.9990).

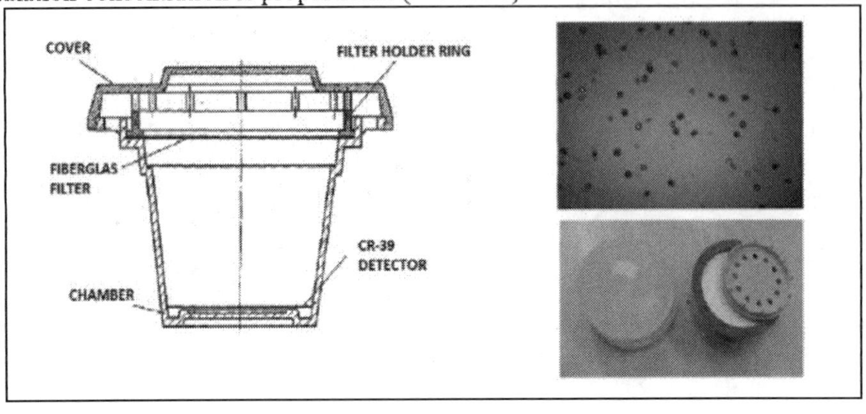

Fig. 1 Karlsruhe-type passive time integrating radon detector

Next important definitions are Radon Potential of a building site (RP) and Radon Index (RI). RP is the value expressing the radon index of the building site, calculated according to the formula:

$$RP= (C_{Rn} - 1)/(- \log k - 10) \quad (1)$$

where C_{Rn} is the radon concentration in soil gas (kBq/m^3) and k is permeability of the soil expressed in (m^2). If RP < 10, then radon index is assumed low, if $10 \leq RP < 35$, then radon index is assumed medium, and high if RP $\geq$ 35. RI indicates the level of risk of radon release from bedrock, and can be expressed numerically as the radon potential of a building site. The categories are low, medium, and high.

4.2 Workplaces

Important change in new regulations is that doses related to radon are included to exposure. As a consequence, exposed workers should be categorized. To match the methods of exposure assessment to the anticipated exposure level , two categories of workers were established, depending on the exposure level. The Category A was chosen for workers who may be exposed to an annual effective dose exceeding 6 mSv and category B, if such doses are greater than 1 mSv and not greater than 6 mSv.

In compliance with the new version of Atomic law employers must make sure, that an appropriate measurement strategy and mitigation of radon exposure was implemented. They are obliged to:

- ensure radiation protection employee exposed to radon,
- make notification in case exceeding the limit 300 Bq/m^3 in workplaces,
- make sure, that a good measurement strategies and mitigation of radon exposure was implemented,
- in case of workplaces located in:
  - ground floor or basements levels, on the areas where high level of radon concentration is possible,
  - underground workplaces,
  - places related to the treatment of water extracted from underground sources

  ensure measurements of radon concentration in air or potential alpha energy concentration
- optimize the exposure of workers working in the mentioned above workplaces,
- inform workers in writing about:
  - increased exposure to radon,
  - results of measurements of radon concentration in air or potential alpha energy concentration,
  - the ionizing radiation dose they received,
  - action taken to reduce radon exposure in the workplace.

If there is a possibility of exceeding the reference level of 300 Bq/m^3 or obtain dose higher than 1 mSv per year, the employer shall take measures to reduce the exposure of workers to radon.

4.3 Homeowners

In compliance with new regulations the seller or lessor of the building shall at the request of the purchaser or tenant of such building provide information on the average annual radon concentration in air in the building. This information shall include:

- average indoor radon concentration in the air,
- comparison measurements results to the reference value of the annual radon concentration in air - 300 Bq/m^3,
- the actual basis for drawing up the information, in particular the results of the measurements and the entity that carried them out.

The above obligation does not apply to non-business renters. When new administrative buildings (hospitals, schools) are planned investor must specify the radon index of the building site. If radon index is medium or high, technical measures shall be taken to prevent radon entry into the building.

If an investor in the construction project has provided measures or technical solutions that prevent radon penetration into the building the above obligation are not required.

4.4 Responsibilities and competences

In new regulations responsibilities and competences of the Chief Sanitary Inspector and minister responsible for construction are similar. Among others they are responsible for organize campaigns promoting the use of measures to reduce concentration of radon, provides information, education and training activities on available measures to limit indoor radon concentrations, monitors the application of measures to limit the average annual concentration of radon, cooperates with the professional councils of architects and construction engineers and other professional organizations and associations, local government bodies and specialists in the field of protection against ionizing radiation, publishes information on instruments used to finance measures aimed at reducing the average annual concentration of radon. Main difference is, that activity field of minister responsible for construction concerns only on new building and Chief Sanitary Inspector's all other buildings designed to accommodate people.

4.5 National Action Plan for Radon Exposure

The ministry responsible for health in cooperation with the Chief Sanitary Inspector will be responsible for developing "National Action Plan for Radon Exposure" an action plan for long-term hazards arising from exposure to radon in dwellings and workplaces. The goal of the action plan is to ensure appropriate long-range strategy for eliminating avoidable risks from radon exposures in dwellings, buildings with public access and workplaces for any source of radon ingress, whether from soil, building materials or water. In line with the new regulations, the National Action Plan should specify:

- strategy for conducting surveys of indoor radon concentrations or soil gas concentrations for the purpose of estimating the distribution of indoor radon concentrations, for the management of measurement data and for the establishment of other relevant parameters (such as soil and rock types, permeability and radium-226 content of rock or soil),
- approach, data and criteria used for the delineation of areas or for the definition of other parameters that can be used as specific indicators of situations with potentially high exposure to radon,
- identification of types of workplaces and buildings with public access, such as schools, underground workplaces, and those in certain areas, where measurements are required, on the basis of a risk assessment, considering for instance occupancy hours,
- the basis for the establishment of reference levels for dwellings and workplaces,
- assignment of responsibilities (governmental and non-governmental), coordination mechanisms and available resources for implementation of the action plan,

 https://doi.org/10.1051/e3sconf/20182801041

- strategy for reducing radon exposure in dwellings,
- strategies for facilitating post construction remedial action,
- strategy, including methods and tools, for preventing radon ingress in new buildings, including identification of building materials with significant radon exhalation,
- strategy for communication to increase public awareness and inform local decision makers, employers and employees of the risks of radon, including in relation to smoking,
- guidance on methods and tools for measurements and remedial measures,
- where appropriate, provision of financial support for radon surveys and for remedial measures, in particular for private dwellings with very high radon concentrations,
- long-term goals in terms of reducing lung cancer risk attributable to radon exposure (for smokers and non- smokers),

The National Action Plan should be review and update at least once every four years.

5 Conclusions

The impact of the changes in Polish Atomic Law will be important in many areas of life, since it implies for the first time an obligation to develop a regulatory frame to actively work on reducing the radon exposure not only of workers, but also of the general public and lowering the reference level for the annual average activity concentration in air to a maximum value of 300 Bq/m^3.

References

1. World Health Organization *Handbook on Indoor Radon: A Public Health Perspective*, (2009)
2. Council Directive 2013/59/EURATOM of 5 December 2013 *laying down basic safety standards for protection against the dangers arising from exposure to ionising radiation, and repealing Directives 89/618/Euratom, 90/641/Euratom, 96/29/Euratom, 97/43/Euratom and 2003/122/Euratom*. (2013)
3. M. Biernacka, J. Henschke, J. Jagielak, A. Koczyński, K. Mamont-Cieśla , Postępy Fiz.Med. **26**,1–2, (1991)
4. *The activity of the President of the National Atomic Energy Agency and the assessment of the state of nuclear safety and radiological protection in Poland in 2016*, **56**, (2017)
5. Decree of the Council of Ministers of 18 January 2005 *on dose limits of ionising radiation*. The Polish Law Journal, no 20, pos. 168, (2005)

Determination of water-soluble elements in PM2.5, PM10, and PM2.5-10 collected in the surroundings of power plants

Elwira Zajusz-Zubek[1,*], *Anna Mainka*[1] *and Konrad Kaczmarek*[2]

[1]Department of Air Protection, Silesian University of Technology, 22B Konarskiego St., 44-100 Gliwice, Poland
[2]Institute of Mathematics, Silesian University of Technology, 23 Kaszubska St., 44-100 Gliwice, Poland; E-Mail: Konrad.Kaczmarek@polsl.pl

Abstract. The analysis reported in this study was performed to characterize the concentrations and water-soluble content of trace elements (As, Cd, Co, Cr, Hg, Mn, Ni, Pb, Sb and Se) in PM2.5, PM10 and PM2.5-10 samples collected in the surroundings of power plants in southern Poland. The solubility of trace elements bound to PM2.5 and PM10 was higher than for PM2.5-10, and in most cases, significant differences were revealed in the relative percentage concentrations of the water-soluble fractions. The occurrence of Cd, Cr, Mn, Ni, Pb and Se in first PCA (Principal Component Analysis) factor (PC1) – indicate coal combustion processes as the potential source of these elements. Other factors indicate two further anthropogenic sources: the resuspension of road dust due to vehicular activities and waste burning in domestic sources – factor (PC2), and, soil dust sources affected by fugitive dust from the mining processes and unpaved roads, as well as transportation and deposition of coal –factor (PC3).

1 Introduction

The concentration and composition of airborne particulate matter (APM) at a specific location depends on a large number of natural and anthropogenic factors, such as local and regional PM sources, meteorological conditions and geographical situation [1,2]. Particles with diameters of below 10 μm have various sources. They mainly originate from the re-suspension of road dust, coal and oil combustion, and industrial processes [3,4].

In the evaluation of the environmental impact of coal-fired power plants, PM is an important factor. Its hazardous potential results from the ability of dust to penetrate into an organism via the respiratory tracts, and from the presence of dangerous substances contained in the dust particles. Among the dangerous substances loaded on the PM, many studies have referred to the total content of hazardous air pollutants (HAPs), especially trace elements (As, Be, Cl, Cr, Cd, Co, Hg, Mn, Ni, Pb, Sb and Se) [5–7]. However, there is a shortage of data on the chemical forms of trace elements (chemical speciation) which

* Corresponding author: elwira.zajusz-zubek@polsl.pl

occur in APM [8–10]. In developing countries, energy processes based on coal — as well as coal pyrolysis — are the most significant sources of HAPs [11,12].

The evaluation of the potential toxicities of trace elements is based on the distributions of their chemical forms, as divided into four groups (highly mobile, mobile, less mobile and not mobile). Mobile elements are particularly bioavailable to the environment. The amount of potentially bioavailable trace elements in PM can be estimated from the water-soluble fraction and from the fraction obtained with a dilute salt by extraction [13–16]. Thus, investigations of the soluble or extractable elements in PM have attracted far greater attention in this field of study.

In our study, chemical fractionation was applied in order to characterize (1) the mobility of trace elements in respirable (PM2.5), inhalable (PM10) and coarse fractions (PM2.5–10), collected from areas near coal-fired power plants; and (2) principal component analysis (PCA), to identify the possible sources of these elements in this area.

2 Material and Methods

2.1 Sampling Sites

Due to the influence of the particular heating season, especially on domestic and municipal emissions, the sampling campaigns were only carried out during the summer season. Samples were collected from 28 May to 23 September 2014 (16 weeks), in the surroundings of four working power plants fired with hard coal. Each sampling campaign was divided into four sessions for each location. In order to collect a sufficient amount of PM (with diameters <1, 1–2.5, and 2.5–10 μm), each session was conducted continuously for seven days. In total, there were 48 samples and 32 blanks (stored in the sampling area but not exposed).

All of the sampling points were located to the northeast of the selected power plants, in accordance with the dominant wind direction in the region. The distance of the points from their corresponding power plants was approx. 2 km. The choice of a specific location was dependent on the possibility of locating the measuring apparatus on a particular estate with the permission of the owners.

The sampling points were all located in southern Poland. Three of them (P1, P3, and P4) were situated in the industrial macro-region of Upper Silesia, while P2 was situated approx. 70 km to the west of this macro-region. Figure 1 shows the study area and the sampling locations in the surroundings of the four selected power plants.

In each studied area, the amount of APM is generally associated with local pollutant sources. For all the sampling sites, the major source was the power plants, including the transportation and deposition of coal and fly ash. Other sources included industrial activities, the combustion of coal and wood for domestic cooking purposes, and vehicle transport. Additionally, points P1 and P2 were impacted by soil resuspension from agriculture activities and vehicle traffic on unpaved roads. Meanwhile, at points P3 and P4, there was an impact from the urban traffic from the neighboring cities (Katowice and Jaworzno).

https://doi.org/10.1051/e3sconf/20182801042

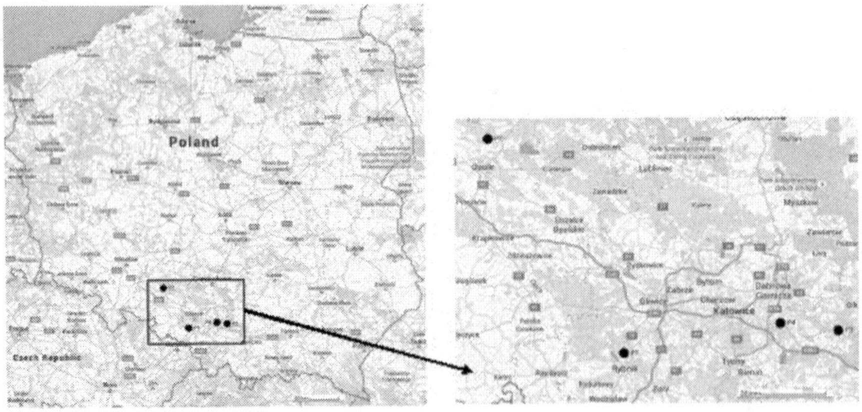

Fig. 1. Location of the sampling sites in southern Poland (Map data: ©2017 Google, ORION-ME).

2.2 Sampling Method

The methodology adopted here follows that of [13]. The samples were collected at sites P1–P4, near four working power plants, using a three-stage impactor: a Dekati® PM10 impactor (Finland) with a flow rate of 30 dm^3/min. The impactor differentiates particles into four size fractions with cut-off diameters of <1, 1–2.5, 2.5–10 and >10 μm, respectively. The particles with diameters >1 μm were collected on polycarbonate membranes (Nuclepore, Whatman International Ltd., Maidstone, UK; diameter 25 mm), while the backup filter material was made of Teflon (Pall International Ltd., New York, NY, USA; diameter 47 mm). The average volume of air aspired by the filters was approx. 300 m^3. Blank filters were stored in the sampling area. According to the manufacturer's information, the impactor is characterized by uncertainties below 2.8%. The inlet tube was installed 1.5 m above the ground, which is important when considering human exposure. The membranes and filters were conditioned before and after sampling (temperature 20±1°C; relative humidity 50±5%) for 48 h, and weighted with a microbalance precision of 1 μg (MXA 5/1, RADWAG, Radom, Poland).

2.3 Chemical Analysis and Quality Control

Three PM fractions (with cut-off diameters of 2.5 and 10 μm), collected in the surroundings of four working power plants, applied the sequential extraction procedure following [13]. The evaluation of the chemical forms of the trace elements underwent the procedure of leaching, based on the scheme described by Tessier et al. [17], and modified by Fernández Espinosa et al. [18], Sanchez-Rodas et al. [19] and Schleicher et al. [20]. The fractionation of PM (2.5, 10 and 2.5–10 μm) separated the bound elements into four fractions. The first fraction (F1) covers soluble and exchangeable elements. The second fraction (F2) includes carbonates, oxides, and reducible elements. The third fraction (F3) contains elements bound to organic matter, oxidizable and sulphidic elements. The last fraction (F4) encloses residual elements.

For every fraction, the As, Cd, Co, Cr, Hg, Mn, Ni, Pb, Sb and Se content was determined. The total content of each trace element was calculated as the sum of the four extraction steps [18,21].

Determining the concentration of the various trace elements for each sample was carried out by inductively coupled plasma mass spectrometry (ICP-MS) (NexION 300D, PerkinElmer, Inc., Waltham, MA, USA). The instrument was equipped with a multiplier collector and a radio frequency power of 3 kW was applied to the plasma.

All of the statistical analyses, including a univariate and multivariate analysis, as well as a principal component analysis, were performed using the statistical package, Statistica 10 (StatSoft, Tulsa, OK, USA). Non-parametric tests were undertaken to confirm the parametric results — that is, the corresponding non-parametric tests led to the same conclusions of significance/non-significance as the parametric tests. Throughout the study, a *p*-value of <0.05 was considered to indicate statistical significance.

3 Results and Discussion

3.1 APM and Trace Element Concentrations

In our study, the concentrations of trace elements in PM2.5, PM10 and PM2.5-10 (coarse fraction with particles of an aerodynamic diameter between 2.5 and 10 μm) were examined at the four locations (P1, P2, P3 and P4) in the surroundings of four working power plants (PP1, PP2, PP3 and PP4). To compare the trace element concentrations, the results showing their total content in PM2.5, PM10, and PM2.5-10 collected at all sites are presented in Table 2. The mean PM2.5, PM10 and PM2.5-10 concentrations determined at the four sampling sites P1–P4 were 13.10 μg/m^3 (range 6.86–19.65 μg/m^3), 16.33 μg/m^3 (range 10.39–22.08 μg/m^3) and 3.23 μg/m^3 (range 2.27–4.56 μg/m^3), respectively.

Table 2. The PM mass concentrations (μg/m^3) and trace element concentrations (ng/m^3) of PM2.5, PM10 and PM2.5-10.

PM2.5, μg/m^3		Elements concentration, ng/m^3									
		As	**Cd**	**Co**	**Cr**	**Hg**	**Mn**	**Ni**	**Pb**	**Sb**	**Se**
Average	13.10	0.85	0.78	0.18	29.85	0.56	8.08	3.01	17.00	1.13	7.95
Min	6.86	0.37	0.27	0.04	16.21	0.04	4.58	0.38	9.70	0.28	1.09
Max	19.65	1.34	1.07	0.32	43.51	1.10	10.99	7.13	32.73	2.47	14.77
SD	3.54	0.29	0.21	0.14	13.58	0.51	1.91	2.27	6.54	0.71	6.51
PM10, μg/m^3											
Average	16.33	0.89	0.99	0.27	44.64	0.83	11.73	4.12	21.21	1.29	11.39
Min	10.39	0.40	0.36	0.06	24.19	0.07	6.88	0.52	11.76	0.32	1.27
Max	22.08	1.40	1.32	0.48	65.22	1.61	15.84	8.19	35.34	2.66	21.46
SD	3.18	0.30	0.27	0.21	20.42	0.77	2.72	3.02	7.25	0.73	9.92
PM2.5-10, μg/m^3											
Average	3.23	0.04	0.21	0.09	14.80	0.27	3.65	1.11	4.21	0.16	3.44
Min	2.27	0.03	0.06	0.02	7.98	0.02	2.09	0.14	2.05	0.05	0.15
Max	4.56	0.06	0.38	0.16	21.71	0.53	5.18	3.06	12.62	0.24	6.83
SD	0.63	0.01	0.12	0.07	6.84	0.26	1.01	1.05	3.02	0.08	3.42

Generally, among the trace elements, the highest concentrations in all the determined PM fractions were revealed to be of Cr and Pb. The similar distribution of trace elements in all PM fractions can be explained by the high share of PM2.5 in PM10 (79.04±7.60%). Considering the total concentrations of different trace elements in PM2.5 and PM10, there were no significant differences ($p>0.05$) for As, Co, Hg, Ni, Pb, Sb, and Se, while the total concentrations of Cd, Cr and Mn differed significantly ($p=0.001$–0.05). Comparing the total concentrations of trace elements in PM2.5-10 with PM2.5 or PM10 generally, significant differences ($p<0.05$) were observed. Only Hg revealed similar concentration in PM2.5 and in the coarse fraction (PM2.5-10).

3.2 Soluble Fraction in APM

There is actually a widespread agreement that information about the mass and total content of the trace elements in APM is necessary but insufficient, in order to evaluate the overall pollution and hazard levels. Figure 2 illustrates the distribution of the water-soluble fraction in the total concentrations of As, Cd, Co, Cr, Hg, Mn, Ni, Pb, Sb, and Se extracted from PM2.5, PM10 and PM2.5-10. Generally, the solubility was higher for PM2.5 and PM10 than for PM2.5-10, and in some cases, there were significant differences in the relative percentage concentrations for the water-soluble fractions. The fractionation data indicate high average substantial percentages of As (95.4%), Sb (57.0%), Cd (50.1%), Mn (38.2%) and Pb (35.2%) in PM2.5, and As (93.9%), Sb (51.3%), Cd (44.3%), Mn (34.2%) and Pb (28.9%) in PM10. In PM2.5-10, the highest contribution was also of As (62.1%) and the following was Mn (22.7%). The concentrations of the water-soluble fractions of trace elements in PM2.5 and PM10 were only significantly different ($p<0.05$) for Co and Hg, whereas the water-soluble concentrations of trace elements in PM2.5-10 in comparison with PM2.5 and PM10 were significantly different ($p<0.05$) for all the elements except Hg between coarse (PM2.5-10) and fine (PM2.5) fractions.

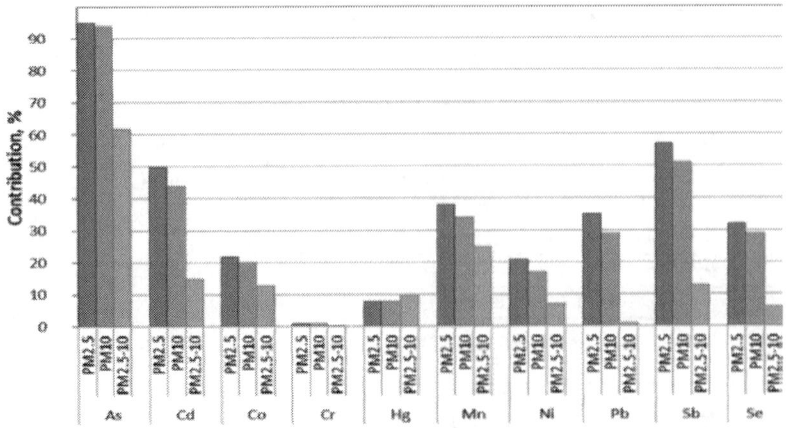

Fig. 2. The contribution of water-soluble trace elements in PM2.5, PM10 and PM2.5-10 samples collected from areas near the coal-fired power plants.

Similar studies on the elemental composition of the water-soluble fraction of PM2.5 samples collected in Greek cities were conducted by Manousakas et al. [8]. A substantial percentage (≥30%) of the Cd, V, Zn, Cu, Pb, and Ni displayed water-soluble characteristics in the samples collected in a coastal city, and the same was true of the As, Cd, Mn, Ni and Pd in samples collected near lignite power plants and mines. In our study, the water-soluble fraction of Cd, Mn, and Pb also contributed considerably (>30%) to the total concentrations, in contrast to Ni, for which the contribution of the water-soluble fraction was far lower (7.4%). However, the Greek results cannot be directly compared with our study (of hard coal-fired power plants), due to the different emission profiles of lignite-fired power plants.

A significant percentage (>30%) of water-soluble elements — Cd, Mn, Pb, and Sb — was also reported by Sarti et al. [9] in PM2.5 collected near municipal waste incinerators (in the Bologna area). At the same time, the authors state that among the various combustion processes, waste incineration didn't seem to be the main source of these soluble metals in PM2.5. During a five-year field study in the peri-urban area of Ferrara, in northern Italy, Canepari et al. [22] studied the concentration and solubility of trace elements contained in fine (PM2.5) and coarse (PM2.5-10) fractions. The results show similar extraction percentages of the following elements in the fine (PM2.5) fraction: As, Cd, Mn, Pb, and Sb. They emphasize that the solubility for PM2.5 was higher than that for PM2.5-10 and that in some cases there were significant differences in the relative percentage concentrations for the extracted and the residue fractions [22,23].

3.3 Principal Component Analysis (PCA)

In order to determine the associations between the elements generated by diverse emission sources, PCA was applied to identify the possible sources of water-soluble trace elements in the PM2.5-10 samples. By using chemical fractionation data instead of total element content, PCA can evaluate emission sources with higher selectivity [23]. The relationships between the 10 trace elements are readily seen as a 3-D plot of the PCA loadings (Fig. 3).

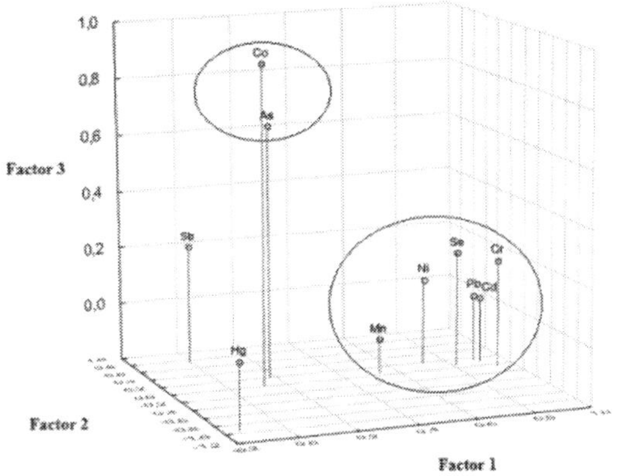

Figure 3. PCA loading 3-D plot (PC1 vs. PC2 vs. PC3) for 10 trace elements.

Three factors were obtained, accounting for 70.9% of the total variance:

(1) PC1 with high loadings of Cd, Cr, Mn, Ni, Pb, and Se;
(2) PC2 with high loadings of Hg and Sb; and
(3) PC3, which includes As and Co.

The first component (PC1) contributed 40.6% of the total variance. It is interesting to note that two groups of elements — Se–Cr and Cd–Pb — are linked to each other at a higher level than other elements. This is compatible with results from previous studies the results [24,25], demonstrating that Cd and Pb can be reliably linked to fossil fuel burning (oil and coal). Moreover, in the case of Cd and Pb, the link between these elements can be attributed to an association that extends far beyond direct industrial combustion [25].

The second component (PC2) contributed 16.0% of the variance, and included Hg and Sb. However, the lack of association between Hg and Sb (or any other elements), observed in the opposite signs of PC2 loadings, indicate separate emission sources. Hg can be distributed from primary and secondary anthropogenic sources. The primary sources are mining, extraction and the burning of fossil fuels that contain Hg as a trace contaminant. The secondary anthropogenic sources are those in which emissions occur through the intentional use of mercury, including mercury use in industrial processes and waste incineration. The largest single source category of anthropogenic mercury emission into the air is the stationary combustion of coal, associated with energy or heat production in major power plants [26]. Nevertheless, it might indicate the influence of sources other than coal combustion. Bearing in mind the general problem in the region whereby low-quality fuels (biomass, culm or even refuse) are procured and burned for domestic purposes, the main source of water-soluble Hg in the coarse fraction could be waste incineration. In general, Sb is naturally present in soil [27]; hence, anthropogenic sources can be attributed to the resuspension of road dust, due to vehicular activities like brake wear [28,29]. In the region, Sb is also emitted from long-term mining and smelting activities [30]. Moreover, Sb revealed an intermediate degree of similarity to As and Co.

The third component (PC3) contributed about 14.3% of the variance. It essentially constitutes a loading of Co and As (to a lesser extent). As is regarded as a marker element of coal combustion and coke-making processes [31,32]; however, the PCA doesn't confirm this source. It would seem that the As in the water-soluble fraction originated from a source other than coal combustion processes. The presence of Co in soil may originate from anthropogenic activities, such as the application of cobalt-containing sludge or phosphate fertilizers, the disposal of cobalt-containing wastes, and the atmospheric deposition from activities such as mining, smelting, refining, and combustion [33]. Dai et al. [32] matched Co to soil dust. In addition, the PC3 component might point to soil dust sources affected by fugitive dust from the mining processes and unpaved roads, as well as transportation and the deposition of coal.

4 Conclusions

A chemical extraction procedure coupled with ICP-MS was used for the characterization of the water-soluble elements contained in PM2.5, PM10 and PM2.5-10 samples, collected in the surroundings of hard coal-fired power plants in southern Poland. The results included the concentration, trace element composition and water solubility of the APM. The average concentration levels of PM2.5, PM10 and PM2.5-10 were 13.10 $\mu g/m^3$, 16.33 $\mu g/m^3$ and 3.23 $\mu g/m^3$, respectively.

The solubility of trace elements bound to PM2.5 and PM10 was higher than for PM2.5-10, and in most cases revealed significant differences in the relative percentage concentrations of the water-soluble fractions.

Principal component analysis (PCA) was performed on the water-soluble elements to identify the possible sources of these elements in this area. The occurrence of Cd, Cr, Mn, Ni, Pb and Se in one PCA factor (PC1), indicated coal combustion processes as a potential source of these elements.

Second (PC2) and third (PC3) components were loaded with Hg and Sb, and As and Co, respectively. Therefore, they can be attributed to other anthropogenic sources. PC2 could have been contributed to the resuspension of road dust due to vehicular activities, and waste burning in domestic sources. Meanwhile, PC3 points to soil dust sources affected by fugitive dust from the mining processes and unpaved roads, as well as the transportation and the deposition of coal.

In our opinion, to better interpret the results, it is worth to use chemical fractionation data instead of total element content, so the PCA is able to evaluate emission sources with a higher selectivity.

Acknowledgements: this work was supported by the Faculty of Power and Environmental Engineering, Silesian University of Technology (statutory research). All the results used in this paper have received funding from the National Science Centre Poland under grant 2012/05/B/ST10/00524.

References

1. A. Mues, A. Manders, M. Schaap, L. H. van Ulft, E. van Meijgaard, P. Builtjes, Atmos. Environ. **80**, 232 (2013)
2. G. Titos, H. Lyamani, M. Pandolfi, A. Alastuey, L. Alados-Arboledas, Atmos. Environ. **89**, 593 (2014)

3. Y. Cheng, S. Lee, Z. Gu, K. Ho, Y. Zhang, Y. Huang, J. C. Chow, J. G. Watson, J. Cao, R. Zhang, Particuology **18**, 96 (2015)
4. B. Srimuruganandam, S. M. Shiva Nagendra, Sci. Total Environ. **433**, 8 (2012)
5. A. Mainka, E. Zajusz-Zubek, K. Kaczmarek, Int. J. Environ. Res. Public Health **12**, 7990 (2015)
6. K. Mohiuddin, V. Strezov, P. F. Nelson, E. Stelcer, Atmos. Environ. **83**, 72 (2014)
7. C. Petaloti, A. Triantafyllou, T. Kouimtzis, C. Samara, Chemosphere **65**, 2233 (2006)
8. M. Manousakas, H. Papaefthymiou, K. Eleftheriadis, K. Katsanou, Sci. Total Environ. **493**, 694 (2014)
9. E. Sarti, L. Pasti, M. Rossi, M. Ascanelli, A. Pagnoni, M. Trombini, M. Remelli, Atmos. Pollut. Res. **6**, 708 (2015)
10. E. Zajusz-Zubek, A. Mainka, Z. Korban, J. S. Pastuszka, Atmos. Pollut. Res. **6**, 1 (2015)
11. J. Konieczyński, E. Zajusz-Zubek, Arch. Environ. Prot. **37**, 3 (2011)
12. J. Konieczyński, E. Zajusz-Zubek, M. Jabłońska, Sci. World J. **2012**, 1 (2012)
13. E. Zajusz-Zubek, K. Kaczmarek, A. Mainka, Int. J. Environ. Res. Public Health **12**, 13085 (2015)
14. E. Zajusz-Zubek, *Evaluation of Forms in Which Occur Selected Trace Elements in Suspended Dust (PM10) and Respirable Fraction (PM2.5) in the Surroundings of Coal-Fired Power Plants and Coking Plants in the Non-Heating Season* (Publishing House of Silesian University of Technology, Gliwice, 2016)
15. S. Yadav, P. G. Satsangi, Environ. Monit. Assess. **185**, 7365 (2013)
16. P. Sipos, C. Choi, Z. May, Chemie Der Erde - Geochemistry **76**, 481 (2016)
17. A. Tessier, P. G. C. Campbell, M. Bisson, Anal. Chem. **51**, 844 (1979)
18. A. J. Fernández Espinosa, M. Ternero Rodríguez, F. J. Barragán De La Rosa, J. C. Jiménez Sánchez, Atmos. Environ. **36**, 773 (2002)
19. D. Sanchez-Rodas, A. Sanchez De La Campa, V. Oliveira, J. De La Rosa, J. Inorg. Biochem. **108**, 112 (2012)
20. N. J. Schleicher, S. Norra, F. Chai, Y. Chen, S. Wang, K. Cen, Y. Yu, D. Stüben, Atmos. Environ. **45**, 7248 (2011)
21. P. Richter, P. Griño, I. Ahumada, A. Giordano, Atmos. Environ. **41**, 6729 (2007)
22. S. Canepari, M. L. Astolfi, C. Farao, M. Maretto, D. Frasca, M. Marcoccia, C. Perrino, Environ. Sci. Pollut. Res. **21**, 4010 (2014)
23. S. Canepari, A. Pietrodangelo, C. Perrino, M. L. Astolfi, M. L. Marzo, Atmos. Environ. **43**, 4754 (2009)
24. Y. Zhai, X. Liu, H. Chen, B. Xu, L. Zhu, C. Li, G. Zeng, Sci. Total Environ. **493**, 109 (2014)
25. S. M. Enamorado-Báez, J. M. Gómez-Guzmán, E. Chamizo, J. M. Abril, Atmos. Res. **155**, 118 (2015)
26. E. G. Pacyna, J. M. Pacyna, K. Sundseth, J. Munthe, K. Kindbom, S. Wilson, F. Steenhuisen, P. Maxson, Atmos. Environ. **44**, 2487 (2010)
27. C. A. Tylenda, D. W. Sullivan, *Handbook on the Toxicology of Metals* (Elsevier, 2015)
28. P. Pant, R. M. Harrison, Atmos. Environ. **77**, 78 (2013)
29. F. Fujiwara, R. J. Rebagliati, J. Marrero, D. Gómez, P. Smichowski, Microchem. J. **97**, 62 (2011)
30. W. Rogula-Kozłowska, K. Klejnowski, P. Rogula-Kopiec, B. Mathews, S. Szopa, Bull. Environ. Contam. Toxicol. **88**, 722 (2012)
31. A. B. Vicente, T. Sanfeliu, M. M. Jordan, J. Environ. Manage. **108**, 92 (2012)
32. Q.-L. Dai, Aerosol Air Qual. Res. **2015**, 875 (2015)
33. D. Lison, *Handbook on the Toxicology of Metals* (Elsevier, 2015)

AUTHOR INDEX